GALILEO'S COMMANDMENT

GALILEO'S
COMMANDMENT
An Anthology of Great
Science Writing

Edited by

Edmund Blair Bolles

W. H. Freeman
New York

For Laura

DESIGN: Victoria Tomaselli

Library of Congress Cataloging-in-Publication Data

Galileo's commandment : an anthology of great science writing / edited by
  Edmund Blair Bolles.
        p.   cm.
    Includes index.
    ISBN 0-7167-3035-9
    1. Science—Popular works.   2. Science.   3. Science news.
  4. Technical writing.   I. Bolles, Edmund Blair, 1942–   .
  Q162.G22   1997   97-5997
  500–dc21   CIP

Printed in the United States of America

First printing, 1997

Contents

CHAPTER SEVEN

"Those Who Would Judge the Book Must Read It" 245

CHAPTER EIGHT

"Somehow the Wave Had to Exist" 301

Contents Arranged by Date

Contents Arranged by Science

Introduction

I love great science writing for the same reason I enjoy splendid auto-biography or classic letters and journals. It puts me in direct contact with an active, probing mind. Scientific theories change; new discoveries become old hat; and stirring mysteries become trite facts, but the excitement of being in the presence of a thinking brain is forever fresh. It is the presence of a living imagination that keeps great science writing alive, just as it does other forms of writing, yet the nature of this literary form remains mostly unremarked and unexamined.

Scientists look on their texts as sources of information, and for them, any sense of the author's humanity is a secondary effect. True, some scientists admit to the importance of the human side of what they read. I recall, for example, the head of a research lab who worried that papers in modern science journals had become so abstract that they almost never gave readers any sense of an author who had wrestled with ideas. He feared that young scientists just coming into his lab would think that their own difficulties were unusual, a mark of inferiority. Some published scientists, as well, are dissatisfied with this aloofness. Once I interviewed a neuroscientist who kept interrupting his account to tell me, "I can explain it so glibly now, but we had to think hard for every word." Yet however much scientist-readers may appreciate the human presence in science writing and no matter how badly scientist-writers want others to value their own struggles, the official doctrine holds that ultimately personalities do not matter. Knowledge and progress are what matter. Scientists are not used to thinking of even their finest writing as literature.

The aim behind this collection is to show readers that science writing can be great writing in precisely the same sense that other genres are great: it has something important to say; it says it by presenting readers with unique imaginations; and readers in turn are inspired to think in ways that, by themselves, they never could. Scientists have been wrong to devalue this side of their literature, and nonscientists have been wrong to think there was nothing in this writing for them.

Every generation since Voltaire has had an energetic minority who recognized that participating in the ideas of one's age and civilization required keeping up with the sciences. Voltaire saw that point clearly enough to present Isaac Newton's ideas to French readers. He realized that the critical ingredient in the new science was a free imagination and that hiding from science meant hiding from one's own humanity. Yet Voltaire's view has had to compete with the widespread belief that science is a threat, an insult to humanity.

The duel between science and the arts has been a recurring literary theme for centuries. By the end of the 1700s, poets knew they could not stop science, but they did manage to persuade much of the literary world to despise it. Anti-Newtonian poets like Goethe and William Blake made grand arguments, while lesser lights created the Romantic idea of the successful-yet-mad scientist. Earlier writers like Jonathan Swift had portrayed crackpot scientists who pursued impossible ambitions, but the new crackpot, like Dr. Frankenstein, was crazy in another way. He was mad because he was successful and could unleash previously unknown powers against humanity. Since then we have seen an endless stream of inquiring Dr. Jekylls, Dr. Moreaus, and Dr. Strangeloves—lunatic scientists who poison society by wondering too much about matter and too little about the human heart.

I hope that by the end of this anthology, readers will feel the absurdity of that antiscience stereotype. I have chosen the pieces in the spirit of a curator selecting works for an exhibition by an underrated artist. Just as a collection of paintings by, say, Edvard Munch can inspire an increased respect for the artist by simply showing what is there, I hope that this book's readers will see that science writing can be fresh, pleasurable, and not at all like cold toast.

Galileo's Commandment

To make my case, I have focused on straight science writing, not philosophical writing about science. I have included no essays here about the beauty of science, the role of science in society, the relation between science and technology, and the like. The literary potential of such prose has already been recognized. Instead, this book presents writing that makes a contribution to science.

In Bertolt Brecht's play *The Life of Galileo*, Galileo (the character) says, "Science knows only one commandment: contribute to science." In Galileo (the man)'s own case, the contribution was heroic. It advanced across many borders of the imagination. Most scientists cannot make contributions on Galileo's level, just as most dramatists cannot approach Shakespeare. Yet by concentrating on nature and trying to understand nature on its own terms, many people have contributed to science, and any writer who has fulfilled this one commandment has been eligible for inclusion in this anthology.

Many scientists, both famous and forgotten, professional or not, can be found here. The oldest selection in this volume comes from Herodotus, the Greek historian and traveler who wrote a splendid natural history of the Nile Valley. In this selection, Herodotus contributed to science because he was trying to understand what he saw in natural terms. He wrote in about 444 B.C., which to us seems a long time ago, but his visit to Egypt was at about the midway point between the founding of the first royal dynasty in a united Egypt (ca. 2850 B.C.) and today. By Herodotus's time, more than two thousand years' worth of cultured people had sought to understand the Nile, but mostly they looked to supernatural and occult explanations of what they saw. Herodotus, however, looked to the natural actions of the Nile River itself, and in so doing he made a contribution to science.

This book also contains selections by trained scientists who devoted their careers to writing for nonspecialists. "Mere popularizers," some professionals might sneer, but these writers too are obeying Galileo's commandment. Consider, for example, the selection about ice by Louise Young. She was trained in physics, and during World War II she worked at MIT designing radar antennas for the navy, but she has devoted most of her professional career to writing for nonscientists. Her ability to organize knowledge so that the world becomes intelligible is not a trivial feat. Most people (no matter how many facts are at their fingertips) cannot do it. Writers like Louise Young have used their literary skills to make important contributions to science.

Scientific Imaginations

Initially, I conceived of this anthology simply as a collection of great science writing. Because science writing is not taken seriously as literature, even the most important pieces are hard to find. Yes, books by Darwin are still in demand, and Galileo still sits on the back shelves in college bookstores, but Lavoisier, Wallace, Mach, Maxwell,

Rutherford, Helmholtz, and many others are almost impossible to find outside the archives of great research libraries. Plato is still available; Lavoisier should still be available. So I went to work reading these works. As the number of possible selections mounted, I saw that this anthology could be more than just a collection of science writing. Placed together, the works began to show unexpected properties. They began to glow and interact with one another, just as the radium in Marie Curie's lab began shining brighter as more of it was amassed in one place. My standards for inclusion became tougher, and the works responded by shining even more fiercely in one another's company. I tried to select only pieces produced by robust imaginations and disciplined by a ruthless honesty.

Mention of a "robust imagination" might strike some people as a contradiction. Science looks for truth; imagination makes up lies. But that is a very Philistine way of looking at the world. Imagination is the faculty of getting beyond our ignorance of the here and now. The obvious examples of imaginative science include Alfred Wallace's lying sick in the East Indies and conceiving of a way that evolution could proceed through a process of natural selection. Einstein pictured the world as seen from a railway car traveling at almost the speed of light. Newton noticed the unexpected regularity in the way light moved through a prism, and he was inspired to imagine particles of light moving in straight lines. Insights like these unquestionably spring from imagination, but scientific imagination means more than occasionally having a big new idea.

A novelist—say Joseph Heller, the author of *Catch-22*—does not have just one or two big ideas that provide the imaginative side of the story. Rather, imagination is active throughout the work, in every paragraph and line of dialogue. The same holds true for great science writing. For example, in Wallace's essay on evolution, his imagination is always present as he describes the annual mortality of birds, the ancestors of giraffes in times of stress, and the habits of the passenger pigeon.

Imagination's steady presence also is apparent in Galileo's account of his first look through a telescope. In this case, Galileo did not even have a grand organizing idea such as Wallace enjoyed. Yet his considerable imagination let him recognize mountains on the moon, moons around Jupiter, and an immensity of stars that expanded the size of the universe by a fantastic magnitude.

Just as other literary imaginations take many forms, scientific imaginations follow many drummers. This book is organized to show off the diversity of scientific imagination, the engine that propels science, just as it propels all other creative activity. In great science writing, of course, the imagination is disciplined by a ruthless honesty. Again, it is the same in other forms of literature. Poets struggle over every word to get past the clichés and to say something worth saying. They must be ruthless with themselves in this struggle; every great poet knows how hard it is to overcome the trite and conventional. Scientists too must be ruthless with themselves. Although they test their ideas by means of experiment, they must also search for alternative explanations and counterexamples. Francis Bacon helped launch the scientific era by providing a long list of ways in which people can go too easy on themselves and their ideas. Galileo wrote dialogues because he believed it was vital to test his own theories through argument rather than simple assertion. Indeed, it was said in Galileo's day that he could argue better for the old earth-centered picture of the universe than any of the old believers could. This same ruthlessness is visible in the most recent science writing. George Smoot, writing in 1994, describes how he spent years compiling a list of all the things that might fool him as he prepared to search the heavens for evidence that the Big Bang had happened.

Scientific Voices

By its nature, disciplined, imaginative writing produces a distinct literary voice. It has to be that way because it is the clichés that make so many of us sound alike. Any person who abandons the clichés must either fall silent or develop a unique voice. Inevitably then, imaginative writings have voices that put the writers on the page alongside their ideas.

A writer's voice is what people sometimes mean by style, although technically style refers to the words a writer chooses. Style is the objective data on the printed page that a literary critic studies. Not many science writers are great stylists. Lucretius, Rachel Carson, Loren Eiseley, and a few others wrote beautifully in the stylistic sense of the term, but they are the exceptions. Voice, however, is different. Like stage presence in actors, voice emerges from a writer's being. The writer is often unconscious of it and does not always try to create it. Often too, the untrained reader responds to it without knowing just what it is. Voice inspires a subjective feeling in readers about a writer and is what editors commonly look for when they are considering pieces for a general audience. Usually, it is the writer's voice that keeps a reader reading. Grabby openers and astonishing facts can get readers out of the starting gate; strong subject matter helps them stay the course; but it is the writer's voice that pulls readers forward and makes them care.

Science is about the world, not the scientist, yet the more that scientists contribute to science, the more of themselves ends up in their writing. Perhaps the best illustration of voice's place is in the writing of Charles Darwin and Alfred Wallace. Each independently developed a theory of evolution through natural selection, and their theories were so similar that Darwin said that his chapter headings had turned up in Wallace's essay. Yet for all the similarity of their theories, they were very different people, and no reader is likely to confuse their voices. In this book's selection by Darwin, the voice is calm, detailed, and observant. In the selection by Wallace, his voice is enthusiastic, expansive, and theoretical. Even if the story of Darwin and Wallace were not one of the most famous in science history, an alert reader could quickly guess from their voices which writer was likely to nurse his theory for years and which would write up his idea almost immediately after conceiving it.

Even the famous Dr. Einstein could give himself away. His subject matter often seemed remote and his style so Olympian that you have to be in on the joke to realize how humbly he was speaking. When he says, "Since the introduction of the special principle of relativity has been justified," a more boastful man might have written "after my special principle of relativity was proved right." The cynic may suspect a false humility. We all know who it was that introduced and justified relativity theory. But Einstein's voice continues, "Every intellect which strives after generalization must feel the temptation to venture the step towards the general principle of relativity." Why not say more personally, "I wanted to find a general principle of relativity"? But that would be my voice, or maybe your voice, certainly not

Einstein's voice. He thinks so naturally in terms of generalizations that he even generalizes about his own mind. And why does he call the search for a second theory a "temptation" instead of simply a desire? He immediately explains: "But a simple and apparently quite reliable consideration seems to suggest that, for the present at any rate, there is little hope of success in such an attempt." Something blocks his way. There is a fact, a "reliable consideration," that gets in the way and gives even the great Einstein pause. That pause, of course, was the integrity that disciplines even the most daring scientific imaginations.

Since his death, Einstein's brain has been preserved in the hope that one day we will know enough about brains to know what made Einstein so special. Maybe one day that project will succeed. In the meantime, Einstein stands exposed in his prose.

Complex people have many voices. Anthologies, by their nature, tend to let one voice stand for the whole of an author. The only writer with multiple selections in this volume is Galileo, and by comparing the voices in these pieces, readers can see a few of the many different Galileos—Galileo the enthusiast reporting the worlds seen through a telescope; Galileo the skeptical inquirer doubting the Aristotelian account of gravity; and Galileo the confident teacher proving how bodies fall. Again, the rule is straightforward—the more contributions to science made, the more voices a science writer is likely to display.

Science Reading

The rewards of reading science writing are two sided. First is the contact with voices, imaginations, and inspirations that can be made only through reading. The second is a new understanding of nature that is clear enough to bring delight. Understanding often comes only after a struggle, but then success eases the spirit and the labor becomes a source of pride. Yet the fact that understanding requires an effort makes a special demand on readers. Literature always poses a challenge to readers by asking if they are honest enough to follow the author's lead. Even though the author has been ruthless enough to break with clichés, that effort does not spare readers. To join in the author's understanding, readers must be equally ruthless with themselves.

Please note that I have promised understanding, not truth. Scientists, of course, want the truth, and science writers try to report facts while recognizing opinion, but the essays in this collection are amalgams of insight and error. Even when they think scientifically, people are likely to get it wrong. Herodotus thought the Nile Valley used to be a lake. It was not. Isaac Newton denied that light travels in waves. Robert Hooke did not believe that light came in particles. Both were wrong. Ernest Rutherford maintained that a chain reaction could not produce an atomic explosion. Fred Hoyle wrote equations that described fictitious ways that the universe spawns galaxies. Rachel Carson's wonderful account of the seafloor left out the fact that the floor was expanding. Error is inherent in this genre.

It turns out, though, that understanding can grow even when it is planted in a garden of errors. The idea sounds surprising, but a glance at history shows what I mean. Herodotus was wrong about the lake, but he was correct that rivers can effect great changes. His was an idea to build on, and other essays in this book like Saussure's on glaciers and Carson's on the seafloor are variations on Herodotus's theme of natural movement and environmental change. Newton and Hook were wrong to deny the other's conclusion, but between them they worked out the facts about light. The essays in this collection about quantum physics could not have been written without Newton's and Hook's errors.

So even without a final truth, there is such a thing as scientific progress and growth in understanding the natural world. By now, progress has accelerated to such a speed that great changes in understanding take far less time than a human life. Just since the middle of the twentieth century, we have discovered how life can reproduce itself, why Africa and South America have coasts that fit together like pieces of a jigsaw puzzle, what the earliest eons of the universe were like, and why opium is enjoyable. Viewed more closely, each of these recent achievements includes dubious points that must be either clarified or revised or maybe even rewritten, yet our understanding has grown.

Organization of This Book

I have organized the book to show how understanding progresses from bafflement to mastery. The book has three parts: an opening section in which scientists try to understand science, a long middle section in which readers can see the many ways that scientific imaginations try to understand nature, and a third section in which writers transform scientific efforts into literary achievements.

The essays in Part One ("The Scientific Imagination Examined") are self-conscious about science. The authors had scientific arguments to make, but they also wanted to consider how science itself prospers.

Part Two ("The Scientific Imagination in Action") contains about two-thirds of the book's selections and is the heart of this anthology. The pieces have been placed along a spectrum that displays the breadth of the scientific imagination. At one end of this range are the observers who saw much but understood little, and at the other end are the great generalizers, the superstars of science, who could explain things on an abstract level.

The third part ("Style in the Scientific Imagination") contains the most self-consciously literary writing, but it is still science. These writers present narratives about how the world works, and they contribute to science by teaching.

A Note on Cuts

Readers will find bracketed ellipses [. . .] scattered throughout the selections, which mean that words have been cut from the original. (If you see just . . . without the brackets, the ellipsis was in the original.) The main reason for the cuts was space. Sometimes an author made a reference to a remark not found in this anthology. For example, a selection from a book might say, "Atoms, as we saw in chapter 5, are small." But chapter 5 has not been included in this volume, so the selection reads, "Atoms [. . .] are small." And sometimes cuts were made because the author was just too long-winded, telling us more about frogs or such than most people care to know.

GALILEO'S COMMANDMENT

Science knows only one commandment: contribute to science.
— Bertolt Brecht, *The Life of Galileo*

The Scientific Imagination Examined

"This vain presumption of understanding everything can have no other basis than never understanding anything. For anyone who had experienced just once the perfect understanding of one single thing, and had truly tasted how knowledge is accomplished, would recognize that infinity of other truths of which he understands nothing."

—GALILEO, *THE TWO CHIEF WORLD SYSTEMS*

The selections in this first part are the most philosophical in the book, although they are not philosophical essays. The contributors to this section wanted to understand science itself, so they examined how science succeeds and where it spins its wheels. Perhaps the section's archetypal essay is the piece by the Swiss psychologist Jean Piaget, because it is simultaneously an account of scientific research and a meditation on one aspect of the scientific imagination. Piaget is trying to understand why children do not see the absurdity of certain remarks, but at the same time he knows that his findings have implications for the kind of intellectual playfulness that scientists need if they are to think freely.

Instead of asking Socrates' questions about what knowledge is and why knowing is possible, Piaget looks at the more practical issue of identifying barriers to understanding. All the other

contributors to this part also discuss the routes to understanding and
the sources of misunderstanding. They are interested in concrete success and failure, not in finding the best definitions of abstractions.
Even the selection by Karl Popper, who was a philosopher by trade,
focuses on the actions and personalities of scientists rather than on an
abstract theory of knowledge.

Taken together, the essays in this part draw a picture of the human
nature of scientific contributions. Courage, persistence, and enthusiasm are present, but so are self-deceit, prejudice, and vanity. I have put
this section at the beginning of the book to illustrate that when scientific imaginations are examined, they prove to be as complex and contradictory as anything else produced by human creativity. Yet its
humanity seems to bring strength. Scientific achievement has provided
the props that let us live our daily lives, and it does so by showing its
human origins, not by defying our natures.

The part is divided into three chapters. The first contains enthusiastic pieces about science. They ring with the confidence and the tri-
umph that Mach expressed when he said that every real problem
could be solved through close observation and searching thought.
The science in these stories seems like pure pie on earth. Ask the
question; get the answer. Even in Isaac Asimov's essay, in which a
search lasts for centuries, it is centuries of steady progress.

Chapter 2 finds the highway to understanding is a bit bumpier.
Herbert Butterfield's story of chemistry's struggle has a different fla-
vor from Asimov's account of chemical progress. The selection by
John McPhee is playful but includes a serious side in his look at how
technical language simultaneously appalls people while allowing scientists to speak precisely to one another. According to this chapter,
progress is not inevitable. The chapter's sources of scientific success—
imaginative flexibility and intellectual courage—are thus not just
repetitions of Mach's praise for observation and thought. Instead,
they show how much of the whole person is responsible for scientific
achievement.

The writers in the final chapter view their subjects with alarm. By
looking at the barriers to success, these authors make exactly the opposite contribution of those shouts of hurrah in the first chapter. The
enemy in chapter 3's accounts is a rigidity and a dishonesty that goes
beyond what Chomsky called the limits of actual knowledge. It is

ironic that the great essay on intellectual honesty was written by Francis Bacon, a politician whose career was suddenly cut short when he was convicted of bribery and locked in the Tower of London. Adding to the sauce in this story was Bacon's defense that, yes, he took the money, but, no, he never let it influence his decisions. It was the most commonplace of self-deceptions, yet for all that, Bacon knew exactly how to describe dishonesty in scientific thought.

➤⊷⊙⊶◄

"Every Real Problem Can and Will Be Solved"

Ernst Mach

ISAAC ASIMOV

Death in the Laboratory, from *The Magazine of Fantasy and Science Fiction* (1965)

Isaac Asimov (1920–1992) was one of the century's most prolific writers, with more than 500 books to his credit. They include popular science fiction novels, studies of Shakespeare, and collections of limericks, but mostly they are stories of science. Asimov is reputed to have worked on three projects at a time, using a different typewriter for each and scooting from one typewriter to another on a roller chair. How he found time to learn the facts needed for his prose was never explained, but the basis of his drive is clear enough: a passion for science as the force of reason in the face of superstition. For him, the success of science was another victory for the good guys.

In the following amusing essay, Asimov reminds us of something all too easily overlooked about science—that pursuing it can call for physical as well as intellectual courage. Besides the chemists that Asimov describes, there are famous examples like Charles Darwin's going off on an around-the-world voyage and Benjamin Franklin's risking death by electrocution to establish the nature of lightning. Alfred Wegner, a contributor to this volume, died during a scientific expedition in Greenland, and Marie Curie, another contributor, died of cancer caused by her long studies of radium.

I'm a great one for iconoclasm. Given half a chance, I love to say something shattering about some revered institution, and wax sarcastically cynical about Mother's Day or apple pie or baseball. Naturally, though, I draw the line at having people say nasty things about institutions I personally revere.

Like Science and Scientists, for instance. (Capital S, you'll notice.)

Scientists have their faults, of course. They can be stodgy and authoritarian and theories can get fixed in place and resist dislodging [. . .] but not as often in science (I like to think) as in any other form of human endeavor. [. . .]

Occasionally someone treats the discovery of xenon fluoride as an example of the manner in which stodgy theories actually inhibit experimentation.

I can hear them say it: "Stupid lazy chemists just got the idea into their heads that the noble gases formed no compounds so no one bothered to try to see if they *could* form compounds. After all, if everyone knows that something can't be done, why try to do it? And yet, if, at any time, any chemist had simply bothered to mix xenon and fluorine in a nickel container—"

It does sound very stupid of a chemist not to stumble on something that easy, doesn't it? Just mix a little xenon and fluorine in a nickel container, and astonish the world, and maybe win a Nobel Prize.

But do you know what would have happened if the average chemist in the average laboratory had tired to mix a little xenon (very rare and quite expensive, by the way) with a little fluorine? A bad case of poisoning, very likely, and, quite possibly, death.

If you think I'm exaggerating, let's consider the history of fluorine. That history does not begin with fluorine itself—a pale yellow-green assassin never seen by human eyes until eighty years ago—but with an odd mineral used by German miners about five hundred years ago.

The substance is mentioned by the first great mineralogist of modern times, George Agricola. In 1529, he described its use by German miners. The mineral melted easily (for a mineral) and when added to ore being smelted, the entire mixture melted more easily, thus bringing about a valuable saving of fuel and time.

Something which is liquid flows, and the Latin word for "to flow" is *fluere*, from which we get "fluid." [. . .] From the same root comes the word Agricola used for the mineral that liquefied and flowed so easily. That word was *fluores*.

In later years, [. . .] when it became customary to add the suffix "-ite" to the names of minerals, a new alternate name was "fluorite." [. . .]

Fluorite is still used today as a flux (or liquefier) in the making of steel. The centuries pass but a useful property remains a useful property.

In 1670, a German glass-cutter, Heinrich Schwanhard, was working with fluorite and exposing it, for some reason, to the action of strong acids. A vapor was given off and Schwanhard bent close to watch. His spectacles clouded and, presumably, he may have thought the vapor had condensed upon them.

The cloud did not disappear, however, and on close examination, the spectacles proved to have been etched. The glass had actually been partly dissolved and its smooth surface roughened.

This was very unusual, for few chemicals attack glass, which is one of the reasons chemists use glassware for their equipment. Schwanhard saw a Good Thing is this. He learned to cover portions of glass objects with wax (which protected those portions against the vapors) and etched the rest of the glass. In this way, he formed all sorts of delicate figures in clear glass against a cloudy background. He got himself patronized by the Emperor and did very well, indeed.

But he kept his process secret and it wasn't until 1725 that chemists, generally, learned of this interesting vapor.

Through the eighteenth century, there were occasional reports on fluorite. A German chemist, Andreas Sigismund Marggraf, showed, in 1768, that fluorite did not contain sulfur. He also found that fluorite, treated with acid, produced a vapor that chewed actual holes in his glassware.

However, it was a Swedish chemist, Carl Wilhelm Scheele, who really put the glass-chewing gas on the map about 1780. He, too, acidified fluorite and etched glass. He studied the vapors more thoroughly than any predecessor and maintained the gas to be an acid. Because of this, Scheele is commonly given the credit for having discovered this "fluoric acid" (as it was termed for about a quarter of a century).

The discovery, unfortunately, did Scheele's health no good. He isolated a large number of substances and it was his habit to smell and taste all the new chemicals he obtained, in order that this might serve as part of the routine characterization. Since in addition to the dangerous "fluoric acid," he also isolated such nasty items as hydrogen sulfide (the highly poisonous rotten-egg gas we commonly associate with school chemistry laboratories) and hydrogen cyanide (used in gas-chamber executions), the wonder is that he didn't die with the stuff in his mouth.

His survival wasn't total, though, for he died at the early age of forty-three, after some years of invalidship. There is no question in my mind but that his habit of sniffing and sipping unknown chemicals drastically shortened his life.

While most chemists are very careful about tasting, by the way, much more careful than poor Scheele ever was, this cannot be said about smelling, even today. Chemists may not deliberately go about sniffing at things, but the air in laboratories is usually loaded with gases and vapors and chemists often take a kind of perverse pleasure in tolerating this, and in reacting with a kind of superior professional amusement at the nonchemists who make alarmed faces and say "phew."

This may account for the alleged shortened life expectancy of chemists generally. I am not speaking of this shortened life expectancy as an established fact, please note, since I don't know that it is. I say "alleged." Still, there was a letter recently in a chemical journal by someone who had been following obituaries and who claimed that chemists died at a considerably younger age, on the average, than did scientists who were not chemists. This could be so.

There were also speculations some years back that a number of chemists showed mental aberrations in later years through the insidious long-term effects of mercury poisoning. This came about through the constant presence of mercury vapor in the laboratory, vapor that ascended from disregarded mercury droplets in cracks and corners. (All chemists spill mercury now and then.)

To avoid creating alarm and despondency, however, I might mention that some chemists lived long and active lives. The prize specimen is the French chemist Michel Eugène Chevreul, who was born in 1786 and died in 1889 at the glorious age of one hundred and three! What's more, he was active into advanced old age, for in his nineties he was making useful studies on gerontology (the study of the effect of old age on living organisms) using himself (who else) as a subject.

He attended the elaborate celebration of the centennial of his own birth and was exuberantly hailed as the "Nestor of science." [. . .]

Of course, Chevreul worked with such nondangerous substances as waxes, soaps, fats, and so on, but consider then the German chemist Robert Wilhelm Bunsen. As a young man he worked with organic compounds of arsenic and poisoned himself nearly to the point of death. At the age of twenty-five, one of those compounds exploded and caused him to lose the sight of one eye. He survived, however, and went on to attain the respectable age of eighty-eight.

Yet it remains a fact that many of the chemists, in the century after Scheele, who did major work in "fluoric acid" died comparatively young.

Once Scheele had established the gas produced from acidified fluorite to be an acid, a misconception at once arose as to its structure. The great French chemist Antoine Laurent Lavoisier had decided at just about that time that all acids contained oxygen and, indeed, the word "oxygen" is from the Greek phrase meaning "acid producer."

It is true that many acids contain oxygen (sulfuric acid and nitric acid are examples) but some do not. Consider, for instance, a compound called "muriatic acid," from a Latin word meaning "brine" because the acid could be obtained by treating brine with sulfuric acid.

It was supposed, following Lavoisier's dictum, that muriatic acid contained oxygen, was perhaps a compound of oxygen with an as-yet-unknown element called "murium." Scheele found that on treating muriatic acid with certain oxygen-containing compounds, a greenish gas was obtained. He assumed that muriatic acid had added on additional oxygen and named the gas "oxymuriatic acid."

The English chemist Humphry Davy, however, after careful work with muriatic acid was able to show that the acid did not contain oxygen. Rather it contained hydrogen and was probably a compound of hydrogen and an as-yet-unknown element. Furthermore, if oxygen combined with muriatic acid, the chances were that it combined with the hydrogen, pulling it away and leaving the as-yet-unknown element in isolation. The greenish gas which Scheele had called oxymuriatic acid was, Davy decided, that element, and in 1810 he renamed it "chlorine" from the Greek word for "green" because of its color.

Since muriatic acid is a compound of hydrogen and chlorine, it came to be known as "hydrogen chloride" (in gaseous form) or "hydrochloric acid" (in water solution).

Other acids were also found to be free of oxygen. Hydrogen sulfide and hydrogen cyanide are examples. (They are very weak acids, to be sure, but the oxygen-in-acid proponents could not fall back on the assumption that oxygen is required for strong acids, since hydrochloric acid, though not containing oxygen, is, nevertheless, a strong acid.)

Davy went on to show that fluoric acid was another example of an acid without oxygen. Furthermore, fluoric acid had certain properties that were quite reminiscent of hydrogen chloride. It occurred to a French physicist, André Marie Ampère, therefore, that fluoric acid might well be a compound of hydrogen with an element very like chlorine. He said as much to Davy, who agreed.

By 1813, Ampère and Davy were giving the new element (not yet isolated or studied) the same suffix as that possessed by chlorine in order to emphasize the similarity. The stem of the name would come from fluorite, of course, and the new element was "fluorine," a name that has been accepted ever since. Fluoric acid became "hydrogen fluoride" and fluorite became "calcium fluoride."

The problem now arose of isolating fluorine so that it might be studied. This proved to be a problem of the first magnitude. Chlorine could be isolated from hydrochloric acid by having oxygen, so to speak, snatch the hydrogen from chlorine's grip, leaving the latter isolated as the element. Oxygen was more active than chlorine, you see, and pulled more strongly at hydrogen than chlorine could.

The same procedure could not, however, be applied to hydrogen fluoride. Oxygen could not, under any conditions, snatch hydrogen from the grip of fluorine. (It was found, many years later, that elementary fluorine could, instead, snatch hydrogen from oxygen. Fluorine, in reacting with water—a compound of hydrogen and oxygen—snatches at the hydrogen with such force that the oxygen is liberated in the unusually energetic form of ozone.)

The conclusion was inescapable that fluorine was more active than chlorine and oxygen. In fact, there seemed reason to suspect that fluorine might be the most active element in existence (a deduction that later chemists amply confirmed) and that no simple chemical reaction could liberate fluorine from compounds such as hydrogen fluoride or calcium fluoride, since no other element could force hydrogen or calcium out of the strong grip of fluorine.

But then, who says it is necessary to restrict one's self to chemical reactions. In 1800, the electric battery was invented and within weeks, it had been found that an electric current passing through a compound could split it apart ("electrolysis") where ordinary chemical reactions might be able to perform that task only under extreme conditions. Water, for instance, was broken up to hydrogen and oxygen. Hydrogen (and various metals) can be made to appear at the negative electrode, while oxygen (and other nonmetals) can be made to appear at the positive electrode. [. . .]

There was no reason, it seemed to Davy, that the same technique might not work with calcium fluoride. Here the calcium would appear at the negative electrode and fluorine at the positive. He tried it and got nowhere. Oh, he might have isolated fluorine at the positive electrode, but as soon as it was formed, it attacked whatever was in sight: water, glass, even silver or platinum, which Davy had used as his container. In no time at all, Davy had fluorine compounds on his hands, but no fluorine.

It was a losing proposition in another way, too, for Davy managed to be severely poisoned during his work on fluorine compounds, through breathing small quantities of hydrogen fluoride. It didn't kill him, but it undoubtedly contributed to the fact that he died at the age of fifty after some years of invalidship.

Others were less lucky than Davy, even. In the 1830s, two English brothers, Thomas and George Knox, decided not to take it for granted that fluorine could not be liberated by chemical means (scientists are not as stodgy as their critics like to pretend). They tried to coax chlorine into reacting with mercury fluoride, accepting the mercury and liberating the fluorine. They failed, and both underwent long and agonizing sieges of hydrogen fluoride poisoning.

A Belgian chemist, P. Louyet, who had followed the attempts of the Knox brothers closely, tried to repeat their work and failed even more spectacularly. He was entirely killed by hydrogen fluoride.

One of Louyet's assistants was the French chemist Edmond Frémy. He had watched some of Louyet's experiments and decided that trying to isolate fluorine by chemical reactions got one nothing but a ticket to the morgue. He returned to Davy's electrolytic method and worked with the most gingerly caution. His reward was that he lived to be eighty.

In 1885, he repeated Davy's attempt to electrolyze calcium fluoride with the same results—any fluorine that developed tackled everything in reach and was gone at once.

He next decided to work with hydrogen fluoride itself. Hydrogen fluoride is a liquid at slightly less than room temperatures and can be more easily dealt with. It needn't be kept red-hot during the electrolysis as calcium fluoride has to be.

Unfortunately, hydrogen fluoride in Frémy's day was always obtained in water solution. To try to electrolyze a water solution of hydrogen flouride meant that two different elements could come off at the positive electrode, oxygen or fluorine. Since oxygen was less active and more easily pulled away from hydrogen, only oxygen appeared at the electrode if there was even a small quantity of water present in the hydrogen fluoride.

Frémy therefore worked out methods for producing completely water-free hydrogen fluoride: "anhydrous hydrogen fluoride." He was the first to do so. Unfortunately, he found himself stymied. Anhydrous hydrogen fluoride would not pass an electric current. If he added some water, an electric current would pass—but only oxygen would be produced.

In the end, he, too, gave up and as the 1880s dawned, fluorine was still victor. It had defeated the best efforts of many first-class chemists for three-quarters of a century; had invalided some and killed others outright.

Frémy had a student, the French chemist Ferdinand Frédéric Henri Moissan, who took up the battle, and proceeded to attack the fluorine problem with bulldog tenacity.

He went back to chemical methods once again. He decided he must begin with a fluorine compound that was relatively unstable. The more stable a compound after all, the more tightly fluorine is holding the other atoms and the more difficult it is to pry that fluorine loose.

In 1884, Moissan came to the conclusion that phosphorus fluoride was comparatively unstable (for a fluoride). This seemed particularly hopeful since phosphorus happened to be unusually avid in its tendency to combine with oxygen. Perhaps in this case, oxygen could pull atoms away from fluorine. Moissan tried and succeeded only partially. The oxygen grabbed at the phosphorus all right but the fluorine did not let go and Moissan ended with a compound in which phosphorus was combined with both oxygen and fluorine.

Moissan tried another tack. Platinum is an extremely inert metal; even fluorine attacks it only with difficulty. Hot platinum, however, does seem to have the ability to combine easily with phosphorus. If he passed phosphorus fluoride over hot platinum, would the platinum perhaps combine with phosphorus rather than with fluorine, and set the fluorine free?

No such luck. Both phosphorus and fluorine combined with the platinum and in a matter of minutes a lot of expensive platinum was ruined for nothing. (Fortunately for Moissan, he had a rich father-in-law, who subsidized him generously.)

Moissan, like Frémy before him, decided to back away from straight chemistry and try electrolysis.

He began with arsenic fluoride and after fiddling with that, unsuccessfully, he decided to abandon that line of investigation because he was beginning to suffer from arsenic poisoning. So he turned to hydrogen fluoride (and underwent four different episodes of hydrogen fluoride poisoning, which eventually helped bring him to his death at the age of fifty-four).

Moissan remembered perfectly well that Frémy's anhydrous hydrogen fluoride would not carry an electric current. Something had to be added to make it do so, but not something that would offer an alternate element for production at the positive electrode. Why not another fluoride? Moissan dissolved potassium hydrogen fluoride in the anhydrous hydrogen fluoride and had a mixture which could pass a current and which could produce only fluorine at the positive electrode.

Furthermore, he made use of equipment built up out of an alloy of platinum and iridium, an alloy that was even more resistant to fluorine than platinum itself was.

Finally, he brought his entire apparatus to −50° C. All chemical reactions are slowed as temperature decreases and at −50° C even fluorine's savagery ought to be subdued.

Moissan turned on the current and hydrogen bubbled off the negative electrode like fury, but nothing showed at the positive electrode. He stopped to think. The positive electrode was inserted into the platinum-iridium vessel through a stopper. The stopper had to be an insulator so it couldn't be platinum or any metal; and that stopper had been eaten up by fluorine. No wonder he hadn't gotten any gas.

Moissan needed a stopper made of something that would not carry an electric current and would be untouched by fluorine. It occurred to him that the mineral fluorite already had all the fluorine it could carry and would not be attacked further. He therefore carefully carved stoppers out of fluorite and repeated the experiment.

On June 6, 1886, he obtained a pale yellow-green gas about his positive electrode. Fluorine had finally been isolated, and when Moissan later repeated the experiment in public, his old teacher Frémy watched.

Moissan went on, in 1899, to discover a less expensive way of producing fluorine. He made use of copper vessels. Fluorine attacked copper violently, but after the copper was overlaid with copper fluoride, no further attacks need be expected. In 1906, the year before his death, Moissan received the Nobel Prize in chemistry for his feat.

Even so, fluorine remained the bad boy of the table of elements for another generation. It could be isolated and used; but not easily and not often. Most of all, it couldn't be handled with anything but supreme caution for it was even more poisonous than hydrogen fluoride.

Meanwhile the noble gases were discovered in the 1890s and although they were recognized as being extremely inert, chemists tried over and over to force them into some kind of compound formation. (Don't believe the myth that chemists were so sure that the noble gases wouldn't react that they never tried to test the fact. Dozens of compounds were reported in the literature—but the reports, until quite recently, always proved to be mistaken.)

It wasn't until the early 1930s that chemical theory had been developed to the point where one need not tackle the noble gases at random in an effort to form compounds. The American chemist Linus Pauling, in 1933, was able to show, through logical arguments, that xenon ought to be able to form compounds with fluorine. Almost at once two chemists at Pauling's school, the California Institute of Technology, took up the challenge. They were Donald M. Yost and Albert L. Kaye.

All the xenon they could get hold of was 100 cc worth at normal air pressure and they could get hold of no fluorine. They had to rig up a device of their own to prepare fluorine; and it worked only intermittently. Doing the best they could, they found they could obtain no clear signs of any compound. Neither were they completely certain that no compound had been formed. The results were inconclusive.

There was no immediate follow-up. The results didn't warrant it. Chemists knew the murderous history of fluorine, and enthusiasm for such experiments ran low.

During World War II, fluorine was needed in connection with atomic bomb research. Under that kind of pressure, methods for the production of fluorine in quantity, and in *reasonable security* were developed.

By the 1950s, it was finally possible to run nonmilitary experiments involving fluorine without much risk of suicide. Even then, there were only a few laboratories equipped for such work and those had a great many things to do with fluorine other than mixing them with noble gases.

"Just mix xenon and fluoride in a nickel container" indeed. It could not have been done, in reasonable safety, and with reasonable hopes of success, any more than ten years before it actually was done in 1962; and, under the circumstances, the ten years' delay was a remarkably reasonable one and reflects no discredit whatever on Science.

ARTHUR S. EDDINGTON

The Story of Algol, from
Stars and Atoms (1927)

S ir Arthur Stanley Eddington (1882–1944) was a physicist, astronomer, and popular science writer. In 1919 he turned Albert
Einstein into a celebrity by measuring the bending of starlight during
a solar eclipse. Einstein had said that gravity works as a kind of lens,
changing the direction in which light travels, and he specified with
mathematical exactness just how far the light would be bent. The
irony of the research was that Eddington had conducted the experiment as a nonbeliever, expecting that his measurements would prove
Einstein wrong. Instead, he sent a telegram telling Einstein that the
effect was precisely as he had predicted, and Einstein was suddenly
known around the world as the man who had overthrown Newton.

In his own work, Eddington was best known for having described the evolution and composition of the stars, establishing
information that only a few decades earlier had been considered
unknowable.

Eddington was also a great writer with a talent for creating
images that make it easy for general readers to grasp abstract data.
In this selection he finds the reality behind a flickering light, and in
so doing, he shows how scientific success builds on itself. The science described in this essay—calculating the details of an inter-
twined set of stars—depended on an implicit trust in established
theories of gravity, motion, and so on. In the popular imagination,
scientific success often means overthrowing established ideas, but
here Eddington shows how success can come through the clever use
of existing knowledge. For most of us, the light coming from existing stars seems hopelessly simple, but Eddington explains that by
measuring details like the stars' mix of wavelengths (their spectra),
any pulses, and their "redshift" (distortions of wavelengths), a
wealth of information can be decoded.

This is a detective story, which we might call 'The Missing Word
and the False Clue.'

In astronomy, unlike many sciences, we cannot handle and probe
the objects of our study; we have to wait passively and receive and

decode the messages that they send to us. The whole of our information about the stars comes to us along rays of light; we watch and try to understand their signals. There are some stars which seem to be sending us a regular series of dots and dashes—like the intermittent light from a lighthouse. We cannot translate this as a morse code; nevertheless, by careful measurement we disentangle a great deal of information from the messages. The star Algol is the most famous of these 'variable stars.' We learn from the signals that it is really two stars revolving round each other. Sometimes the brighter of the two stars is hidden, giving a deep eclipse or 'dash;' sometimes the faint star is hidden, giving a 'dot.' This recurs in a period of 2 days 21 hours—the period of revolution of the two stars.

There was a great deal more information in the message, but it was rather tantalizing. There was, so to speak just one word missing. If we could supply that word the message would give full and accurate particulars as to the size of the system—the diameters and masses of the two components, their absolute brightness, the distance between them, their distance from the sun. Lacking the word the message told us nothing really definite about any of these things.

In these circumstances astronomers would scarcely have been human if they had not tried to guess the missing word. The word should have told us how much bigger the bright star was than the fainter, that is to say, the ratio of the masses of the two stars. Some of the less famous variable stars give us complete messages. (These could accordingly be used for testing the relation of mass and absolute brightness, and are represented by triangles in the figure.) The difficulty about Algol arose from the brightness of the bright component which swamped and made illegible the more delicate signals from the faint component. From the other systems we could find the most usual value of the mass ratio, and base on that a guess as to its probable value for Algol. Different authorities preferred slightly different estimates, but the general judgement was that in systems like Algol the bright component is twice as massive as the faint component. And so the missing word was assumed to be 'two;' on this assumption the various dimensions of the system were worked out and came to be generally accepted as near the truth. That was sixteen years ago.

In this way the sense of the message was made out to be that the brighter star had a radius of 1,100,000 kilometers (one and a half times the sun's radius), that it had half the mass of the sun, and thirty

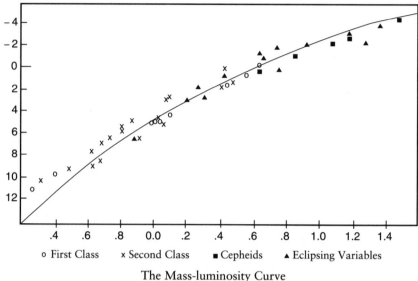

o First Class   x Second Class   ■ Cepheids   ▲ Eclipsing Variables

The Mass-luminosity Curve

times the sun's light power, &c. It will be seen at once that this will
not fit our curve in the figure; a star of half the sun's mass ought to be
very much fainter than the sun. It was rather disconcerting to find so
famous a star protesting against the theory; but after all the theory is
to be tested by comparison with facts and not with guesses, and the
theory might well have a sounder basis than the conjecture as to the
missing word. Moreover, the spectral[1] type of Algol is one that is not
usually associated with low mass, and this cast some suspicion on the
accepted results. [. . .]

We can try in succession various guesses [as to a better "missing
word"] until we reach one that gives the bright component a mass
and luminosity agreeing with the curve in the figure. The guess 'two'
gives, as we have seen, a point which falls a long way from the
curve. Alter the guess to 'three' and recalculate the mass and brightness on this assumption; the corresponding point is now somewhat
nearer to the curve. Continue with 'four', 'five', &c.; if the point
crosses the curve we know that we have gone too far and must take
an intermediate value in order to reach the desired agreement. This
was done in November 1925, and it appeared that the missing word

---

[1] *Spectral type. For a charming discussion of what can be learned and said about stars,
based on the spectra they create, see the essay by Anne Cannon.*

must be 'five', not 'two'—a rather startling change. And now the message ran—

Radius of bright component = 2,140,000 kilometres.
Mass of bright component = 4.3 × sun's mass.

If you compare these with the original figures you will see that there is a great alteration. The star is now assigned a large mass much more appropriate to a B-type[2] star. It also turns out that Algol is more than a hundred times as bright as the sun; and its parallax[3] is 0.028″—twice the distance previously supposed.

At the time there seemed little likelihood that these conclusions could be tested. Possibly the prediction as to the parallax might be proved or disproved by a trigonometrical determination; but it is so small as to be almost out of range of reasonably accurate measurement. We could only adopt a 'take it or leave it' attitude—'If you accept the theory, this is what Algol is like; if you distrust the theory, these results are of no interest to you.'

But meanwhile two astronomers at Ann Arbor Observatory had been making a search for the missing word by a remarkable new method. They had in fact found the word and published it a year before, but it had not become widely known. If a star is rotating, one edge or 'limb' is coming towards us and the other going away from us. We can measure speeds towards us or away from us by means of the Doppler effect on the spectrum,[4] obtaining a definite result in miles per second. Thus we can and do measure the equatorial speed of rotation of the sun by observing first the east limb then the west limb and taking the difference of velocity shown. That is all very well on the sun, where you can cover up the disk except the special part that you want to observe; but how can you cover up part of a star when a star is a mere point of light? *You* cannot; but in Algol the covering up is done for you. The faint component is your screen. As it passes in front of the bright star there is a moment when it leaves a

---

[2] *B-type stars are the hottest ones.*

[3] *Parallax. In this context, the term refers to the distance between the two stars in question, as seen from the earth.*

[4] *The "Doppler effect on the spectrum" refers to the distortions in wavelength measurement that arise from the expansion of the universe.*

thin crescent showing on the east and another moment when a thin crescent on the west is uncovered. Of course, the star is too far away for you actually to see the crescent shape, but at these moments you receive light from the crescents only, the rest of the disk being hidden. By seizing these moments you can make the measurements just as though you had manipulated the screen yourself. Fortunately the speed of rotation of Algol is large and so can be measured with relatively small error. Now multiply the equatorial velocity by the period of rotation; that will give you the circumference of Algol. Divide by 6.28, and you have the radius.[5]

That was the method developed by Rossiter and McLaughlin. The latter who applied it to Algol found the radius of the bright component to be 2,180,000 kilometres.

So far as can be judged his result has considerable accuracy; indeed it is probable that the radius is now better known than that of any other star except the sun. If you will now turn back [to p. 19] and compare it with the value found from the theory you will see that there is cause for satisfaction. McLaughlin evaluated the other constants and dimensions of the system; these agree equally well, but that follows automatically because there was only one missing word to be supplied. In both determinations the missing word or mass ratio turned out to be 5.0.

This is not quite the end of the story. Why had the first guess at the mass ratio gone so badly wrong? We understand by now that a disparity in mass is closely associated with a disparity in brightness of the two stars. The disparity in brightness was given in Algol's original message; it informed us that the faint component gives about one-thirteenth of the light of the bright one. (At least that was how we interpreted it.) According to our curve this corresponds to a mass ratio 2½, which is not much improvement on the original guess 2. For a mass ratio 5 the companion ought to have been much fainter—in fact its light should have been undetectable. Although considerations like these could not have had much influence on the original guess, they seemed at first to reassure us that there was not very much wrong with it.

---

[5] 6.28 is 2 times π. Eddington is using a formula so basic that even today it is still taught in schools: The circumference of a circle equals the radius times 2π.

Let us call the bright component Algol A and the faint component Algol B. Some years ago a new discovery was made, namely Algol C. It was found that Algol A and B together travel in an orbit round a third star in a period of just under two years—at least they are travelling round in this period, and we must suppose that there is something present for them to revolve around. Hitherto we had believed that when Algol A was nearly hidden at the time of deepest eclipse all the remaining light must come from Algol B; but now it is clear that it belongs to Algol C, which is always shining without interference. Consequently the mass ratio 2½ is that of Algol A to Algol C. The light from Algol B is inappreciable as it should be for a mass ratio 5.

The message from Algol A and B was confused, not only on account of the missing word, but because a word or two of another message from Algol C had got mixed up with it; so that even when the missing word was found to be 'five' and confirmed in two ways, the message was not quite coherent. In another place the message seemed to waver and read 'two-and-a-half'. The finishing step is the discovery that 'two-and-a-half' belongs to a different message from a previously unsuspected star, Algol C. And so it all ends happily.

The best detective is not infallible. In this story our astronomical detective made a reasonable but unsuccessful guess near the beginning of the case. He might have seen his error earlier, only there was a false clue dropped by a third party who happened to be present at the crime, which seemed to confirm the guess. This was very unlucky. But it makes all the better detective story of it.

⊱──⊰

ERNST MACH

A New Sense, from
Popular Scientific Lectures (1897)

E rnst Mach (1838–1916) was driven by a desire to bring physics
and psychology into harmony. He often complained that when
physicists talked about nature, they spoke about mechanics and
absolutes, but the moment they began talking about people, their
language changed completely. He wanted to change the way we talk
about both nature and ourselves, so that people would refer to the
same world when they discussed either physics or humanity.

Mach was influential in both fields of study. The American
psychologist William James visited him in Prague, where Mach
spent most of his career, and considered him a leading psychologist. Einstein thought of him as a great physicist who had pio-
neered relativistic thinking. Mach did much to develop an understanding of sensory perception. His research measuring the speed
of sound, giving us the Mach number, was a typical Mach project
that linked the pure physics of traveling energy with the purely psychological sensation of sound.

Mach was a popular lecturer and a great proponent of the idea
that science was the one way to understanding. His lectures were
widely published and stand as a testament to sharp scientific thinking. The English translation remained in print for decades and went
through many editions. In this selection from his lectures, Mach
links his knowledge of physics—the forces at play during motion—
with his curiosity about perception. His research identified a sense
that Aristotle had missed, the sense of being oriented in space. As
Mach says, several other people made the discovery at the same
time. His point in this lecture is not to claim priority (a claim he did
not make) but to assert his confidence in science's ultimate success
at explaining nature. (Note: Mach's translator did not do him justice, and this translation from the German is slightly revised from
the version published a century ago.)

You know the peculiar sensation that one has in falling, as when
one jumps from a high springboard into the water, and that is also
experienced in some measure at the beginning of the descent of eleva-

tors and swings. The reciprocal gravitational pressure on the different parts of our body, which is certainly felt in some manner, vanishes in free descent, or, in the case of the elevator, is diminished on the beginning of the descent. A similar sensation would be experienced if we were suddenly transported to the moon where the acceleration of gravity is much less than upon the earth. [. . .]

Whether we move by means of our legs, or we are moved by a vehicle or a boat, at first only a part of our body is directly moved and the rest of it is afterwards set in motion by the first part. We see that pressures, pulls, and tensions are always produced between the parts of the body in this action, which pressures, pulls, and tensions give rise to sensations by which the forward or rotary movements in which we are engaged are made perceptible. But it is quite natural that sensations so familiar should be little noticed and that attention should be drawn to them only under special circumstances when they occur unexpectedly or with unusual strength.

Thus my attention was drawn to this point by the sensation of falling and subsequently by another singular occurrence. I was rounding a sharp railway curve once when I suddenly saw all the trees, houses, and factory chimneys along the track swerve from the vertical and assume a strikingly inclined position. What had hitherto appeared to me perfectly natural, namely, the fact that we distinguish the vertical so perfectly and sharply from every other direction, now struck me as enigmatical. Why is it that the same direction can now appear vertical to me and now cannot? By what is the vertical distinguished for us? (Compare [the figure below.])

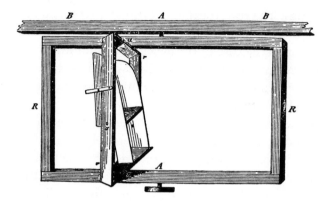

The rails are raised on the [. . .] outward side of the track to insure the stability of the carriage [. . .] against the action of the centrifugal force, the whole being so arranged that the combination of the force of gravity with the centrifugal force of the train shall give rise to a force perpendicular to the plane of the rails.

Let us assume, now, that under all circumstances we somehow sense the direction of the total resultant mass-acceleration, whencesoever it may arise, as the vertical. Then, both the ordinary and the extraordinary phenomena will be alike rendered intelligible.

I now wanted to put the view I had reached to a more convenient and exact test than was possible on a railway journey where one has no control over the determining circumstances and cannot alter them at will. I accordingly had the simple apparatus constructed that is represented in [the figure above].

In a large frame, which is fastened to the walls, rotates about a vertical axis a second frame, and within the latter a third one, which can be set at any distance and position from the axis, made stationary or movable, and is provided with a chair for the observer.

The observer takes his seat in the chair and to prevent disturbances of judgment is enclosed in a paper box. If the observer, together with the [inner] frame, is then set in uniform rotation, he will feel and see the beginning of the rotation both as to direction and amount very distinctly although every outward visible or tangible point of reference is wanting. If the motion is continued uniformly, the sensation of rotation will gradually cease entirely and the observer will imagine himself at rest. But if [the inner frame] be placed outside the axis of rotation, at once on the rotation beginning, a strikingly apparent, palpable, actually visible inclination of the entire paper box is produced, slight when the rotation is slow, strong when the rotation is rapid, and continuing as long as the rotation lasts. It is absolutely impossible for the observer to escape perceiving the inclination, although here also all outward points of reference are wanting. If the observer, for example, is seated so as to look towards the axis, he will feel the box strongly tipped backwards, as it necessarily must be if the direction of the total resultant force is perceived as the vertical. For other positions of the observer the situation is similar.

Once, while performing one of these experiments and after rotating so long that I was no longer conscious of the movement, I suddenly caused the apparatus to be stopped, whereupon I immediately felt and saw myself with the whole box rapidly flung round in rotation in the opposite direction, although I knew that the whole apparatus was at rest and every outward point of reference for the perception of motion was wanting. Everyone who disbelieves in sensations of movement should be made acquainted with these phenomena. Had Newton known them and had he ever observed how we may actually imagine ourselves turned and displaced in space without the assistance of stationary bodies as points of reference, he would certainly have been confirmed more than ever in his unfortunate speculations regarding absolute space.

The sensation of rotation in the opposite direction after the apparatus has been stopped, slowly and gradually ceases. But on accidentally inclining my head once during this occurrence, the axis of apparent rotation was also observed to incline in exactly the same manner both as to direction and to amount. It is accordingly clear that the acceleration or retardation of rotation is felt. The acceleration operates as a stimulus. The sensation, however, like almost all sensations, though it gradually decreases, lasts perceptibly longer than the stimulus. Hence the long continued apparent rotation after the stopping of the apparatus. The organ, however, which causes the persistence of this sensation must have its seat in the *head*, since otherwise the axis of apparent rotation could not assume the same motion as the head.

If I were to say, now, that a light had flashed upon me in making these last observations, the expression would be a feeble one. I ought to say I experienced a perfect illumination. My juvenile experiences of vertigo occurred to me. I remembered Flourens's[6] experiments relative to the section of the semi-circular canals of the labyrinths of doves and rabbits, where this inquirer had observed phenomena similar to vertigo, but which he preferred to interpret, from his bias to the acoustic theory of the labyrinth, as the expression of painful auditory disturbances. I saw that Goltz[7] had nearly, but not quite, hit the bull's eye with his theory of the semi-circular canals. This inquirer, who from his happy habit of following his own natural thoughts without regard for tradition, has cleared up so much in science, spoke, as early as 1870, on the ground of experiments, as follows: "It is uncertain whether the semi-circular canals are auditory organs or not. In any event they form an apparatus which serves for the preservation of equilibrium. They are, so to speak, the sense-organs of equilibrium of the head and indirectly of the whole body." I remembered the galvanic dizziness which had been observed by [Johan] Ritter and [Jan] Purkinje on the passage of a current through the head, when the persons experimented upon imagined they were falling towards the cathode.[8] The experiment was immediately repeated and sometime later (1874) I was able to demonstrate the same objectively with fishes, all of which placed themselves sidewise and in the same direction in the field of the current as if at command. [. . .]

Let us picture to ourselves the labyrinth of the ear with its three semi-circular canals lying in three mutually perpendicular planes (Fig. 3), the mysterious position of which inquirers have endeavored to explain in every possible and impossible way. Let us conceive the nerves of the ampullae, or the dilated extensions of the semi-circular canals, equipped with a capacity for responding to every imaginable stimulus with a sensation of rotation just as the nerves of the retina of the eye when excited by pressures, by electrical or chemical stimuli always respond with the sensation of light; let us picture to ourselves,

---

[6] *Marie-Jean Pierre Flourens (1794–1867), a French physiologist whose pioneering research on pigeons established basic brain anatomy (1824). He demonstrated that damage to the semicircular canals of the inner ear had no effect on hearing.*

[7] *Friedrich Goltz (1834–1902), a German physiologist who proposed the "hydrostatic concept" that the semicircular canals are stimulated by the weight of their fluid. Mach, Josef Breuer, and Crum Brown, each working independently, proposed the "hydrodynamic concept" that the semicircular canals are stimulated by the motion of the fluid in them.*

[8] *Falling toward the source of the electric current.*

further, that the unusual excitation of the ampullae nerves is produced by the inertia of the contents of the semi-circular canals. [. . .] It will be seen that on this supposition all the single facts which, without the theory, appear as so many different individual phenomena, become from this single point of view clear and intelligible.

I had the satisfaction, immediately after the communication in which I set for this idea,[9] of seeing a paper by [Josef] Breuer[10] appear in which this author had arrived by entirely different methods at results that agreed in all essential points with my own. A few weeks later appeared the researches of Crum Brown of Edinburgh, whose methods were even still nearer mine.[11] [. . .]

Aristotle has said that "The sweetest of all things is knowledge." And he is right. But if you were to suppose that the *publication* of a new view were productive of unbounded sweetness, you would be mightily mistaken. No one disturbs his fellow men with a new view unpunished. Nor should the fact be made a subject of reproach to these fellow men. To presume to revolutionize the current way of thinking with regard to any question is no pleasant task, and above all not an easy one. They who have advanced new views know best what serious difficulties stand in their way. With honest and praise-

---

[9] *Mach published his paper in November 1873.*

[10] *The same Breuer who later worked with Sigmund Freud.*

[11] *The matching results of Mach, Breuer, and Brown is an excellent example of the shared discoveries of scientific legend in which a timeless mystery is suddenly understood by several independent researchers.*

worthy zeal, men set to work in search of everything that does not suit with them. They seek to discover whether they cannot explain the facts better or as well, or approximately as well, by the traditional views. And that, too, is justified. But at times some extremely artless animadversions are heard that almost nonplus us. "If a sixth sense existed it could not fail to have been discovered thousands of years ago." Indeed; there was a time, then, when only seven planets could have existed! But I do not believe that any one will lay any weight on the philological question of whether the set of phenomena which we have been considering should be called a sense. The phenomena will not disappear when the name disappears. It was further said to me that animals exist that have no labyrinth, but which can yet orient themselves, and that consequently the labyrinth has nothing to do with orientation. We do not walk forsooth with our legs, because snakes propel themselves without them! [. . .]

I should like to close with a reminiscence from the year 1863. Helmholtz's[12] *Sensation of Tone* had just been published and the function of the cochlea[13] now appeared clear to the whole world. In a private conversation that I had with a physician, the latter declared it to be an almost hopeless undertaking to seek to fathom the function of the other parts of the labyrinth, whereas I, in youthful boldness, maintained that the question could hardly fail to be solved, and that very soon, although of course I had then no glimmering of how it was to be done. Ten years later the question was substantially solved.

Today, after having tried my powers frequently and in vain on many questions, I no longer believe that we can make short work of the problems of science. Nevertheless, I should not consider an "ignorabimus" as an expression of modesty, but rather as the opposite. That expression is a suitable one only with regard to problems that are wrongly formulated and that are therefore not problems at all. Every real problem can and will be solved in due course without supernatural divination, entirely by accurate observation and close, searching thought.

---

[12] *Herman von Helmholtz (1821–1894), a German physiologist and physicist. His book on tone worked out the physics of transforming sound waves into nervous signals that the brain can use.*

[13] *The cochlea is the organ in the ear that generates the nervous signals for hearing.*

JOHN B. WATSON

The New Science of Animal Behavior, from
Harper's (1909)

Before John B. Watson (1878–1958) became famous for developing the controversial behaviorist school of psychology, he had
already enjoyed great success in devising a technique for studying the
mental experiences of animals. The selection I have chosen comes
from that prebehaviorist period and is startling in the way it both
anticipates and denies the revolution that was still in Watson's
future. He found a way to determine whether an animal sees colors,
can distinguish among musical pitches, and has other subjective
experiences beyond the direct observation of scientists. The selection
in this volume by Sir Julian Huxley uses some of the methods that
Watson reports here and shows the value of Watson's technique. The
following essay explains how an imagination for technique can contribute to science. Watson managed to obtain data supporting the
idea—previously thought untestable—that there was a real evolutionary continuity between human and animal minds.

This selection was published in 1909, and to this day, researchers interested in the subjective experiences of infants and animals continue to use techniques based on the ideas in this essay.
Four years after the article's publication, however, Watson published an essay entitled "Psychology as the Behaviorist Views It," in
which he launched behaviorism, a theory that denies there is a mind
at all. Specifically, behaviorism says that the subjective aspects of an
event—feeling thirst before drinking, for example—have no effect
on what a person or an animal does and thus are of no interest to
psychology. Watson's brilliant technique for eluding the problem
of subjectivity was turned into a theory that denied there was such
a problem.

A few decades ago we heard much of the new science of experimental psychology. The "new psychology," as it was called, flourished vigorously and supplanted its rival, the older, speculative or
metaphysical type. In its infancy this science had as its province the
experimental analysis of the human mind. As time went on it became
evident that the human mind, like the human body, had passed

through developmental stages before reaching its present relatively high stage of perfection. If this is true—and there can no longer be a doubt of it—it becomes as necessary to study the minds of animals as it is to study the human mind. This new branch of experimental psychology is called animal psychology or animal behavior. The latter term is probably the preferable one, since many biologists are studying the behavior of animals, and some of them, being ignorant of the aims and methods of experimental psychology, object to any naming of the field which would imply that its workers are in any measure psychologists. The subject is large enough, however, for both the psychologist and the biologist. The goal of both is the same—the right understanding of all the factors which enter into the development of human life.

Not later than ten years ago our knowledge of the behavior of animals consisted largely of the chance observations made by naturalists and of the anecdotes which the lovers of animals had recorded about their own pets or the pets of their neighbors. The status of animal psychology at that time was similar to that of physics when the latter science concerned itself with the question as to whether the sun revolved daily around the earth. The older naturalists, by the mistaken way in which they carried on their observations gave us what has aptly been called a supranormal psychology of animals. If a cat, which has been shut up in a room while its mistress is away, goes to the window and turns a button, thereby permitting the window to swing open, what more natural on the part of the mistress when she returns and finds the cat gone and discovers the mode of exit than to assume that the cat understood the relation existing between the button and the window and reasoned that if it turned the button the window would swing open? And when this anecdote comes to the ear of the naturalist, why does he not have the right to generalize upon this single incident and conclude that "reasoning" is a part of the cat's mental equipment? Or, if the squirrel during the time of plenty buries a store of nuts, and when the time of scarcity comes goes and scratches them up, why not assume that the squirrel "remembers" that he buried the nuts in such and such places, and realizes that if he goes to these spots he can again find food? Again, if one of these naturalists were asked whether or not animals have color vision, the reply would be: "Certainly. Is not the bull angered by the flaunting of a red rag? Does not the female bird select her mate by reason of his attractive plumage? Why else, from an evolutionary standpoint, should the males put on their gaudy plumage in the mating season?"

Gradually, in the course of time, after a sufficient number of such observations were at hand, there appeared numerous books, which took as their subject-matter, the mental life of animals. Animals high and low in the zoological scale were accredited with all the sensations which man possesses, and with many which man does not possess. It was affirmed in these book that animals consciously remember their past acts; that they have emotions similar in most respects to those displayed by man. It was even affirmed that the sentiments of justice, charity, religion, and other social virtues are not wholly wanting. These books are of no value to the science. They consist simply of tabulations of clever acts, which, if performed by man, would presumably call for the exercise of conscious processes. It is impossible to find in these publications a single carefully conducted experimental test of the acts in question. As fit companions to this type of book, we have the extravagant creations of the "literary naturalists." Unfortunately, the literary naturalists exist even at the present time. These men spend a day or two in the woods watching the animals at work and at play. They may or may not make written notes of the results of their observations, but when it comes time to write up the story their mental notes are rounded into beautiful literary form. If there arises a conflict between the fact observed and what the author really wants to say, so much the worse for the fact.

It is easy now to point out the mistake which the early naturalists made and which the untrained man makes to-day in observing the behavior of animals. In our example the cat probably did turn the button, but this part of the observation is not the whole story—by what tedious and long-drawn-out process did the animal first stumble upon the right movement? To complete the observation, it would be necessary to take a number of such animals, put them in the same situation, and observe the whole process of learning the act. In other words, we should have to resort to experiment. It was the recognition of the need of experiment which led Principal Lloyd Morgan of Bristol College, England, to separate himself some twenty years ago from the "anecdotal" school and begin an experimental investigation of the behavior of higher animals. Time and laboratory facilities were not at his disposal, but he made many tests upon chicks and upon his pet dog, which usually accompanied him on his morning walks. Morgan's work did not attract many followers until some ten years ago, when Thorndike of Columbia University took up the study in the United States. Since then many other American and Continental students have begun studies upon animals, and to-day, in this country at least, the field of animal behavior is one of the most active in the whole of science. The work is now no longer confined to the study of higher vertebrates. It has been extended until it includes the study of all living organisms.

How do we make a laboratory study of the mind of an animal? It is not possible to get into its mind and see for ourselves the drama of mental events which is taking place there, consequently how is it ever possible to get any clear insight into the working of its mind? At first sight we seem to have here an insuperable obstacle to the study. A little reflection, however, will show that we are forever debarred from studying the mind of our human neighbor in this direct way; yet surely no one in this day would be hardy enough to deny that we can and do get a very definite and scientific notion of the way our neighbor's mind works. We study our human subject in two ways—by watching what he does under given and controllable conditions, and by attending to what he says under those conditions. Still further reflection will show that speech is only a refined and highly organized way of acting or behaving. Instead of reacting with the arm or leg our human subject when speaking reacts with the muscles of his throat. If it is admitted that speech is only a refined mode of behavior (and of this there is no doubt), we are forced to the conclusion that all of our knowledge of the minds of others comes from our observations of what they do. If we control the conditions under which a human subject reacts, and record such reactions, as is done in the psychological laboratory, we get that body of knowledge which is called "human experimental psychology."

In a similar way we take our animal subjects into the laboratory, preferably when they are young (very often at birth), and watch the gradual way in which their instinctive life develops. This gives us a key to what all animals of a particular species naturally and instinctively do—*i.e.,* the acts which they perform without training, tuition, or social contact with their fellow animals. It teaches the psychologist, too, the way to go about the animal's education—*i.e.,* gives him a notion of the problems which the structural peculiarities of the animal will permit him to learn. He would not give a starfish the same problems that he would give the bird, nor the amoeba (the lowest form of animal life) the same problems as the monkey. But it is very desirable before the detailed work of any animal's education is begun to know something about the way his sense organs work. We must know the avenues through which we may appeal to him. Is our animal normal in his color vision? If not, what are the defects? Is he totally color-blind, or only partly so? How keen is his ability to discriminate between two equally illuminated circles when they differ only in size? Can he discriminate between a circle and a square, a square and a triangle, or between a circle and an ellipse? Is he more sensitive to differences in the intensity of white light than a human being, or less so? It sometimes takes months or even years for an investigator to answer accurately a single one of these questions; yet every sense organ—smell, hearing, touch, etc.—should be tested in this careful way before it can be said that we know our animal and are ready to begin his education.

How can we answer the question, for example, whether an animal has color vision? I have already said, "by watching what he does." Let me illustrate, from some of my own experiments upon the color vision of monkeys, how the psychologist makes the animal tell whether it has color vision or not. Two colors (obtained by passing sunlight through a prism), red and green, or yellow and blue, are made to illuminate two small metal food-boxes, set flush with the floor of the apparatus, the whole being placed in an otherwise dark room. In the food-box (closed by a hinged lid which the animal must pull open) illuminated by the red we place a single grape; in the box illuminated by the green we also place a grape, but arrange the conditions so that the animal can open only the box illuminated by the red light. The grape is kept in the other box so that we may be sure that the animal is not being guided by smell. We first train our animal outside the dark room to get food by pulling open these little boxes. What happens when we take him into the dark room and confront him with the two boxes illuminated by the two different lights? Obviously he can choose either of the two boxes; he does not know which one to open, no association has as yet been established between red light and food and green light and no food. As a matter of fact, he is just as likely to go to one as to the other—*i.e.,* make as many wrong choices as right. If we leave the red color on the same side always, the animal will learn to go to that side even though it is color-blind, by merely forming the habit of going to the right or left— a so-called position habit. We must guard against this by having our apparatus so arranged that we can present the red now on the right, now on the left. The animal must learn to follow the red light regardless of its position. If our tests are continued long enough, an association is established between red color and food. It required about twenty-five days for Jimmie, a rhesus monkey of mine, to form this simple association. At the end of this time he would choose the red (color with food) about ninety times out of a hundred on the average. It began to look as though he might have color vision, but our test had not been carried far enough. So far we have said nothing about the difference in the brightness or intensity of the two colors. We know that colors as they are seen in the sunlight spectrum differ to our eyes quite a little in their brightness. A color-blind person could learn to distinguish between red and green in the above test by reason of the fact that they differ so in brightness. The apparatus must be so constructed that we can easily and immediately change the brightness of either the red or the green and the monkey must be able to choose the red always under these several conditions; red very dark, green very bright; both red and green dark; both red and green bright. Many such changes were introduced into my tests in order to confuse the monkey if possible, but without much success. The monkey apparently possesses to a very high degree the ability to choose between colors.

Only a few of the animals have been tested in this careful way. When the results are all in, we will know as much about the animal type of vision as we do about the vision of a human being. The results, so far attained, seem to indicate that monkeys have color vision. The common cat is apparently defective or wholly lacking in it. Opinion is divided about dogs—the evidence seems fair, however, that the Russian hunting-dog is without it. The dancing mice (tests made by Professor Yerkes of Harvard) and white rats are very defective in color vision. We know very little about the color vision of birds, but there is some evidence that they possess it in a high degree.

It can thus be seen that when we have carefully studied the color vision in all the main types of animals we will have a set of facts giving us the complete evolutionary history of color vision. That human color vision has had such an evolutionary history comes out very clearly from the fact that studies on the human cup-shaped retina (the sensitive coat or lining of the eye) have shown it to be almost if not totally color-blind upon the periphery; sensitive only to yellow and blue in the middle region, and sensitive to all the colors of the rainbow only in a very small region near its geometric centre. When we consider this atavistic condition of our own retina the need of such studies upon the color vision of animals is fully apparent.

The students of behavior are as busily engaged in testing the other senses of animals as in testing vision. The dog has been shown to have a wonderful power of analyzing musical chords and of detecting very slight differences in pitch. Other animals have been shown to have extremely defective hearing; even some fairly highly developed animals seem to be almost, if not wholly, lacking in hearing.

Apparatus and methods are at hand for forcing the animal to tell us about the kind of world he lives in. If it is a smell world, we shall find it out. If it is a world of vision in which there are no colors, we shall not long remain ignorant. When all such evidence is in, we shall have an invaluable body of facts which will all but revolutionize the present popular way of looking upon the mind as the proud possession of the human race.

Having found by our previous studies what sense organs of our animal may be appealed to, our next step is to find out how he learns. Learning is the great problem in all human psychology, and any facts we can gather about the way the animal learns will be helpful to us. We can find such facts only by resorting to experiment. We choose some problem for the animal to solve: He must raise a latch, pull a string, slide out a bar, or thread a maze before the goal is reached—i.e., before food can be obtained. The animal, like the human being in this respect, will work at a problem only so long as it compels his interest. We must keep his interest by providing a stimulus. This stimulus may be food, escape from confinement, punishment for wrong action, etc.: So long as we keep the stimulus constant, the animal will work steadily at the task from day to day.

By such experiments we have established the fact that when animals learn to open doors, run mazes, etc., by their own unaided efforts, they achieve the first success in nearly all cases by some happy accident. If a rat is hungry and is confined in a large cage with a small box containing food which it can get access to only by raising a latch, it begins its task by the display of a repertoire of instinctive acts, common to every member of the rat race. It runs around and over the box, gnawing the wires, pushing into every mesh of the wire with its nose, clawing, etc. This random instinctive exercise of energy results early in the knowledge of the fact that the door of the box is the only movable portion. The rat's activity becomes centred here. Since the latch is attached to the small door, the chance has rapidly become better that some movement of the rat, such as butting or clawing may raise the latch from the socket. In a period of time, which may vary from two minutes to twenty or even longer this happy accident will occur, the door will then fall open, and the rat get the food. Will the animal on the second trial run immediately to the latch and raise it? I have hundreds of records, not only upon the rat, but upon other animals as well, which show that such is not the case. The individual animal may take longer on the second trial than upon the first, but the average of the second trial of a number of animals will be shorter than the average of the first. On successive trials the time of success gets shorter and shorter, until finally the animal will open the latch as soon as confronted with the box. This is the type of hundreds of similar experiments which have been made upon animals ranging in development from the monkey to the mouse. Most of the numerous acts of the trained animals on the vaudeville stage are acquired in this laborious way (as the trainer well knows, would he but confess it). Such a mode of learning is the rule, and any higher method (so called "reasoning") is an exception so rare that it is doubtful if it exists.

The question immediately arises, Cannot the animal be aided in its learning? Cannot instruction be given by the experimenter, or by another trained animal? Students of animal behavior are more or less unsettled in their minds about this question. Thorndike was not able to impart instruction to his monkeys, dogs, and cats. They were likewise unable to profit by watching a trained animal perform the act, or by being "put through" the act—*i.e.*, by having the experimenter make their limbs go through the act. My experiments upon monkeys

agree with those of Thorndike. I was not able to get them to imitate
such simple acts as lifting a latch, pushing a banana from a large glass
cylinder by means of a large, light stick, or dragging in grapes beyond
their reach by means of a light wooden rake. In the Harvard psychological laboratory results confirmative of imitation in animals have
been obtained. Berry has found that the white rat has some slight
ability to imitate and that the Manx cat is somewhat better equipped
in this respect. Latterly, Haggerty, whose article appeared in a recent
number of *The Century*, finds a fairly well advanced stage of imitation in monkeys. It is unfortunate in one sense that this conflict of
evidence exists. Fortunate in another, in that interest has been so
aroused that more work will be undertaken upon the problem. Thus
much has been unquestionably established: Animals do not possess
the function of imitating in anything like the degree which they were
formerly supposed to. The function is certainly rudimentary.

The fact that animals learn by such a "trial-and-error method"
as that just described for the rat does not mean that its methods are
far different from those of a small child. Nearly all of the early co-
ordination of infants, from the proper manipulation of the milk-
bottle onward are formed in this laborious way. Imitation appears as
an actual new method of learning only after the proper control of
the various separate muscles of the arms and legs has been learned
by the "trial-and-error method"—at the beginning of the tenth
month; it develops slowly, reaches its maximum in the third and
fourth years, when the elementary social habits, language, personal,
etc., are forming, and then declines. Reasoning in the older use of the
term appears much later. It is a very much over-rated psychological
process, even in adults. The amount of actual reasoning going on in
the minds of most of us, after the pattern of the old conventional
type—

All men are mortal.

Socrates was a man;

Therefore, Socrates was mortal—is just about *nil*. What we do is
to *think*. Past connections and associations will appear in the mind
when we are in a situation which demands some action. If the situation is wholly new, no proper action will take place. It is only when

our memory can supply the separate steps in the act demanded by the present difficult situation that we find the proper (and new) combination springing at first haltingly and then boldly into our mind. We can reason about playing the piano all we care to, but we cannot play the *Spring Song* until we have learned to control each separate movement by the slow and very same process that the animal employs. The point that I would make in all this is that there is no royal road to habit and knowledge. Man gets his first steps in exactly the same way as does the animal. Studies in animal behavior, while not fulfilling the hopes of the early students of evolution in showing that animals have exalted types of intellect, nevertheless are forcing us to reconsider our extravagant notions of the all-sufficiency of the human mind. Continuity between the mind of man and brute, the idea of the early students, will still be shown to exist, not by exalting the mind of the brute, but rather by the reverse process of showing the defects in the human mind.

It is a far cry from such theoretical questions—questions, however, which always have interested the human race and always will—to that of the practical value of animal psychology. It is not a difficult matter to show that there is a practical import to the study. At present the routine of studies in the secondary schools, high schools, and even in the colleges is based upon custom and tradition and not upon experimental studies which show its fitness. If the question were asked why we have four fifty-minute periods each day, each devoted to one topic—for example, to algebra, literature, Latin, physiography, respectively, there would be no answer unless it were that it is customary. No explanation could be vouchsafed as regards the time allotted to each subject, nor as to why these specific subjects were chosen. Experimental studies may sometimes show that quite a different routine is desirable. Certain types of study may be found to be antagonistic, certain others mutually supplementary. Educational systems dealing with that most precious article, the human child, are necessarily conservative, and are slow to introduce changes and to have resort to experiment. Fortunately, there is no such sentiment in regard to the courses of study prescribed for animals. We may vary the course of training *ad libitum*. Suppose it is desired that our animals learn five problems in the shortest time; do we get our best results by forcing them to work intensively on one problem at a time, then, when it is learned by putting them to work upon another, and so on until all five are learned; or shall they be allowed to work at regular intervals on all five each day until all are learned? Which is the more economical as regards time, and which method fixes the problem more firmly in memory? Or again, if the animal has to learn one problem at a time, shall he work upon it once each day for sixty days (assuming that he can learn it perfectly in sixty trials), or four times each day in rapid succession for fifteen days; does he learn it better by doing it a larger number of times each day for a shorter number of days, or by the other method? Which method enables the animal to retain ("remember") the act better? Such questions cannot be answered speculatively, but must be answered by actually carrying out the experiment. That the reply the animal psychologist returns to such and similar questions has an intimate bearing upon instruction in schools can readily be seen. The ability to cope with such problems gives the study of animal behavior its practical *raison d'être*.

"Language of the Sort That Would Have Attracted Gilbert and Sullivan"

John McPhee

KARL POPPER

Heroic Science, from
Replies to My Critics (1974)

Karl Popper (1902–1994) was a philosopher, not a scientist, and he was interested in philosophical questions, such as identifying the nature of truth, rather than in scientific questions. Ordinary science bored him with its mundaneness, just as most scientists are bored by philosophy because of its woolliness. Yet for all the oil-and-water quality of the Popper–science mix, Popper was the twentieth century's most important philospher of science and one of the century's leading contributors to the scientific enterprise. His contribution was unique, for it shaped the way scientists think about what they are doing. Until the middle of this century, scientists insisted that they were seeking truth and that their propositions had to be judged for their truthfulness. Popper rejected this. Who can say what is true? We can only say that we know that some things are false. For a statement to be scientific, it must be falsifiable. There must be some set of conceivable circumstances that, if they existed, would prove a theory false. Many scientists have embraced Popper with great enthusiasm, and these days the charge that a proposition is "unfalsifiable" is as ferocious a blow as science criticism can offer.

There nonetheless are skeptics. Historians of science like to point out that many great changes in science, such as the abandonment of a sun-centered universe or the acceptance of natural selection, did not come through a process of falsification. There also are many philosophical objections to Popper's theory of truth. Logical positivism, for example, so sought to determine what kinds of statements could be "verified." Popper deliberately turned that approach on its head by looking for statements that could be falsified.

Popper's success with scientists sometimes puzzles commentators, but in this selection we can see the strength of his appeal. Popper's portrayal of honest scientific effort is so intense that science looks like the most courageous activity a person can undertake. But it also is a tragic one because, in this picture, you can find out that you are certainly wrong, but you can never know that you are certainly right.

The great scientists, such as Galileo, Kepler, Newton, Einstein, and Bohr (to confine myself to a few of the dead) represent to me a simple but impressive idea of science. Obviously, no such list, however much extended, would *define* scientist or science *in extenso*. But it suggests for me an oversimplification, one from which we can, I think, learn a lot. It is the working of great scientists which I have in my mind as my paradigm for science. Not that I lack respect for the lesser ones; there are hundreds of great men and great scientists who come into the almost heroic category.

But with all respect for the lesser scientists, I wish to convey here a heroic and romantic idea of science and its workers: men who humbly devoted themselves to the search for truth, to the growth of our knowledge; men whose life consisted in an adventure of bold ideas. I am prepared to consider with them many of their less brilliant helpers who were equally devoted to the search for truth—for great truth. But I do not count among them those for whom science is no more than a profession, a technique: those who are not deeply moved by great problems and by the oversimplifications of bold solutions.

It is science in this heroic sense that I wish to study. As a side result I find that we can throw a lot of light even on the more modest workers in applied science.

This, then, for me is science. I do not try to define it, for very good reasons. I only wish to draw a simple picture of the kind of men

I have in mind, and of their activities. And the picture will be an over-simplification: these are men of bold ideas, but highly critical of their own ideas; they try to find whether their ideas are right by trying first to find whether they are not perhaps wrong. They work with bold conjectures and severe attempts at refuting their own conjectures.

My criterion of demarcation between science and nonscience is a simple logical analysis of this picture. How good or bad it is will be shown by its fertility.

Bold ideas are new, daring, hypotheses or conjectures. And severe attempts at refutations are severe critical discussions and severe empirical tests.

When is a conjecture daring and when is it not daring, in the sense here proposed? Answer: it is daring if and only if it takes a great risk of being false—if matters could be otherwise, and seem at the time to be otherwise.

Let us consider a simple example. Copernicus's or Aristarchus's[1] conjecture that the sun rather than the earth rests at the centre of the universe was an incredibly daring one. It was, incidentally, false; nobody accepts today the conjecture that the sun is (in the sense of Aristarchus and Copernicus) at rest in the centre of the universe. But this does not affect the boldness of the conjecture, nor its fertility. And one of its main consequences—that the earth does not rest at the centre of the universe but that it has (at least) a daily and an annual motion—is still fully accepted, in spite of some misunderstandings of relativity.

But it is not the present acceptance of the theory which I wish to discuss, but its boldness. It was bold because it clashed with all then accepted views, *and* with the *prima facie* evidence of the senses. It was bold because it postulated a hitherto unknown hidden reality behind the appearances.

It was not bold in another very important sense: neither Aristarchus nor Copernicus suggested a feasible crucial experiment. In fact, they did not suggest that anything was wrong with the traditional appearances: they let the accepted appearances severely alone; they only reinterpreted them. They were not anxious to stick out their necks by predicting new observable appearances.

---

[1] *Aristarchos of Samos (flourished about 270 B.C.) and Nicolas Copernicus (1473–1543).*

To the degree that this is so, Aristarchus's and Copernicus's theories may be described in my terminology as unscientific or metaphysical. To the degree that Copernicus did make a number of minor predictions, his theory is, in my terminology, scientific. But even as a metaphysical theory it was far from meaningless; and in proposing a new bold view of the universe it made a tremendous contribution to the advent of the new science.

Kepler went much further. He too had a bold metaphysical view, partly based upon the Copernican theory, of the reality of the world. But his view led him to many new detailed predictions of the appearances. At first these predictions did not tally with the observations. He tried to reinterpret the observations in the light of his theories; but his addiction to the search for truth was even greater than his enthusiasm for the metaphysical harmony of the world. Thus he felt forced to give up a number of his favoured theories, one by one, and to replace them by others which fitted the facts. It was a great and a heartrending struggle. The final outcome, his famous and immensely important three laws,[2] he did not really like—except the third. But they stood up to his severest tests—they agreed with the detailed appearances, the observations which he had inherited from Tycho.[3]

Kepler's laws are excellent approximations to what we think today are the true movements of the planets of our solar system. They are even excellent approximations to the movements of the distant binary star systems which have since been discovered. Yet they are merely *approximations* to what seems to be the truth; *they are not true.*

They have been tested in the light of new theories—of Newton's theory and of Einstein's—which predicted small deviations from Kepler's laws. (According to Newton, Kepler's laws are correct only for two-body systems.) Thus the crucial experiments went against Kepler, very slightly, but sufficiently clearly.

---

[2] *Kepler's three laws describe planetary motion. The first says that the planets do not move in circles but in ellipses, with the sun as one of the two focal points; the second is an equation that links the time of the orbit with the space covered; and the third is another equation showing the relation between the time it takes a planet to go around the sun and its distance from the sun.*

[3] *Tycho Brahe (1546–1601), an astronomer who hired Kepler as an assistant.*

Of these three theories—Kepler's, Newton's, and Einstein's—the latest and still the most successful is Einstein's; and it was this theory which led me into the philosophy of science. What impressed me so greatly about Einstein's theory of gravitation were the following points.

1. It was a very bold theory. It greatly deviated in its fundamental outlook from Newton's theory which at that time was utterly successful.

2. From the point of view of Einstein's theory, Newton's theory was an excellent approximation, though false (just as from the point of view of Newton's theory, Kepler's and Galileo's theories were excellent approximations, though false). Thus it is not its truth which decides the scientific character of a theory.

3. Einstein derived from his theory three important predictions of vastly different observable effects, two of which had not been thought of by anybody before him, and all of which contradicted Newton's theory, so far as they could be said to fall within the field of application of this theory at all.

But what impressed me perhaps most were the following two points.

4. Einstein declared that these predictions were crucial: if they did not agree with his precise theoretical calculations, he would regard his theory as refuted.

5. But even if they were observed as predicted, Einstein declared that *his theory was false*: he said that it would be a better approximation to the truth than Newton's, but he gave reasons why he would not, even if all predictions came out right, regard it as a true theory. He sketched a number of demands which a true theory (a unified field theory) would have to satisfy, and declared that his theory was at best an approximation to this so far unattained unified field theory.

(It may be remarked in passing that Einstein, like Kepler, failed to achieve his scientific dream—or his metaphysical dream: it does not matter in this context what label we use. What we call today Kepler's laws or Einstein's theory of gravitation are results which in no way satisfied their creators, who each continued to work on his dream to the end of his life. And even of Newton a similar point can be made: he never believed that a theory of action at a distance could be a finally acceptable explanation of gravity.)

Einstein's theory was first tested by Eddington's famous eclipse experiment of 1919.[4] In spite of his unbelief in the truth of his theory, his belief that it was merely a new important approximation towards the truth, Einstein never doubted the outcome of this experiment; the inner coherence, the inner logic of his theory convinced him that it was a step forward even though he thought that it could not be true. It has since passed a series of further tests, all very successfully. But some people still think the agreement between Einstein's theory and the observations may be the result of (incredibly improbable) accidents. [. . .]

The picture of science at which I have so far only hinted may be sketched as follows.

There is a reality behind the world as it appears to us, possibly a many-layered reality, of which the appearances are the outermost layers. What the great scientist does is boldly to guess, daringly to conjecture, what these inner realities are like. This is akin to myth-making. (Historically we can trace back the ideas of Newton via Anaximander to Hesiod, and the ideas of Einstein via Faraday, Boscovich, Leibniz, and Descartes to Aristotle and Parmenides.) The boldness can be gauged by the distance between the world of appearance and the conjectured reality, the explanatory hypotheses.

But there is another, a special kind of boldness—*the boldness of predicting* aspects of the world of appearance which so far have been overlooked but which it must possess if the conjectured reality is (more or less) right, if the explanatory hypotheses are (approximately) true. It is this more special kind of boldness which I have usually in mind when I speak of bold scientific conjectures. It is the boldness of a conjecture which takes a real risk—the risk of being tested, and refuted; the risk of clashing with reality.

---

[4] *Arthur Eddington (1882–1944) (a contributor to this volume) traveled to West Africa to observe a total eclipse of the sun on May 29, 1919, and to see if—as Einstein's theory of gravity predicted—starlight was deflected by the sun.*

Thus my proposal was, and is, that it is this second boldness, together with the readiness to look out for tests and refutations, which distinguishes "empirical" science from nonscience, and especially from prescientific myths and metaphysics.

I will call this proposal (D): (D) for *"demarcation"*. [. . .]

The difficulties connected with my criterion of demarcation (D) are important, but must not be exaggerated. It is vague, since it is a methodological rule, and since the demarcation between science and nonscience is vague. But it is more than sharp enough to make a distinction between many physical theories on the one hand, and metaphysical theories, such as psychoanalysis, or Marxism (in its present form), on the other. This is, of course, one of my main theses; and nobody who has not understood it can be said to have understood my theory.

The situation with Marxism is, incidentally, very different from that with psychoanalysis. Marxism was once a scientific theory: it predicted that capitalism would lead to increasing misery and, through a more or less mild revolution, to socialism; it predicted that this would happen first in the technically highest developed countries; and it predicted that the technical evolution of the "means of production" would lead to social, political, and ideological developments, rather than the other way round.

But the (so-called) socialist revolution came first in one of the technically backward countries. And instead of the means of production producing a new ideology, it was Lenin's and Stalin's ideology that Russia must push forward with its industrialization ("Socialism is dictatorship of the proletariat plus electrification") which promoted the new development of the means of production.

Thus one might say that Marxism was once a science, but one which was refuted by some of the facts which happened to clash with its predictions (I have here mentioned just a few of these facts).

However, Marxism is no longer a science; for it broke the methodological rule that we must accept falsification, and it immunized itself against the most blatant refutations of its predictions. Ever since then, it can be described only as nonscience—as a metaphysical dream, if you like, married to a cruel reality.

Psychoanalysis is a very different case. It is an interesting psychological metaphysics (and no doubt there is some truth in it, as there is so often in metaphysical ideas), but it never was a science. There may be lots of people who are Freudian or Adlerian cases: Freud himself was clearly a Freudian case, and Adler an Adlerian case. But what prevents their theories from being scientific in the sense here described is, very simply, that they do not exclude any physically possible human behaviour. Whatever anybody may do is, in principle, explicable in Freudian or Adlerian terms. (Adler's break with Freud was more Adlerian than Freudian, but Freud never looked on it as a refutation of his theory.)

The point is very clear. Neither Freud nor Adler excludes any particular person's acting in any particular way, whatever the outward circumstances. Whether a man sacrificed his life to rescue a drowning child (a case of sublimation) or whether he murdered the child by drowning him (a case of repression) could not possibly be predicted or excluded by Freud's theory; *the theory was compatible with everything that could happen—even without any special immunization treatment.*

Thus while Marxism became nonscientific by its adoption of an immunizing strategy, psychoanalysis was immune to start with, and remained so. In contrast, most physical theories are pretty free of immunizing tactics and *highly falsifiable to start with.* As a rule, *they exclude an infinity of conceivable possibilities.*

▻╺╾✦╾╼○╾╼✦╾╼╸◅

JOHN MCPHEE

Naming the Rocks, from
Basin and Range (1981)

J ohn McPhee (1931–) is a writer for *The New Yorker* magazine
who fell in love with geology, despite a background in literature.
In the 1950s he wrote scripts for live television dramas, like *Robert
Montgomery Presents*, and then served his journalism stint at *Time*
magazine. When he came to *The New Yorker*, he seemed to be able
to write about anything. Then he found geology, and his subject
became the many wonders of the natural earth.

In this selection McPhee confronts the barrier that keeps all non-
geologists at arm's length—geology's technical language. Language is
crucial to understanding, and specialized understanding calls for specialized language. Scientific language—be it geological, mathematical, chemical, or psychological—frightens the nonscientist, who believes that understanding is impossible, and draws the contempt of
the skeptic who suspects that behind those alien words are ordinary
ideas. Scientists, on the other hand, like specialized terms, for the
more precise expression they allow and the club membership they
confer. In McPhee's essay we find out how specialized speech can
confuse everybody. A later selection in this book, by Lavoisier, shows
the breakthrough side of the struggle to speak precisely.

I used to sit in class and listen to the terms come floating down the
room like paper airplanes. Geology was called a descriptive science,
and with its pitted outwash plains and drowned rivers, its hanging
tributaries and starved coastlines, it was nothing if not descriptive. It
was a fountain of metaphor—of isostatic adjustments and degraded
channels, of angular unconformities and shifting divides, of rootless
mountains and bitter lakes. Streams eroded headward, digging from
two sides into mountain or hill, avidly struggling toward each other
until the divide between them broke down, and the two rivers that did
the breaking now became confluent (one yielding to the other, giving
up its direction of flow and going the opposite way) to become a
single stream. Stream capture. In the Sierra Nevada, the Yuba had

captured the Bear. The Macho member of a formation in New Mexico
was derived in large part from the solution and collapse of another
formation. There was fatigued rock and incompetent rock and
inequigranular fabric in rock. If you bent or folded rock, the inside of
the curve was in a state of compression, the outside of the curve was
under great tension, and somewhere in the middle was the surface of
no strain. Thrust fault, reverse fault, normal fault—the two sides were
active in every fault. The inclination of a slope on which boulders
would stay put was the angle of repose. There seemed, indeed, to be
more than a little of the humanities in this subject. Geologists communicated in English; and they could name things in a manner that sent
shivers through the bones. They had roof pendants in their discordant
batholiths, mosaic conglomerates in desert pavement. There was ultra-
basic, deep-ocean, mottled green-and-black rock—or serpentine.
There was the slip face of the barchan dune. In 1841, a paleontologist
had decided that the big creatures of the Mesozoic were "fearfully
great lizards," and had therefore named them dinosaurs. There were
festooned crossbeds and limestone sinks, pillow lavas and petrified
trees, incised meanders and defeated streams. There were dike swarms
and slickensides, explosion pits, volcanic bombs. Pulsating glaciers.
Hogbacks. Radiolarian ooze. There was almost enough resonance in
some terms to stir the adolescent groin. The swelling up of mountains
was described as an orogeny. Ontogeny, phylogeny, orogeny—accent
syllable two. The Antler Orogeny, the Avalonian Orogeny, the
Taconic, Acadian, Alleghenian Orogenies. The Laramide Orogeny.
The center of the United States had had a dull geologic history—nothing much being accumulated, nothing much being eroded away. It was
just sitting there conservatively. The East had once been radical—had
been unstable, reformist, revolutionary, in the Paleozoic pulses of three
or four orogenies. Now, for the last hundred and fifty million years,
the East had been stable and conservative. The far-out stuff was in the
Far West of the country—wild, weirdsma, a leather-jacket geology in
mirrored shades, with its welded tuffs and Franciscan mélange (inter-
nally deformed, complex beyond analysis), its strike-slip faults and
falling buildings, its boiling springs and fresh volcanics, its extensional
disassembling of the earth.

There was, to be sure, another side of the page—full of geological language of the sort that would have attracted Gilbert and

Sullivan. Rock that stayed put was called autochthonous, and if it had moved it was allochthonous. "Normal" meant "at right angles." "Normal" also meant a fault with a depressed hanging wall. There was a Green River Basin in Wyoming that was not to be confused with the Green River Basin in Wyoming. One was topographical and was *on* Wyoming. The other was structural and was *under* Wyoming. The Great Basin, which is centered in Utah and Nevada, was not to be confused with the Basin and Range, which is centered in Utah and Nevada. The Great Basin was topographical, and extraordinary in the world as a vastness of land that had no drainage to the sea. The Basin and Range was a realm of related mountains that all but coincided with the Great Basin, spilling over slightly to the north and south. To anyone with a smoothly functioning bifocal mind, there was no lack of clarity about Iowa in the Pennsylvanian, Missouri in the Mississippian, Nevada in Nebraskan, Indiana in Illinoian, Vermont in Kansan, Texas in Wisconsinan time. Meteoric water, with study, turned out to be rain. It ran downhill in consequent, subsequent, obsequent, resequent, and not a few insequent streams.

As years went by, such verbal deposits would thicken. Someone developed enough effrontery to call a piece of our earth an epieugeosyncline. There were those who said interfluve when they meant between two streams, and a perfectly good word like mesopotamian would do. A cactolith, according to the American Geological Institute's *Glossary of Geology and Related Sciences*, was "a quasi-horizontal chonolith composed of anastomosing ductoliths, whose distal ends curl like a harpolith, thin like a sphenolith, or bulge discordantly like an akmolith or ethmolith." The same class of people who called one rock serpentine called another jacupirangite. Clinoptilolite, eclogite, migmatite, tincalconite, szaibelyite, pumpellyite. Meyerhofferite. The same class of people who called one rock paracelsian called another despujolsite. Metakirchheimerite, phlogopite, katzenbuckelite, mboziite, noselite, neighborite, samsonite, pigeonite, muskoxite, pabstite, aenigmatite. Joesmithite. With the X-ray diffractometer and the X-ray fluorescence spectrometer, which came into general use in geology laboratories in the late nineteen-fifties, and then with the electron probe (around 1970), geologists obtained ever closer examinations of the components of rock. What they had long seen through magnifying

lenses as specimens held in the hand—or in thin slices under microscopes—did not always register identically in the eyes of these machines.
Andesite, for example, had been given its name for being the predominant rock of the high mountains of South America. According to the
machines, there is surprisingly little andesite in the Andes. The Sierra
Nevada is renowned throughout the world for its relatively young and
absolutely beautiful granite. There is precious little granite in the
Sierra. Yosemite Falls, Half Dome, El Capitan—for the most part the
"granite" of the Sierra is granodiorite. It has always been difficult
enough to hold in the mind that a magma which hardens in the earth
as granite will—if it should flow out upon the earth—harden as rhyolite, that what hardens within the earth as diorite will harden upon the
earth as andesite, that what hardens within the earth as gabbro will
harden upon the earth as basalt, the difference from pair to pair being
a matter of chemical composition and the differences within each pair
being a matter of texture and of crystalline form, with the darker rock
at the gabbro end and the lighter rock the granite. All of that—not to
mention such wee appendixes as the fact that diabase is a special texture of gabbro—was difficult enough for the layman to remember
before the diffractometers and the spectrometers and the electron
probes came along to present their multiplex cavils. What had previously been described as the granite of the world turned out to be a
large family of rock that included granodiorite, monzonite, syenite,
adamellite, trondhjemite, alaskite, and a modest amount of true granite. A great deal of rhyolite, under scrutiny, became dacite, rhyodacite,
quartz latite. Andesite was found to contain enough silica, potassium,
sodium, and aluminum to be the fraternal twin of granodiorite. These
points are pretty fine. The home terms still apply. The enthusiasm
geologists show for adding new words to their conversation is, if anything, exceeded by their affection for the old. They are not about to
drop granite. They say granodiorite when they are in church and granite the rest of the week.

───◦───

HERBERT BUTTERFIELD

Chemistry Transformed, from
The Origins of Modern Science (1949)

In 1948 Herbert Butterfield (1900–1979), a British historian at Cambridge University, gave a series of lectures on the origins of science. They created enough of a stir to be published in a book that has remained in print ever since. Butterfield argued that "change is brought about, not by new observations or additional evidence but by transpositions taking place in the minds of the scientists themselves." Before Butterfield, the established view was that science was a series of discoveries, the amassing of facts that spoke for themselves. Rather than following an intellectual pursuit like other scholars, scientists were compared to Christopher Columbus and Captain James Cook, explorers whose indisputable discoveries merely had to be marked on a map. Many people, especially working scientists themselves, still hold tightly to the explorer metaphor, but after Butterfield published his lectures, it became difficult to deny that imaginative thinking—what Butterfield called "picking up the stick from the other end"—plays a vital part in scientific achievement. In the following excerpt, Butterfield tells how chemistry was blocked for centuries by men who—despite all their honesty, experiments, and observations—were plagued by misleading ideas and so passed their professional lives baying at an absent moon.

It has often been a matter of surprise that the emergence of modern chemistry should come at so late a stage in the story of scientific progress; and there has been considerable controversy amongst historians concerning the reasons for this. Laboratories and distilleries, the dissolution or the combination of substances and the study of the action of acid and fire—these things had been familiar in the world for a long time. By the sixteenth century there had been remarkable advances on anything that had been achieved in the ancient world in the field of what might be called chemical technology—the smelting and refining of metals, the production and the treatment of glassware, pottery and dyes, the development of such things as explosives, artists' materials and medicinal substances. It would appear that experimentation and even technological progress are insufficient by themselves to provide the basis for the establishment of what we should call a "modern science". Their results need to be related to an adequate intellectual framework which on the one hand embraces the observed data and on the other hand helps to decide at any moment the direction of the next enquiry. [. . .]

When we study the history of science, it is useful to direct our attention to the intellectual obstruction which, at a given moment, is checking the progress of thought—the hurdle which it was then particularly necessary for the mind to surmount. [. . .] In chemistry, [. . .] it would seem that the difficulty in this period[5] lay in certain primary things which are homely and familiar—things which would not trouble a schoolboy in the twentieth century, so that it is not easy for us to see why our predecessors should seem to have been so obtuse. It was necessary in the first place that they should be able to identify the chemical elements, but the simplest examples were perhaps the most difficult of all. For thousands of years, air, water and fire had been wrapped up in a myth somewhat similar to the myth of the special ethereal substance out of which the heavenly bodies and celestial spheres were thought to have been made. Of all the things in the world, air and water seemed most certain to be irreducible elements. [. . .] Even fire seemed to be another element—hidden in many substances, but released during combustion, and visibly making its escape in the form of flame. Bacon and some of his successors in the seventeenth century had conjectured that heat might be a form of motion in microscopic particles of matter. Mixed up with such conjectures, however, we find the view that it was itself a material substance; and this latter view was to prevail in the eighteenth century. Men who had made great advances in metallurgy and had accumulated much knowledge of elaborate and complicated chemical interactions, were as yet unable to straighten out their ideas on these apparently simple topics. It would appear to us today that chemistry could not be established on a proper footing until a satisfactory starting-point could be discovered for the understanding of air and water; and for this to be

---

[5] *The seventeenth and eighteenth centuries.*

achieved it would seem to have been necessary to have a more adequate idea both about the existence of "gases" and about the process of combustion. The whole development depended on the recognition and the weighing of gases; but at the opening of the eighteenth century there was no realisation of the distinctions between gases, no instrument for collecting a gas, and no sufficient consciousness of the fact that the measurements of weight might play the decisive part amongst the data of chemistry. [. . .]

A considerable amount of study had been devoted to the air. [. . .] Earlier in the century, Van Helmont had examined what in those days were regarded as "fumes", but though he discovered and described certain things which we should call "gases", he had regarded these as impurities and exhalations—as earthy matter carried by the air—and for him there was really only one "gas", which itself was only a form assumed by water, water being the basis of all material things. [. . .] The problem of the air was to be elucidated only by a more methodical handling and a more acute examination of the processes of combustion. In this connection the emergence of the phlogiston theory provides a significant moment in the history of chemistry.

This theory, which was to become so fashionable for a time in the eighteenth century, embodied the essential feature of a tradition that went back to the ancient world—namely, the assumption that, when anything burns, something of its substance streams out of it, struggling to escape in the flutter of a flame, and producing a decomposition—the original body being reduced to more elementary ingredients. The entire view was based upon one of those fundamental conclusions of commonsense observation which (like Aristotle's view of motion) may set the whole of men's thinking on the wrong track and block scientific progress for thousands of years. The theory might have represented an advance at the time when it was first put forward; but in future ages no rectification seems to have been possible save by the process of going back to the beginning again. Under the system of the Aristotelians it was the "element" of fire which had been supposed to be released during the combustion of a body. During most of the seventeenth century it was thought to be a sulphurous "element"—not exactly sulphur as we know it, but an idealised or a mystical form of it—materially a different kind of sulphur in the case of the different bodies in which it might appear. A German chemist J. J. Becher, who was contemporary with [Robert] Boyle, said in 1669, that it was *terra pinguis*—an oily kind of earth; and at the opening of the eighteenth century another German chemist, G. E. Stahl, took over this view, elaborating it down to 1731, renaming the *terra pinguis* "phlogiston", and regarding phlogiston as an actual physical substance—solid and fatty, though apparently impossible to secure in isolation. It was given off by bodies in the process of combustion, or by metal in the process of calcination, and it went out in flame to combine with air, or perhaps deposited at least a part of itself in an unusually pure form as soot. [. . .]

It had been realised all the time, and it was known to Stahl, who really developed the phlogiston doctrine, that when burning had taken place or metals had been calcined an actual increase in weight had been discovered in the residue. The fact may have been known to the Arabs; it was realised by some people in the sixteenth century; it was brought to the attention of the Royal Society in London after 1660. In the seventeenth century the view had even been put forward more than once that, in the act of burning, a substance took something out of the air, and that this process of combination accounted for the increase which was observed in the weight. The phlogiston theory—the theory that something was lost to a body in the process of burning—is a remarkable evidence of the fact that at this time the results of weighing and measuring were not the decisive factors in the formation of chemical doctrine. [. . .] [T]he phlogiston theory answered to certain *prima facie* appearances, but stood almost as an inversion of the real truth—a case of picking up the wrong end of the stick. It is remarkable how far people may be carried in the study of a science, even when an hypothesis turns everything upside-down, but there comes a point [. . .] where one cannot escape an anomaly, and the theory has to be tucked and folded, pushed and pinched, in order to make it conform with the observed facts. This happened in the case of the phlogiston theory when the scientist found it impossible to evade the fact of the augmented weight of bodies after combustion or calcination. [. . .]

The phlogiston theory had a further disadvantage in that it carried the implication that nothing which could be burned or calcined could possibly be an element. Combustion implied decomposition. Only after the removal of the phlogiston could you expect to find matter in its elementary forms. If in calcination we today see oxygen combining with a metal, the eighteenth century saw the compound body—the metal—being decomposed and deprived of its phlogiston. If in the reverse process we see the oxygen being removed from a lead oxide to recover the original element, they imagined that they were adding something—restoring the phlogiston—so that the lead which emerged was a chemical compound, a product of synthesis. For men who worked on a system of ideas like this, it was not going to be easy to solve the problem of the nature of chemical elements. [. . .]

When the Swedish chemist, [Carl Wilhelm] Scheele, embarked upon the problem of combustion he found that it would be impossible to arrive at satisfactory answers to his questions until he had dealt with the problem of the air, to which he devoted his attention in the years 1768–73. The fact that the two problems were related and that combustion had even a curious correspondence with respiration had long been realised, and there are hints of it in the ancient world. Certain chemists [. . .] in the seventeenth century had put forward suggestions on this question which were in advance of the views of the phlogistonists. In the seventeenth century, however, the problem had been made more difficult by ideas concerning the purely mechanical operation of the air, or concerning the action of the atmosphere as the mere receptacle for the fumes that were given out on combustion. It was held that if a lighted candle soon went out when enclosed under a container, the increasing pressure of the air loaded with fumes was responsible for extinguishing it. And even after the air-pump had been invented and it could be shown that the candle would not burn in a vacuum, a purely mechanistic theory was still possible—you could argue that the pressure of the air was necessary to force out the fire and flame from the burning substance, so that any rarifying of the atmosphere would rob the flame of its vital impulse. At the time of the phlogiston theory mechanistic ideas still prevailed, for it was eminently the function of the air to absorb the escaping phlogiston, and in time the air became saturated, which accounted for the extinguishing of the candle under a closed container. [. . .]

In 1766 Henry Cavendish carried the story further in some studies of what he described [. . .] as "factitious airs", which, he said, meant "any kind of air which is contained in other bodies in an inelastic sense and is produced from thence by art". Amongst other things he dissolved marble in hydrochloric acid, producing [. . .] "fixed air"; dried the gas and used the device of storing it over mercury, since it was soluble in water; and expanded the description of it, calculating its specific gravity,[6] its solubility in water, etc. He also produced hydrogen by dissolving either zinc or iron or tin in sulphuric or hydrochloric acid, and found that there was no difference in the gas if he used different acids on the metals; and again he calculated its specific gravity. It was clear, therefore, that these two gases had a stable existence, and could be produced with permanent properties—they were not the capricious result of some more inconstant impurities in the air. And though both these gases had actually been discovered much earlier, they had not been separated in the mind from other things of a kindred nature—hydrogen had not been distinguished from other inflammable gases, for example. Even now, however, there was a feeling that in the last resort only one kind of air really existed—common air—and that the varieties were due to the presence or absence of phlogiston. Cavendish was inclined to identify his "inflammable air" with phlogiston, though there were objections to this, since phlogiston had been assumed to be not the burning body itself but a substance that left the burning body—and if the hydrogen was phlogiston how could phlogiston leave itself?

Joseph Priestley further improved the apparatus for collecting gases, and it is possible that, coming as an amateur without the means for any great outlay, he was driven to greater ingenuity in the devising of the requisite instruments. He had actually produced oxygen without realising it by 1771, and long before his time there had been ideas of a specially pure portion, or a specially pure constituent, of the air, which had been recognised as important for breathing and combustion. In August 1774 Priestley isolated oxygen, but at first thought it to be what he called "modified" or "phlogisticated nitrous air", and what we call nitrous oxide. Later, for a moment, after further tests, he decided that it must be common air, but by the middle of March 1775 he realised that it was five or six times more effective than the ordinary atmosphere, and he named it "dephlogisticated air". The discovery had been made a few years earlier by the Swedish

---

[6] *Specific gravity is the ratio of weights between volume of a substance and an equal volume of water.*

chemist, Scheele, who published his results later, but who showed more insight than Priestley by the way in which he recognised the existence of two separate gases in the air. Whoever may deserve the credit, the discovery and isolation of oxygen marks an important date in the history of chemistry.

By this time the position was coming to be complicated and chaotic. You have to remember that a deep prejudice regarded the air as a simple primordial substance, and a deeper prejudice still regarded water as an irreducible element. The balance of opinion, on the other hand, was in favour of regarding the metals as compounds, and if one of these, under the action of an acid, produced hydrogen, it was natural to think that the hydrogen had simply been released from the metal itself. When it was found later than an exploded mixture of hydrogen and oxygen formed water, it was simplest to argue that water was one of the constituents of oxygen, or of both the gases, and had been precipitated in the course of the experiment. When a gas was produced after the combustion of a solid body they gradually sorted out the fact that sometimes it was "fixed air" and sometimes the very different "dephlogisticated air"; but they did not know that the former—carbon dioxide—was a compound, or that the latter—oxygen—was an element. Priestley long thought that "fixed air" was elementary and existed in both common air and in his oxygen—his "dephlogisticated air". Many acids were known, but their components were not recognised and they were often regarded as modifications of one fundamental acid. For the chemist of this time there were all these counters, capable of being shifted and shuffled together, and nobody know how to play with them. So many confusions existed that chemistry was building up strange mythical constitutions for its various substances. It is possible that so long as this anarchy existed any purely doctrinal statement of what a chemical element ought to be [. . .] was bound to be ineffective and beside the point.

At this moment there emerged one of those men who can stand above the whole scene, look at the confused pieces of the jig-saw puzzle and see a way of turning them into a pattern. He was Lavoisier, and it is difficult not to believe that he towers above all the rest and belongs to the small group of giants who have the highest place in the story of the scientific revolution. In 1772, when he was twenty-eight, he surveyed the whole history of the modern study of gases and said that what had hitherto been done was like the separate pieces of a great chain which required a monumental body of directed experiments to bring them into unity. He set out to make a complete study of the air that is liberated from substances and that combines with them; and he declared in advance that this work seemed to him to be "destined to bring about a revolution in physics and in chemistry." Two years later he made a more detailed historical survey of what had been done, and added experiments and arguments of his own to show that when metals are calcined they take an "elastic fluid" out of the air, though he was still confused concerning the question whether the gas which was produced on any given occasion was "fixed air" (carbon dioxide) or oxygen. He came to feel that it was not the whole of the air, but a particular gas in the air, which entered into the processes of combustion and calcination; and that what was called "fixed air" had a complicated origin—when you heated red lead and charcoal together, he said, the gas did not arise from either of the substances alone, but took something from both, and therefore had the character of a chemical compound. On the other hand, he soon came to the conclusion that the red lead when heated in isolation produced a gas which was closely connected with common air.

When he heard that Priestley had isolated a gas in which a candle would burn better than in common air, his mind quickly jumped to the possibility of a grand synthesis. Quite unjustly, he tried to steal the credit for the discovery for himself, but it is true that he was the person who recognised the significance of the achievement and brought out its astonishing implications. In April 1775 he produced a famous paper *On the Nature of the Principle that combines with Metals in Calcination and that increases their Weight*, in which he threw overboard his earlier view that the principle might be "fixed air"—carbon dioxide—and came to the conclusion that it was the purest part of the air we breathe. The idea now came to him that "fixed air" was a compound—a combination of common air with charcoal—and he soon arrived at the thesis that it was charcoal plus the "eminently respirable part of the air." Next, he decided that common air consisted of two "elastic fluids", one of which was this eminently respirable part. Further than this, he decided that all acids were formed by the combination of non-metallic substances with "eminently respirable air", so he described this latter as the acidifying principle, or the *principe oxygine*. As a result of this theory oxygen acquired the name which it now possesses, and in the mind of Lavoisier it ranked as an irreducible element, save that it contained "caloric", which was the principle of heat.

Lavoisier was not one was of those men who are ingenious in experimental devices, but he seized upon the work of his contemporaries and the hints that were scattered over a century of chemical history, and used them to some purpose. Occasionally his experimental results were not as accurate as he pretended, or he put out hunches before he had clinched the proof of them, or he relied on points that had really been established by others. If he used the word "phlogiston", he soon did his structural thinking as though no such thing existed, and he disliked the doctrine before he knew enough to overthrow it. [. . .]

In 1781 Priestley was exploding hydrogen and oxygen [. . .] and noticed that the inside of the glass vessels "became dewy". Scientists had been so accustomed to deposits of moisture from the atmosphere or to collecting gases over water, that this kind of thing had often been observed but had passed without notice. [. . .] Cavendish confirmed the production of dew, and showed that it was plain water, that the gases combined in certain proportions to produce nothing but water, and that no weight had been lost in the course of the proceedings. It was difficult for people to believe at this time that there could not be any transmission or diffusion of weight during such an experiment, but Cavendish denied that any such loss took place. It was still more difficult for anybody to believe that water was not an irreducible element. Cavendish came to the conclusion that hydrogen must be water deprived of its phlogiston and oxygen must be phlogisticated water. Once again Lavoisier was the first to understand the situation, after learning of Cavendish's experiment, and once again he pretended to have made the actual discovery. In November 1783 he showed that water was not, properly speaking, an element, but could be decomposed and recombined, and this gave him new weapons against the phlogiston theory. He himself might have discovered the composition of water earlier than the others, but he had been unable in these years to escape from the tyranny of a preconception of his own—the view that oxygen was the great acidifying principle—which led him to look for an acid product while burning hydrogen at a jet.

He was remarkable in other ways. Finding that organic substances gave mainly fixed air and water when they burned, and knowing that fixed air was a compound of carbon and oxygen, he decided that organic substances must be largely composed of carbon, hydrogen and oxygen, and he did much toward their analysis so far as these ingredients were concerned. Already another Frenchman, de Morveau, had been striving for a revision of chemical nomenclature, and from 1782 Lavoisier worked in co-operation with him, producing a new language of chemistry which is still the basis of the language used today. The chemical revolution which he had set out to achieve was incorporated in the new terminology, as well as in a new treatise on chemistry which he wrote. [. . .] Over a broad field, therefore, he made good his victory, so that he stands as the founder of the modern science.

JEAN PIAGET

Learning to See Through Another's Eyes, from
Judgment and Reasoning in the Child (1928)

Jean Piaget (1896–1980) was one of the most remarkable scientists of the twentieth century, for both the originality of his work and the complexity of his interests. As a novelist sometimes does, he traced two stories at once. First, he used the material he had gathered about children's thinking to report how and when it changes. Second, he wanted to understand the nature of rational thought itself. When he completes the first task, he sounds like a typical scientist who gathers and organizes facts. But in his second quest, he sounds like a typical logician who defines and refines his propositions. Inevitably, most readers prefer one side or the other, even though the two sides cannot be separated. Piaget was a writer who requires his readers to be smarter and more broadly interested than they usually are.

This selection shows Piaget's surprising ability to reveal, at the same time, unexpected facts about children and about thought. His interest in why children cannot solve certain puzzles on an IQ test is classic Piaget. It was his observation that children of certain ages tended to make the same mistakes on IQ tests that set him to thinking about the relation between age and intellect, and he arrived at an unexpected generalization about scientific thinking: For all its seriousness of purpose, it requires a certain kind of game-playing skill, an imaginative flexibility that scientists take for granted—the ability to consider and follow ideas that one does not really believe.

Some readers may find that this piece poses exactly the challenge it analyzes, and they may have to put their own skepticism on hold. Piaget's statement that children younger than 11 cannot contemplate propositions they do not believe may sound questionable to people who have listened to 7-year-olds talk about pretending. But readers who can consider Piaget's report on its own terms will feel that he is on to something. Science is not just knowing the facts and thinking rationally. Think, for example, of Alfred Wegener noticing, and doubting, that South America and Africa look as though they had once been joined or of Richard Feynman playing with the idea that time might change from future to past, or of Carl Sagan calculating the odds that the earth is regularly visited by alien life-forms.

Our readers will remember the five absurd sentences of the Binet-Simon test:[7]—

1. A poor cyclist had his head smashed and died on the spot; he was taken to hospital and it is feared that he will not recover.
2. I have three brothers: Paul, Ernest, and myself.
3. The body of a poor young girl was found yesterday, cut into 18 pieces. It is thought that she must have killed herself.
4. There was a railway accident yesterday, but it was not very serious. The number of deaths was only 48.
5. Someone said: If ever I kill myself from despair I won't choose a Friday, because Friday is a bad day and would bring me ill luck. [. . .]

The question arises why some of these apparently glaring absurdities are not discovered by the child until he reaches the age of 10 (or even 11, as some have claimed). [. . .]

To [learn why], we set out to interpret the answers of some 40 schoolboys of Geneva between the ages of 9 and 11–12 in the following manner. The child is first of all examined by means of the Binet-Simon technique, then, having obtained an answer, we make the child repeat the absurd phrase by heart. The phrase is generally deformed by the child in a significant manner. We then read him the exact text so as to eliminate all factors due to inattention or forgetfulness. Finally, we ask the child to arrange the sentence himself in such a way that "there should no longer be anything silly in it." It is also advisable, in the question of the three brothers for example, to take illustrations from the child's own life. In this way one finally comes to understand more or less what he is saying.

According to our results the order of difficulty of the tests was as follows: the questions of the *three brothers* and of *Friday* were the most difficult, the questions of *accident* by far the easiest. Out of 44 children between 9 and 12 (and 3 of 14), 33 solved the question of the young girl cut into 18 pieces, and 35 that of the railway accident,

---

[7] *An early form of the IQ test. Very few readers are likely to "remember" the questions that Piaget is referring to.*

as against only 13 who understood that of the 3 brothers and 10 that of Friday. [. . .]

Why is it that the accident tests are easier than the others? It is because they appeal directly to the child's sense of reality without any presuppositions about the data. Whereas to avoid killing oneself on Friday is absurd only to anyone who believes in the unlucky character of Friday. In order to discover this absurdity the child has therefore to place himself at the point of view of the person who lays down the premises. [. . .] Similarly, in the question about the brothers, the child is obliged to place himself at a point of view which is not his: the family [. . .] consists of three brothers, and he is expected to place himself at the point of view of one of them so as to count the latter's brothers. [. . .] On the other hand, to judge that a woman cut into 18 pieces has not killed herself or that a dead cyclist cannot be resuscitated is a direct judgment of observation. It does not presuppose any preliminary change of point of view, but simply a certain sense of reality. [. . .] Finally, to describe as 'serious' an accident in which there are 48 deaths might seem to involve a formal kind of reasoning if one started from a given definition of the word 'serious.' But this question does not occur to the child; here again his judgment is immediate and absolute, very different therefore from that involved by the tests of Friday and the three brothers. [. . .]

Let us try, quite shortly, to justify these statements with regard to the Friday test. The wrong answers given to this test give us the clue to the child's difficulty. Most of them show an inability to accept the premisses as such and to reason from these premisses in a purely deductive manner.

"*People can kill themselves every day*, says Bai (age 9; 6), *they don't need to kill themselves on a Friday.*" "*Friday is not unlucky,*" says Van (age 9; 10). "*He doesn't know if it will bring him ill luck*" (Berg, 11; 2). "*Perhaps Friday will bring him good luck*" (Arn, 10; 7), etc.

The children all refuse to admit the premisses and do not see that that is not the point. What is required is to accept the premisses, then to reason correctly, *i.e.* to avoid the contradictions of the test.

But the subjects do not see the contradiction because they do not attempt to reason from the point of view of the person who is speaking. They do not get out of their own point of view, and consequently stick at the premises, which they refuse to admit even in the capacity of data.

Campa (10; 3) and Ped (9; 6) are in exactly the same predicament, but they try to justify the premises: "*It is a day when you must never eat meat.*" There is therefore nothing absurd to them in the test, but again this is because they do not reason from the data but judge the latter from their own point of view.

But from the moment that the child admits the premises as given, without justifying or invalidating them, he is ready to resolve the test correctly. Some of them are of opinion that "*he would do better to kill himself on Friday, since it is an unlucky day,*" and then the correct answer comes: "*Since he would be dead it couldn't bring him ill luck*" (Bled, 10; 10).

Thus the difficulty of reasoning formally (*i.e.* of admitting a datum as such and deducing what follows from it) is the real difficulty of the test. That is why this test is, in our opinion, better suited to the age of 11 or even 12 than to that of 10. Indeed, there was an interval of at least a year between the success of this test and that of the accident tests.

We are now in a position to understand what formal reasoning really consists in, and how its structure may be influenced by social factors such as ego-centrism and the socialization of thought.

The first deductive operation of which the mind is capable consists either in foreseeing what will happen when such and such conditions are given, or in reconstructing what has happened when such and such results are given. This is a step forward in intellectual growth which the child is able to take very early. [. . .] But all reasoning at this stage is still limited by one essential qualification: deduction bears only upon the beliefs which the child has adopted himself, in other words, it deals only with his personal conception of reality. The child will be able to say: "Half 9 is not 4, because 4 and 4 make 8," or [when he has failed to find an object in one box and is pointing to the next one] "*Then it must be in that box, anyhow!*" etc., because in these cases deduction deals with propositions which he admits personally, or rejects personally. But if we say to a child: "Let us admit, for example, that dogs have six heads. How many heads will there be in a yard where there are 15 dogs?" the child will refuse to give an answer, because he will not 'assume' the hypothesis. We, on the contrary, even though we admit that the premises are absurd, will be perfectly able to reason from them and conclude that there will be 90 heads in that yard. This is because we distinguish between real or empirical necessity (dogs cannot have six heads) and formal or logical necessity (if dogs had six heads there would necessarily be 90 heads, etc.). [. . .]

Moreover, we are here in presence of a fact which anyone can observe for himself by questioning a child. Up to a certain age it is almost impossible to make a child assume a suggested hypothesis unless one forces him to believe it and thus changes it into an affirmation. In the experiments on air which we shall publish shortly we find children from 8 to 10 who know that there is air everywhere, particularly in this room. We say to them: "If there were no air would this [an object suspended by a string which we swing round rapidly] make a draught?—*Yes.*—Why?—*Because there is always air in the room.*—But in a room where all the air had been taken away would it make any?—*Yes, it would.*—Why?—*Because there would be some air left,*" etc. Or with younger children in an enquiry into animism: "If you could touch the sun, would he feel it?—*You can't touch him.*—Yes, but if you could manage to, would he feel it?—*He is too high up.*—Yes, but if . . . .,*" etc.

This shows us what formal deduction really is; it consists in drawing conclusions, not from a fact given in immediate observation, nor from a judgment which one holds to be true without any qualifications (and thus incorporates into reality such as one conceives it), but in a judgment which one simply assumes, *i.e.* which one admits without believing it, just to see what it will lead to. This is the form of deduction which we have placed round about the age of 11–12 as opposed to the simpler forms of inference which appear first.

These tests of ours will perhaps give the impression that formal deduction is a very special form of thought and of very little use to children. But this is far from being the case. In the first place all mathematical reasoning is formal or, as the logicians put it, hypothetico-deductive. Whenever we say to a child: "Let this be a triangle," or "A piece of cloth costs 12 francs," we are forcing him to reason *in conformity with premises that are simply given,* which means disregarding reality and even suppressing any memories or real observations which might block the way to the process of reasoning. Reasoning of this kind is done from pure hypotheses. Even if the subject-matter be of a concrete nature and the problems given to the child contain measurements and actual observations, the sort of reasoning required is none the less formal in the sense that the child will have to remember a number of rules and definitions which are independent of his own observation. Either a mathematical problem is presented to a child as a purely empirical problem, in which case he will be kept in ignorance of the deductive power of even elementary arithmetic; or he will be compelled to reason strictly, but then this process, in so far as it appeals to fixed rules and to previously admitted propositions, will be a process of formal reasoning.

Nay, more; all deduction, even when it deals with reality given in direct observation, is formal to the extent that it claims to be strict. For deduction that is concerned with objects such as they appear to us in direct observation can obviously not be strict, but only probable, or based on analogy. From the fact that water boils at different temperatures according to pressure, nothing precise can be deduced. In order to deduce strictly one must (1) bring about certain ideal conditions such as could not be realized in immediate experience, and in this way reach the knowledge of laws which will perhaps never be verified but will remain mere constructions of the mind; (2) one must deal with ideal objects, *i.e.* objects defined clearly and in such a manner as to prevent their being confused with the variable objects presented in observation. (The chemical definition of water, H_2O or H_2O_2 stands for a body that is never found pure in nature). So that the form of deduction which is a necessary condition for arriving at general laws or mathematical relations is such that the stricter it is, the more formal it becomes; that is to say it will have to presuppose ideal definitions and hypotheses which cannot be directly verified.

Now theoretical as these conditions may appear, they are none the less indispensable to the psychology of the child's mind. When boys of 9–10 years old spontaneously give weight as an explanation of why bodies float, one often feels that they have some intuition of density. They will say, for instance, that given equal volumes (*i.e.* when they are shown two equal volumes, one of wood, the other of water) wood is lighter than water. Before the age of 9, on the contrary, children who have just declared that wood is light, think that it is heavier than water. What is this very marked progress shown at the age of 9 but an attempt—implicit as yet—to replace immediate reality (absolute weight without any reference to volume) by a ratio (weight/ volume), *i.e.* to replace the real object by a slightly more ideal object.

By following this mental orientation the child will inevitably come to apply formal deduction to nature herself, and he will do so to the extent that he substitutes ratios, laws and fixed (*i.e.* ideal) definitions for simple empirical observation.

We have expatiated at some length upon this subject in order to show that formal deduction, *i.e.* reasoning from premises that are merely assumed and not supplied by immediate belief is of fundamental importance, not only in mathematics, but in every kind of reflection about nature. Let us now return to the origins of formal thinking. The above analysis has shown that two factors were particularly necessary for the right functioning of any formal reasoning: (1) a sort of detachment from one's own point of view or from the point of view of the moment, enabling one to place oneself at that of others and to reason first from premises admitted by them, then more generally from every kind of purely hypothetical proposition; (2) owing to the mere fact of having placed oneself inside the beliefs of others, or more generally inside a hypothesis, one must, in order to reason formally, be able to remain on the plane of mere assumption without surreptitiously returning to one's private point of view or to that of the reality of the moment. To be formal, deduction must detach itself from reality and take up its stand upon the plane of the purely possible, which is by definition the domain of hypothesis. In a word, formal thought presupposes two factors, one social (the possibility of placing oneself at every point of view and of abandoning one's own), the other, which is connected with the psychology of belief (the possibility of assuming alongside of empirical reality a purely possible world which shall be the province of logical deduction).

"The Actual Limits of What Is Known"

Noam Chomsky

STEPHEN JAY GOULD

The Misuse of Darwin, from
The Mismeasure of Man (1981)

Stephen Jay Gould (1941–) is probably America's most admired science writer. His name, along with Carl Sagan's, was mentioned every time I sought recommendations for this book. His regular essays in *Natural History* magazine have explored the whole range of biological evolution and its subtleties. Any one of them might easily have been selected. But Gould has written other works as well, and this excerpt is from one of those. Many science writers survey success, but Gould has also surveyed delusions. His stated ambition is to warn of the ways in which supposedly honest scientists can fool themselves. His position is that the frauds who make up their data are less worrisome than the self-deceivers who take their shaky data more seriously than they merit.

Gould's book *The Mismeasure of Man* examines more than a century of work by people who believed they were doing scientific research and who were accepted as scientists but whose work was profoundly antiscientific. These researchers were not really trying to understand things that puzzled them. Instead, they argued like lawyers, marshaling evidence that supported positions they already held. In this case, they wanted to collect scientific evidence showing that northern European white men were superior to women and other peoples. As long as one form of data could be used to argue that point, they gathered it eagerly. The crisis came when the data and the conclusion no longer matched, for then they had to choose between acting like scientists trying to understand why their expectations were wrong or like loyal ideologues sticking by their assumptions no matter what the evidence says.

Once the fact of evolution had been established, nineteenth-century naturalists devoted themselves to tracing the actual pathways that evolution had followed. They sought, in other words, to reconstruct the tree of life. Fossils might have provided the evidence, for only they could record the actual ancestors of modern forms. But the fossil record is extremely imperfect, and the major trunks and branches of life's tree all grew before the evolution of hard parts permitted the preservation of a fossil record at all. Some indirect criterion had to be found. Ernst Haeckel, the great German zoologist, refurbished an old theory of creationist biology and suggested that the tree of life might be read directly from the embryological development of higher forms. He proclaimed that "ontogeny recapitulates phylogeny" or, to explicate this mellifluous tongue-twister, that an individual, in its own growth, passes through a series of stages representing *adult* ancestral forms in their correct order—an individual, in short, climbs its own family tree.

Recapitulation ranks among the most influential ideas of late nineteenth-century science. It dominated the work of several professions, including embryology, comparative morphology, and paleontology. All these disciplines were obsessed with the idea of reconstructing evolutionary lineages, and all regarded recapitulation as the key to this quest. The gill slits of an early human embryo represented an ancestral adult fish; at a later stage, the temporary tail revealed a reptilian or mammalian ancestor. [. . .]

Recapitulation also provided an irresistible criterion for any scientist who wanted to rank human groups as higher and lower. The *adults* of *inferior* groups must be like *children* of *superior* groups, for the child represents a primitive adult ancestor. If adult blacks and women are like white male children, then they are living representatives of an ancestral stage in the evolution of white males. An anatomical theory for ranking races [. . .] had been found.

Recapitulation served as a general theory of biological determinism. All "inferior" groups—races, sexes, and classes—were compared with the children of white males. E. D. Cope, the celebrated American paleontologist who elucidated the mechanism of recapitulation [. . .] , identified four groups of lower human forms on this criterion: non-white races, all women, southern as opposed to northern European whites, and lower classes within superior races. [. . .] Cope preached the doctrine of Nordic supremacy and agitated to curtail the immigration of Jews and southern Europeans to America. To explain the inferiority of southern Europeans in recapitulatory terms, he argued that warmer climates impose an earlier maturation. Since maturation signals the slowdown and cessation of bodily development, southern Europeans are caught in a more childlike, hence primitive, state as adults. Superior northerners move on to higher stages before a later maturation cuts off their development:

> There can be little doubt that in the Indo-European race maturity in some respects appears earlier in tropical than in northern regions; and though subject to many exceptions, this is sufficiently general to be looked upon as a rule. Accordingly, we find in that race—at least in the warmer regions of Europe and America—a larger proportion of certain qualities which are more universal in women, as greater activity of the emotional nature when compared with the judgment. . . . Perhaps the more northern type left all that behind in its youth.

[. . .] Cope also focused upon the skull, particularly upon "those important elements of beauty, a well-developed nose and beard", [. . .] but he also derided the deficient calf musculature of blacks:

> Two of the most prominent characters of the negro are those of immature stages of the Indo-European race in its characteristic types. The deficient calf is the character of infants at a very early stage; but, what is more important, the flattened bridge of the nose and shortened nasal cartilages are universally immature conditions of the same parts in the Indo-European. . . . In some races—e.g., the Slavic—this undeveloped character persists later than in some others. The

Greek nose, with its elevated bridge, coincides not only with aesthetic beauty, but with developmental perfection.

In 1890 American anthropologist D. G. Brinton summarized the argument with a paean of praise for measurement:

> The adult who retains the more numerous fetal, infantile or simian traits, is unquestionably inferior to him whose development has progressed beyond them. . . . Measured by these criteria, the European or white race stands at the head of the list, the African or negro at its foot. . . . All parts of the body have been minutely scanned, measured and weighed, in order to erect a science of the comparative anatomy of the races. [. . .]

If anatomy built the hard argument of recapitulation, psychic development offered a rich field for corroboration. Didn't everyone know that savages and women are emotionally like children? Despised groups had been compared with children before, but the theory of recapitulation gave this old chestnut the respectability of main-line scientific theory. "They're like children" was no longer just a metaphor of bigotry; it now embodied a theoretical claim that inferior people were literally mired in an ancestral stage of superior groups.

G. Stanley Hall, then America's leading psychologist, stated the general argument in 1904: "Most savages in most respects are children, or, because of sexual maturity, more properly, adolescents of adult size". [. . .] A. F. Chamberlain, his chief disciple, opted for the paternalistic mode: "Without primitive peoples, the world at large would be much what in small it is without the blessing of children."

The recapitulationists extended their argument to an astonishing array of human capacities. Cope compared prehistoric art with the sketches of children and living "primitives" [. . .] : "We find that the efforts of the earliest races of which we have any knowledge were quite similar to those which the untaught hand of infancy traces on its slate or the savage depicts on the rocky faces of cliffs." James Sully, a leading English psychologist, compared the aesthetic senses of children and savages, but gave the edge to children [. . .] :

> In much of this first crude utterance of the aesthetic sense of the child we have points of contact with the first manifesta-

tions of taste in the race. Delight in bright, glistening things, in gay things, in strong contrasts of color, as well as in certain forms of movement, as that of feathers—the favorite personal adornment—this is known to be characteristic of the savage and gives to his taste in the eyes of civilized man the look of childishness. On the other hand, it is doubtful whether the savage attains to the sentiment of the child for the beauty of flowers.

Herbert Spencer, the apostle of social Darwinism, offered a pithy summary [. . .] : "The intellectual traits of the uncivilized . . . are traits recurring in the children of the civilized."

Since recapitulation became a focus for the general theory of biological determinism, many male scientists extended the argument to women. E. D. Cope claimed that the "metaphysical characteristics" of women were

> . . . very similar in essential nature to those which men exhibit at an early stage of development. . . . The gentler sex is characterized by a greater impressibility; . . . warmth of emotion, submission to its influence rather than that of logic; timidity and irregularity of action in the outer world. All these qualities belong to the male sex, as a general rule, at some period of life, though different individuals lose them at very various periods. . . . Probably most men can recollect some early period of their lives when the emotional nature predominated—a time when emotion at the sight of suffering was more easily stirred than in maturer years. . . . Perhaps all men can recall a period of youth when they were hero-worshippers— when they felt the need of a stronger arm, and loved to look up to the powerful friend who could sympathize with and aid them. This is the "woman stage" of character.

In what must be the most absurd statement in the annals of biological determinism, G. Stanley Hall—again, I remind you, not a crackpot, but America's premier psychologist—invoked the suicide rates of women as a sign of their primitive evolutionary status [. . .] :

> This is one expression of a profound psychic difference between the sexes. Woman's body and soul is phyletically older

and more primitive, while man is more modern, variable, and less conservative. Women are always inclined to preserve old customs and ways of thinking. Women prefer passive methods; to give themselves up to the power of elemental forces, as gravity, when they throw themselves from heights or take poison, in which methods of suicide they surpass man. Havelock Ellis thinks drowning is becoming more frequent, and that therein women are becoming more womanly.

As a justification for imperialism, recapitulation offered too much promise to remain sequestered in academic pronouncements. I have already cited Carl Vogt's low opinion of African blacks, based on his comparison of their brains with those of white children. B. Kidd extended the argument to justify colonial expansion into tropical Africa. [. . .] We are, he wrote, "dealing with peoples who represent the same stage in the history of the development of the race that the child does in the history of the development of the individual. The tropics will not, therefore, be developed by the natives themselves."

In the course of a debate about our right to annex the Philippines, Rev. Josiah Strong, a leading American imperialist, piously declared that "our policy should be determined not by national ambition, nor by commercial considerations, but by our duty to the world in general and to the Filipinos in particular". [. . .] His opponents, citing Henry Clay's contention that the Lord would not create a people incapable of self-government, argued against the need for our benevolent tutelage. But Clay had spoken in the bad old days before evolutionary theory and recapitulation:

> Clay's conception was formed . . . before modern science had shown that races develop in the course of centuries as individuals do in years, and that an undeveloped race, which is incapable of self-government, is no more of a reflection on the Almighty than is an undeveloped child who is incapable of self-government. The opinions of men who in this enlightened day believe that the Filipinos are capable of self-government because everybody is, are not worth considering.

[. . .] And so the story might stand, a testimony to nineteenth-century folly and prejudice, if an interesting twist had not been added during our own century. By 1920 the theory of recapitulation had collapsed. [. . .] Not long after, the Dutch anatomist Louis Bolk proposed a theory of exactly opposite meaning. Recapitulation required that adult traits of ancestors develop more rapidly in descendants to become juvenile features—hence, traits of modern children are primitive characters of ancestral adults. But suppose that the reverse process occurs as it often does in evolution. Suppose that juvenile traits of ancestors develop so slowly in descendants that they become adult features. This phenomenon of retarded development is common in nature; it is called neoteny (literally, "holding on to youth"). Bolk argued that humans are essentially neotenous. He listed an impressive set of features shared by adult humans and fetal or juvenile apes, but lost in adult apes: vaulted cranium and large brain in relation to body size; small face; hair confined largely to head, armpits, and pubic regions; unrotated big toe. [. . .]

Now consider the implications of neoteny for the ranking of human groups. Under recapitulation, adults of inferior races are like children of superior races. But neoteny reverses the argument. In the context of neoteny, it is "good"—that is, advanced or superior—to retain the traits of childhood, to develop more slowly. Thus, superior groups retain their childlike characters as adults, while inferior groups pass through the higher phase of childhood and then degenerate toward apishness. Now consider the conventional prejudice of white scientists: whites are superior, blacks inferior. Under recapitulation, black adults should be like white children. But under neoteny, white adults should be like black children.

For seventy years, under the sway of recapitulation, scientists had collected reams of objective data all loudly proclaiming the same message: adult blacks, women, and lower-class whites are like white upper-class male children. With neoteny now in vogue, these hard data could mean only one thing: upper-class adult males are inferior because they lose, while other groups retain, the superior traits of childhood. There is no escaping it.

At least one scientist, Havelock Ellis, did bow to the clear implication and admit the superiority of women, though he wriggled out of a similar confession for blacks. He even compared rural with urban men, found that men of the city were developing womanly anatomy, and proclaimed the superiority of urban life [. . .] : "The large-headed, delicate-faced, small-boned man of urban civilization is much nearer to the typical woman than is the savage. Not only by his large brain, but by his large pelvis, the modern man is following a path first marked out by woman." But Ellis was iconoclastic and controversial (he wrote one of the first systematic studies of sexuality), and his application of neoteny to sexual differences never made much impact. Meanwhile, with respect to racial differences, supporters of human neoteny adopted another, more common, tactic: they simply abandoned their seventy years of hard data and sought new and opposite information to confirm the inferiority of blacks.

Louis Bolk, chief defender of human neoteny, declared that the most strongly neotenized races are superior. In retaining more juvenile features, they have kept further away from "the pithecoid ancestor of man". [. . .] "From this point of view, the division of mankind into higher and lower races is fully justified. [. . .] It is obvious that I am, on the basis of my theory, a convinced believer in the inequality of races". Bolk reached into his anatomical grab-bag and extracted some traits indicating a greater departure for black adults from the advantageous proportions of childhood. Led by these new facts to an old and comfortable conclusion, Bolk proclaimed [. . .] : "The white race appears to be the most progressive, as being the most retarded." Bolk, who viewed himself as a "liberal" man, declined to relegate blacks to permanent ineptitude. He hoped that evolution would be benevolent to them in the future:

> It is possible for all other races to reach the zenith of development now occupied by the white race. The only thing required is continued progressive action in these races of the biological principle of anthropogenesis [i.e., neoteny]. In his fetal development the negro passes through a stage that has already become the final stage for the white man. Well then, when retardation continues in the negro too, what is still a transitional stage may for this race also become a final one. [. . .]

Bolk's argument verged on the dishonest for two reasons. First, he conveniently forgot all the features—like the Grecian nose and full beard so admired by Cope—that recapitulationists had stoutly emphasized because they placed whites *far* from the conditions of childhood.

Secondly, he sidestepped a pressing and embarrassing issue: Orientals, not whites, are clearly the most neotenous of human races. [. . .] Women, moreover, are more neotenous than men. I trust that I will not be seen as a vulgar white apologist if I decline to press the superiority of Oriental women and declare instead that the whole enterprise of ranking groups by degree of neoteny is fundamentally unjustified. Just as Anatole France and Walt Whitman could write as well as Turgenev with brains about half the weight of his, I would be more than mildly surprised if the small differences in degree of neoteny among races bear any relationship to mental ability or moral worth.

><+>-O-<+><

NOAM CHOMSKY
The Case Against B. F. Skinner (1981)

Before he was 30, Noam Chomsky (1928–) had revolutionized the study of linguistics by seeking mathematical rules that described sentences in English so precisely that even a machine could produce them simply by following the rules. Chomsky believes in the importance of mind and of mental rules in accounting for human activity, especially speech. Although that position may not seem exceptional to ordinary readers, it was not the dominant idea among psychologists of the 1950s. Indeed, many psychologists deny that Chomsky is even a scientist. Rather, they see him more as a philosopher and mathematician because he does not conduct experiments and does not observe nature closely. Nonetheless, he has made important contributions to psychology by effectively challenging its lack of interest in the rules that support thought.

Chomsky's interest in the mind led to a famous clash with B. F. Skinner, the dominant American psychologist during the 1940s and 1950s. Skinner's behaviorism denied that anything subjective or even internal shaped human activity. In 1957 Skinner published a book called *Verbal Behavior* that sought to explain all language as habits shaped by the environment. He maintained that no internalized rules or meanings were needed to explain speech. Although Chomsky needed two years to write it, eventually he published a review of Skinner's book that was so devastating that it helped break Skinner's hold on academic psychology.

Many years later, Skinner published a popular book called *Beyond Freedom and Dignity*. Although psychology was by then deeply interested in mental rules, the popular press still saw Skinner as the giant of American psychology, and the first reviews of his book were quite respectful. Then, in the *New York Review of Books*, Chomsky published a disrespectful review. It repeated, in abridged form, the arguments made in his review of *Verbal Behavior*.

If a physical scientist were to assure us that we need not concern ourselves over the world's sources of energy because he has demonstrated in his laboratory that windmills will surely suffice for all future needs, he would be expected to produce some evidence, or other scientists would expose this pernicious nonsense. The situation is different in the behavioral sciences. A person who claims that he has a behavioral technology that will solve the world's problems and a science of behavior that both supports it and reveals the factors determining human behavior is required to demonstrate nothing. One waits in vain for psychologists to make clear to the general public the actual limits of what is known. In view of the prestige of science and technology, this is an unfortunate situation.

Let us now turn to the evidence that Skinner provides for his extraordinary claims: e.g., that "an analysis of behavior" reveals that the achievements of artists, writers, statesmen, and scientists can be explained almost entirely according to environmental contingencies (p. 44);[1] and that it is the environment that makes a person wise or compassionate (p. 171); that "all these questions about purposes, feelings, knowledge, and so on can be restated in terms of the environment to which a person has been exposed" and that "what a person intends to do depends on what he has done in the past and what has then happened" (p. 72) and so on.

According to Skinner, apart from genetic endowment, behavior is determined entirely by "reinforcement." To a hungry organism, food is a positive reinforcer. This means that "anything the organism does that is followed by the receipt of food is more likely to be done again whenever the organism is hungry" (p. 27), but "Food is reinforcing only in a state of deprivation" (p. 37). A negative reinforcer is a stimulus that increases the probability of behavior that reduces the intensity of that stimulus; it is "aversive," and roughly speaking, constitutes a threat (p. 27). A stimulus can become a conditioned reinforcer by association with other reinforcers. Thus money is "reinforcing only after it has been exchanged for reinforcing things" (p. 33). The same is generally true of approval and affection. (The reader may attempt something that Skinner always avoids, namely, to characterize the "stimuli" that constitute "approval.")

Behavior is shaped and maintained by the arrangement of such reinforcers. Thus, "We change the relative strengths of responses by differential reinforcement of alternative courses of action" (pp. 94–95). One's repertoire of behavior is determined by "the contingencies of reinforcement to which he is exposed as an individual" (p. 127). An

---

[1] *These page numbers refer to pages in B. F. Skinner's* Beyond Freedom and Dignity *(hardcover).*

"organism will range between vigorous activity and complete quiescence depending upon the schedules on which it has been reinforced" (p. 186). As Skinner realizes (though some of his defenders do not) meticulous control is necessary to shape behavior in highly specific ways. Thus, "The culture . . . teaches a person to make fine discriminations by making differential reinforcement more precise," (p. 194) a fact that causes problems when "the verbal community cannot arrange the subtle contingencies necessary to teach fine distinctions among stimuli which are inaccessible to it." "As a result the language of emotion is not precise" (p. 106).

The problem in "design of a culture" is to "make the social environment as free as possible of aversive stimuli" (p. 42), "to make life less punishing and in doing so to release for more reinforcing activities the time and energy consumed in the avoidance of punishment" (p. 81). It is an engineering problem, and we could get on with it if only we could overcome the irrational concern for freedom and dignity. What we require is the more effective use of the available technology, more and better controls. In fact, "A technology of behavior is available which would more successfully reduce the aversive consequences of behavior, proximate or deferred, and maximize the achievements of which the human organism is capable" (p. 125). But "the defenders of freedom oppose its use," thus contributing to social malaise and human suffering. It is this irrationality that Skinner hopes to persuade us to overcome.

At this point an annoying, though obvious, question intrudes. If Skinner's thesis is false, then there is no point in his having written the book or our reading it. But if his thesis is true, then there is also no point in his having written the book or our reading it. For the only point could be to modify behavior, and behavior according to the thesis, is entirely controlled by arrangement of reinforcers. Therefore reading the book can modify behavior only if it is a reinforcer, that is, if reading the book will increase the probability of the behavior that led to reading the book (assuming an appropriate state of deprivation). At this point, we seem to be reduced to gibberish.

A counterargument might be made that even if the thesis is false, there is a point to writing and reading the book, since certain false theses are illuminating and provocative. But this escape is hardly available. In this case, the thesis is elementary and not of much interest in itself. Its only value lies in its possible truth. But if the thesis is true, then reading or writing the book would appear to be an entire waste of time, since it reinforces no behavior.

Skinner would surely argue that reading the book, or perhaps the book itself, is a "reinforcer" in some other sense. He wants us to be persuaded by the book, and not to our surprise, he refers to persuasion as a form of behavioral control, albeit a weak and ineffective form. Skinner hopes to persuade us to allow greater scope to the behavioral technologists, and apparently believes that reading this book will increase the probability of our behaving in such a way as to permit them great scope (freedom?). Thus reading the book, he might claim, reinforces this behavior. It will change our behavior with respect to the science of behavior (p. 24).

Let us overlook the problem, insuperable in his terms, of clarifying the notion of "behavior that gives greater scope to behavioral technologists," and consider the claim that reading the book might reinforce such behavior. Unfortunately, the claim is clearly false, if we use the term "reinforce" with anything like its technical meaning. Recall that reading the book reinforces the desired behavior only if it is a consequence of the behavior. Obviously putting our fate in the hands of behavioral technologists is not behavior that led to (and hence can be reinforced by) our reading Skinner's book. Therefore, the claim can be true only if we deprive the term "reinforce" of its technical meaning. Combining these observations, we see that there can be some point to reading the book or to Skinner's having written it only if the thesis of the book is divorced from the "science of behavior" on which it allegedly rests.

Let us consider further the matter of "persuasion." [. . .] Skinner [claims] that "we sample and change verbal behavior, not opinions" as, he says, behavioral analysis reveals (p. 95). Taken literally, this means that if, under a credible threat of torture, I force someone to say, repeatedly, that the earth stands still, then I have changed his opinion. Comment is unnecessary.

Skinner claims that persuasion is a weak method of control, and he asserts that "changing a mind is condoned by the defenders of freedom and dignity because it is an ineffective way of changing behavior, and the changer of minds can therefore escape from the charge that he is controlling people" (p. 97). Suppose that your doctor gives you a very persuasive argument to the effect that if you continue to smoke, you will die a horrible death from lung cancer. Is it necessarily the case that this argument will be less effective in modifying your behavior than any arrangement of true reinforcers?

In fact, whether persuasion is effective or not depends on the content of the argument (for a rational person), a factor that Skinner cannot begin to describe. The problem becomes still worse if we consider other forms of "changing minds." Suppose that a description of a napalm raid on a foreign village induces someone in an American audience to carry out an act of sabotage. In this case, the "effective stimulus" is not a reinforcer, but the mode of changing behavior may be quite effective and, furthermore, the act that is performed (the behavior "reinforced") is entirely new (not in the "repertoire") and may not even have been hinted at in the "stimulus" that induced the change of behavior. In every possible respect, then, Skinner's account is simply incoherent.

Since his William James Lectures of 1947, Skinner has been sparring with these and related problems. The results are nil. It remains impossible for Skinner to formulate questions of the kind just raised in his own terms, let alone investigate them. What is more, no serious scientific hypotheses with supporting evidence have been produced to substantiate the extravagant claims to which he is addicted. Furthermore, this record of failure was predictable from the start, from an analysis of the problems and the means proposed to deal with them.

It must be stressed that "verbal behavior" is the only aspect of human behavior that Skinner has attempted to investigate in any detail. To his credit, he recognized early that only through a successful analysis of language could he hope to deal with human behavior. By comparing the results that have been achieved in this period with the claims that are still advanced, we gain a good insight into the nature of Skinner's science of behavior. My impression is, in fact, that the claims are becoming more extreme and more strident as the inability to support them and the reasons for this failure become increasingly obvious.

It is unnecessary to labor the point any further. Evidently Skinner has no way of dealing with the factors involved in persuading someone or changing his mind. The attempt to invoke "reinforcement" merely leads to incoherence. The point is crucial. Skinner's discussion of persuasion and "changing minds" is one of the few instances in which he tries to come to terms with what he calls the "literature of freedom and dignity." The libertarian whom he condemns distinguishes between persuasion and certain forms of control. He advocates persuasion and objects to control. In response, Skinner claims that persuasion is itself a (weak) form of control and that by using weak methods of control we simply shift control to other environmental conditions, not to the person himself (pp. 97 and 99).

Thus, Skinner claims, the advocate of freedom and dignity is deluding himself in his belief that persuasion leaves the matter of choice to "autonomous man," and furthermore he poses a danger to society because he stands in the way of more effective controls. As we see, however, Skinner's argument against the "literature of freedom and dignity" is without force. Persuasion is no form of control at all, in Skinner's sense; in fact, he is unable to deal with the concept. But there is little doubt that persuasion can "change minds" and affect behavior, on occasion quite effectively.

Since persuasion cannot be coherently described as the result of arrangement of reinforcers, it follows that behavior is not entirely determined by the specific contingencies to which Skinner arbitrarily restricts his attention, and that the major thesis of the book is false. Skinner can escape this conclusion only by claiming that persuasion is a matter of arranging reinforcing stimuli, but this claim is tenable only if the term "reinforcement" is deprived of its technical meaning and used as a mere substitute for the detailed and specific terminology of ordinary language. In any event, Skinner's "science of behavior" is irrelevant: the thesis of the book is either false (if we use terminology in its technical sense) or empty (if we do not). And the argument against the libertarian collapses entirely.

Not only is Skinner unable to uphold his claim that persuasion is a form of control, but he also offers not a particle of evidence to support his claim that the use of "weak methods of control" simply shifts the mode of control to some obscure environmental factor rather than to the mind of autonomous man. Of course, from the thesis that all behavior is controlled by the environment, it follows that reliance on weak rather than strong controls shifts control to other aspects of the environment. But the thesis, insofar as it is at all clear, is without empirical support, and in fact may even be empty, as we have seen in discussing "probability of response" and persuasion. Skinner is left with no coherent criticism of the "literature of freedom and dignity."

The emptiness of Skinner's system is revealed when he discusses more peripheral matters. He claims (p. 112) that the statement, "You should (you ought) to read *David Copperfield*" may be translated, "You will be reinforced if you read *David Copperfield*." No matter how we try to interpret Skinner's suggestion, giving the term "reinforce" its literal sense, we fall into utter confusion. Probably what Skinner has in mind when he says that it is "reinforcing" to read *David Copperfield* is that the reader will like it or enjoy it, and thus be "reinforced."

But this gives the game away. We are now using "reinforce" in a sense quite different from that of the laboratory theory of behavior. It would make no sense at all to try to apply results about "scheduling" of reinforcement, for example, to this situation. Furthermore, it is no wonder that we can "explain" behavior by using the non-technical term "reinforce" with just the meaning of "like" or "enjoy" or "learn something from" or whatever. Similarly, when Skinner tells us that a fascinating hobby is "reinforcing" (p. 36), he is surely not claiming that the behavior that leads to indulging in this hobby will be increased in probability. Rather, he means that we enjoy the hobby. A literal interpretation of such remarks yields gibberish, and a metaphorical interpretation merely replaces an ordinary term by a homonym of a technical term with no gain in precision. [. . .]

Or consider the claim that "we are likely to admire behavior more as we understand it less" (p. 3). In a strong sense of "explain," it follows that we admire virtually all behavior, since we can explain virtually none. In a looser sense, Skinner is claiming that if Eichmann is incomprehensible to us, but we understand why the Vietnamese fight on, then we are likely to admire Eichmann but not the Vietnamese resistance. Similarly, Skinner asserts, "Except when physically restrained, a person is least free or dignified when he is under the threat of punishment" (p. 60). Thus someone who refuses to bend to authority in the face of severe threat has lost his dignity.

The real content of Skinner's system can be appreciated only by examining such cases, point by point. The careful reader will discover that in each case a literal interpretation of Skinner's statements, where terminology is understood in something like the technical sense, yields obvious falsehood, and that a loose metaphorical interpretation does permit the translation of the familiar descriptive and evaluative vocabulary of ordinary discourse into Skinner's terms, of course with a loss of precision and clarity, in view of the poverty of his system.

We can get a taste of the explanatory force of Skinner's theory from such (typical) examples as these: a pianist learns to play a scale smoothly because "smoothly played scales are reinforcing" (p. 204); "A person can know what it is to fight for a cause only after a long history during which he has learned to perceive and to know that state of affairs called fighting for a cause" (p. 90); and so on.

Similarly we can perceive the power of Skinner's behavioral technology by considering the useful observations and advice he offers. "Punishable behavior can be minimized by creating circumstances in which it is not likely to occur" (p. 64). If a person "is strongly reinforced when he sees other people enjoying themselves, . . . he will design an environment in which children are happy" (p. 150). If overpopulation, nuclear war, pollution, and depletion of resources are a problem, "we may then change practices to induce people to have fewer children, spend less on nuclear weapons, stop polluting the environment, and consume resources at a lower rate, respectively" (p. 152).

The reader may search for more profound thoughts than these. He may seek, but he will not find.

Skinner alludes more frequently in this book to the role of genetic endowment than he did in his earlier speculations about human behavior and society. One would think that this would lead to some modification in his conclusions, or to new conclusions. It does not, however. The reason is that Skinner is as vague and uninformative about genetic endowment as he is about control by contingencies of reinforcement. Unfortunately, zero plus zero still equals zero.

FRANCIS BACON

Idols of the Tribe, from
Novum Organum (1620)

Francis Bacon (1561–1626) was a man of action on one side and a philosopher and scholar on the other. As a man of action, he was a civil servant and a faithful, hardworking king's counselor. As a philosopher, he produced a steady flow of essays, treatises, inquiries, and outlines of proposed works. For Bacon, the area where action and philosophy met was in the natural sciences. He knew that science required the deep learning and knowledge suited to his philosophical side but that it also suited his active side because he believed he could use science to lessen humanity's miseries and increase its freedom. Typical of the kind of practical science Bacon most valued was the question that led to his death: he thought that by packing a chicken in snow, he might preserve the meat. While conducting an experiment to test the idea, he caught cold and died. Had he lived, perhaps some early form of frozen food would have been introduced into England during the 1600s.

Bacon's own scientific research is little remembered today and even in his own time was not considered major, yet he holds a central position in the history of science. His most important contribution was his reconsideration of the idea of causes. According to Aristotle, causes were abstract essences that could be found through logical deduction. For example, if all men are mortal and Socrates is a man, then we can deduce that part of Socrates' essence is mortality. Bacon saw cause as a mechanical relationship between two things, and causes were to be learned "inductively." For example, Socrates died from drinking hemlock. If we were to draw up a list of hemlock eaters and note the details of how they died and also any consumers who did not die, we might eventually have enough information to determine how hemlock kills people. Then we shall have attained new knowledge. It was this redirection of attention from abstract to mechanical causes that made Bacon important.

The following selection is from Bacon's *Novum Organum*, the book in which he gives his fullest account of induction. The passage here, his list of "Idols of the Tribe," criticizes common ways of thinking, particularly those that were natural to the heirs of Aristotle. The selection begins by rejecting the tendency to systematize knowledge on the basis of abstract principles, and it ends with Bacon's insistence that it is better to "dissect [nature] into parts" than to resolve it "into abstractions." This argument for the analysis of causes became the standard for modern science, especially after the Royal Society was founded in 1660 along lines that Bacon specifically defined.

The human understanding is of its own nature prone to suppose the existence of more order and regularity in the world than it finds. And though there be many things in nature which are singular and unmatched, yet it devises for them parallels and conjugates and relatives which do not exist. Hence the fiction that all celestial bodies move in perfect circles; spirals and dragons being (except in name) utterly rejected. Hence too the element of Fire with its orb is brought in, to make up the square with the other three which the sense perceives. Hence also the ratio of density of the so-called elements is arbitrarily fixed at ten to one. And so on of other dreams. And these fancies affect not dogmas only, but simple notions also.

>-·->-·-o-·<·-·-<

The human understanding, when it has once adopted an opinion (either as being the received opinion or as being agreeable to itself) draws all things else to support and agree with it. And though there be a greater number and weight of instances to be found on the other side, yet these it either neglects and despises, or else by some distinction sets aside and rejects; in order that by this great and pernicious predetermination the authority of its former conclusions may remain inviolate. And therefore it was a good answer that was made by one who when they showed him hanging in a temple a picture of those who had paid their vows as having escaped shipwreck, and would have him say whether he did not now acknowledge the power of the gods—'Aye,' asked he again, 'but where are they painted that were drowned after their vows?' And such is the way of all superstition, whether in astrology, dreams, omens, divine judgements, or the like; wherein men, having a delight in such vanities, mark the events where they are fulfilled, but where they fail, though this happen much oftener, neglect and pass them by. But with far more subtlety does this mischief insinuate itself into philosophy and the sciences, in which the first conclusion colours and brings, into conformity with itself all that come after, though far sounder and better. Besides, independently of that delight and vanity which I have described, it is the peculiar and perpetual error of the human intellect to be more moved and excited by affirmatives than by negatives; whereas it ought properly to hold itself indifferently disposed towards both alike. Indeed in the establishment of any true axiom, the negative instance is the more forcible of the two.

The human understanding is moved by those things most which strike and enter the mind simultaneously and suddenly, and so fill the imagination; and then it feigns and supposes all other things to be somehow, though it cannot see how, similar to those few things by which it is surrounded. But for that going to and fro to remote and hetergeneous instances, by which axioms are tried as in the fire, the intellect is altogether slow and unfit, unless it be forced thereto by severe laws and overruling authority.

The human understanding is unquiet; it cannot stop or rest, and still presses onward, but in vain. Therefore it that we cannot conceive of any end or limit to the world, but always as of necessity it occurs to us that there is something beyond. Neither again can it be conceived how eternity has flowed down to the present day, for that distinction which commonly received of infinity in time past and in time to come can by no means hold; for it would thence follow that one infinity is greater than another, and that infinity is wasting away and tending to become finite. The like subtlety arises touching the infinite divisibility of lines, from the same inability of thought to stop. But this inability interferes more mischievously in the discovery of causes: for although the most general principles in nature ought to be held merely positive, as they are discovered, and cannot with truth be referred to a cause; nevertheless the human understanding being unable to rest still seeks something prior in the order of nature. And then it is that in struggling towards that which is further off it falls back upon that which is more nigh at hand, namely, on final causes: which have relation clearly to the nature of man rather than to the nature of the universe, and from this source have strangely defiled philosophy. But he is no less an unskilled and shallow philosopher who seeks causes of that which is most general, than he who in things subordinate and subaltern omits to do so.

The human understanding is no dry light, but receives an infusion from the will and affections, whence proceed sciences which may be called 'sciences as one would.' For what a man had rather were true he more readily believes. Therefore he rejects difficult things from impatience of research; sober things, because they narrow hope; the deeper things of nature, from superstition; the light of experience, from arrogance and pride, lest his mind should seem to be occupied with things mean and transitory; things not commonly believed, out of deference to the opinion of the vulgar. Numberless in short are the ways, and sometimes imperceptible, in which the affections color and infect the understanding.

But by far the greatest hindrance and aberration of the human understanding proceeds from the dullness, incompetency, and deceptions of the senses; in that things which strike the sense outweigh things which do not immediately strike it, though they be more important. Hence it is that speculation commonly ceases where sight ceases, insomuch that of things invisible there is little or no observation. Hence all the working of the spirits inclosed in tangible bodies lies hid and unobserved of men. So also all the more subtle changes of form in the parts of coarser substances (which they commonly call alteration though it is in truth local motion through exceedingly small spaces) is in like manner unobserved. And yet unless these two things just mentioned be searched out and brought to light, nothing great can be achieved in nature, as far as the production of works is concerned. So again the essential nature of our common air, and of all bodies less dense than air (which are very many), is almost unknown. For the sense by itself is a thing infirm and erring; neither can instruments for enlarging or sharpening the senses do much; but all the truer kind of interpretation of nature is effected by instances and experiments fit and apposite; wherein the sense decides touching the experiment only, and the experiment touching the point in nature and the thing itself.

The human understanding is of its own nature prone to abstractions and gives a substance and reality to things which are fleeting. But to resolve nature into abstractions is less to our purpose than to dissect her into parts; as did the school of Democritus, which went further into nature than the rest. Matter rather than forms should be the object of our attention, its configurations and changes of configuration, and simple action, and law of action or motion; for forms are figments of the human mind, unless you will call those laws of action forms.

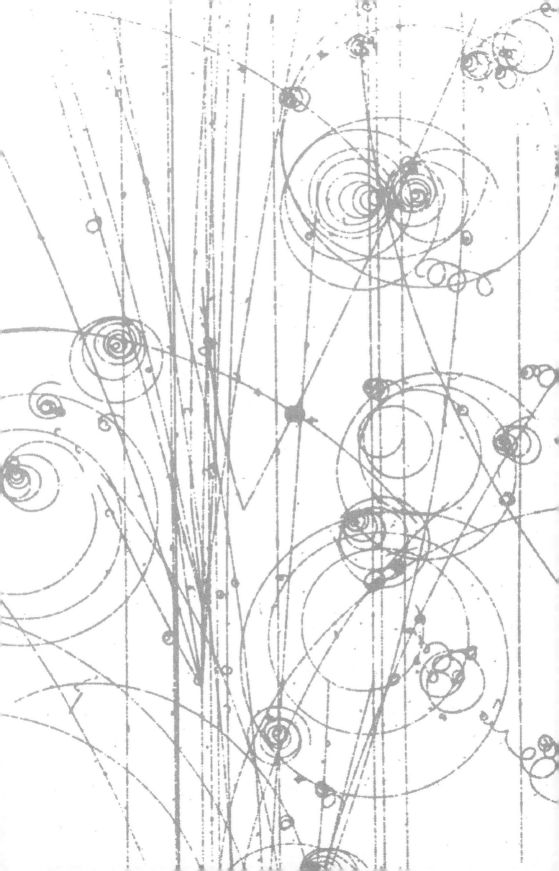

The Scientific Imagination in Action

"It was indeed a laborious task for me to discover how such effects could be accomplished in nature. Yet I finally found something that served me admirably. In a way it is almost unbelievable. I mean that it is astonishing and incredible to us, but not to Nature; for she performs with utmost ease and simplicity things which are even infinitely puzzling to our minds, and what is very difficult for us to comprehend is quite easy for her to perform."

—GALILEO, *THE TWO CHIEF WORLD SYSTEMS*

Although science is supposed to be logical, disciplined, and unified, the imaginations that support it are oddly shaped pegs. For one thing, they tend to be both round and square at the same time. When square, they remain fixed on the concrete experience. They look closely and see what is there, refusing to anticipate even the dot over the *i*. During their round stage, they are great generalizers that speculate abstractly and develop grand simplifications—"oversimplifications" Popper called them. The scientists' shape shifting comes out of an interior freedom to wonder about whatever catches their fancy. Thus Darwin pursued both the tiniest detail and the greatest generality of biological nature, and Lavoisier's chemical investigations led him to think about both language and physical substances. Meanwhile, Richard Feynman was trying to figure out why time changed from past to future and not the other way around. The scientific imaginations in this part seem so varied in taste, so contradictory in outlook, and so lawless in their focus that no well-shaped peg could expect any one person to encompass much of its range. Yet when you look at scientists, you often find much of this complexity and contrariness in single individuals.

This tangle is especially evident in the older science writers. All the ancients and the Renaissance figures in this anthology had extensive imaginations. Figures like Herodotus, Leonardo, Boyle, and Newton appear to have had something to contribute to anything that passed before their eyes. Our own century's scientists have seldom been as broad. Besides having a more specialized focus, their imaginations tend to stay within a narrower band along the possible whole. But taken together, the breadth of imagination has survived. Some of today's scientists are great observers; some are great thinkers; and others are even skilled at uniting ideas that once seemed distinct.

The selections in this part show the range of scientific imaginations needed to progress from wonder to explanation. I confess that the breadth seen here is wider than I had expected when I began collecting essays for this volume. Although I have been reading science prose all my life, I found that I had a somewhat constricted theory of scientific thinking. For example, I had thought that scientific observation without some prior notion to be tested was impossible, and I would have insisted that speculation without data was more metaphysical than scientific. But these assumptions were shown to be false. Science comes less from method than from attitude. In the current argument over so-called creationist science, for example, creationist researchers may follow the prescribed method, but their attitude— seeking to justify a tradition—is distinctly different from a scientist's effort to understand. We see this distinction between method and attitude again in the way that many advertising and marketing firms hunt for data. Their methods may allow them to justify their advertising claims, but their motive is selling rather than understanding.

All the essays in this part reflect this wish to understand. The chapters are arranged to lead readers along the progress of scientific imaginings. Chapter 4 contains the writings of curious observers, people who believe that facts are the means to understand mysteries. Although most of the writers in this chapter are from older times, this open curiosity remains vital to science. The modern contributor in this chapter is George Schaller, a biologist who has spent his adult life in wild places observing animal behavior.

Chapter 5 is just a stride further along the road. Writers are looking beyond the facts toward new theory. Robert Kennedy Duncan, for example, notes the peculiarities of radioactivity, but he also tries to make something new out of what he sees, and the bent of his mind makes him look in the natural world. Many people besides scientists are curious, but scientists want to understand nature in natural terms. That bent does not mean they must be secular or atheistic. James Clerk Maxwell's account of molecules ends with an explicit theological argument, but he knows and he says that at that point, the science has ended.

The next milestone comes in chapter 6, as scientists begin demanding new demonstrative evidence to buttress their explanations of events. Some of the scientists here, like Boyle and Galileo, are skeptical about ideas with long and honored pedigrees. Others have ideas of their own to pursue, but they are united in their demand for a demonstration that will justify an assertion about how nature acts.

By this stage in scientific progress, ideas and theories are being widely discussed. All the pieces in chapter 7 are about the science of others. Thomas Huxley, in the guise of "Darwin's bulldog," contributed much to biology by arguing for Darwin's theory. Voltaire became a sort of Newton's bulldog in Europe, helping make British science an international science.

Chapter 8 shows how much the look of science can differ from the observations of chapter 4. Yet these essays remain scientific because they are efforts to understand and to understand in natural terms. The last three pieces in this chapter are about quantum physics, and each grapples with the same quantum conundrum. They show how science, as it grows more surprising in its conclusions, forces a rethinking of the starting point.

The science of chapters 6, 7, and 8 can move into a loop as the speculations and ideas explored in chapter 8 provoke demands for evidence, as in chapter 6. Depending on how that research goes, more writing and disputation may follow. But eventually, if all goes well, the classic generalizations of the sort included in chapter 9 appear. In popular literature, this kind of ultimate breakthrough is what science is all about. They are transforming achievements.

"Brought Near to That Great Fact—That Mystery of Mysteries"

Charles Darwin

>—<>·-<>—<

GALILEO GALILEI

First Look Through a Telescope, from *The Sidereal Messenger* (1610)

The startling observations reported in this selection hurled Galileo Galilei (1564–1642) onto history's stage. Before 1610, he enjoyed honor among like-minded scholars, but after 1610 he was a famous man, known throughout Europe. It might seem that his celebrity was of the banal sort so common today, which would have come to anybody who first pointed a telescope toward the stars. Readers, however, will quickly discover what is wrong with that cynicism. Galileo did not just see through his telescope; he looked. His writing overflows with excitement and astonishment at the sights he reports, but all the while he was studying what he saw and observing what was new.

It is often said that when Galileo looked through his telescope, he found proof that the earth did go around the sun. But he saw no such thing and made no such claim. What he did report was that the night sky looked very different from the way the old theories insisted that it did. The moon was a place with landmarks like those on earth (and if that fact seems obvious, look at the selection by Kepler); the stars were more abundant than anyone had

dreamed; and Jupiter had planets of its own. All the old ideas, based on naked-eye observations, were suddenly stripped of their prestige and open to the challenge that new observations could be better.

The Moon

Let me review the observations made by me during the two months just past, [. . .] inviting the attention of all who are eager for true philosophy to the beginnings which led to the sight of most important phenomena.

Let me speak first of the surface of the Moon which is turned towards us. For the sake of being understood more easily, I distinguish two parts in it, which I call respectively the brighter and the darker. The brighter part seems to surround and pervade the whole hemisphere; but the darker part, like a sort of cloud, discolours the Moon's surface and makes it appear covered with spots. Now these spots, as they are somewhat dark and of considerable size, are plain to every one, and every age has seen them, wherefore I shall call them *great* or *ancient* spots, to distinguish them from other spots, smaller in size, but so thickly scattered that they sprinkle the whole surface of the Moon, but especially the brighter portion of it. These spots have never been observed by any one before me; and from my observations of them, often repeated, I have been led to that opinion which I have expressed, namely, that I feel sure that the surface of the Moon is not perfectly smooth, free from inequalities and exactly spherical, as a large school of philosophers considers with regard to the Moon and the other heavenly bodies, but that, on the contrary, it is full of inequalities, uneven, full of hollows and protuberances, just like the surface of the Earth itself, which is varied everywhere by lofty mountains and deep valleys.

The appearances from which we may gather these conclusions are of the following nature:—On the fourth or fifth day after new-moon, when the Moon presents itself to us with bright horns, the boundary which divides the part in shadow from the enlightened part does not extend continuously in an ellipse, as would happen in the case of a perfectly spherical body, but it is marked out by an irregular, uneven and very wavy line, as represented in the figure given, for sev-

eral bright excrescences, as they may be called, extend beyond the boundary of light and shadow into the dark part, and on the other hand pieces of shadow encroach upon the light:—nay, even a great quantity of small blackish spots, altogether separated from the dark part, sprinkle everywhere almost the whole space which is at the time flooded with the Sun's light, with the exception of that part alone which is occupied by the great and ancient spots. I have noticed that the small spots just mentioned have this common characteristic always and in every case, that they have the dark part towards the Sun's position, and on the side away from the Sun they have brighter boundaries, as if they were crowned with shining summits. Now we have an appearance quite similar on the Earth about sunrise, when we behold the valleys, not yet flooded with light, but the mountains surrounding them on the side opposite to the Sun already ablaze with the splendour of his beams; and just as the shadows in the hollows of the Earth diminish in size as the Sun rises higher, so also these spots on the Moon lose their blackness as the illuminated part grows larger and larger.

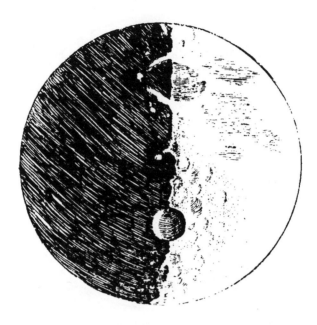

Again, not only are the boundaries of light and shadow in the Moon seen to be uneven and sinuous, but—and this produces still greater astonishment—there appear very many bright points within the darkened portion of the Moon, altogether divided and broken off from the illuminated tract, and separated from it by no inconsiderable interval, which, after a little while, gradually increase in size and brightness, and after an hour or two become joined on to the rest of the bright portion, now become somewhat larger; but in the meantime others, one here and another there, shooting up as if growing, are lighted up within the shaded portion, increase in size, and at last are linked on to the same luminous surface, now still more extended. An example of this is given in the same figure. Now, is it not the case on the Earth before sunrise, that while the level plain is still in shadow, the peaks of the most lofty mountains are illuminated by the Sun's rays? After a little while does not the light spread further, while the middle and larger parts of those mountains are becoming illuminated; and at length, when the Sun has risen, do not the illuminated parts of the plains and hills join together? The grandeur however, of such prominences and depressions in the Moon seems to surpass both in magnitude and extent the ruggedness of the Earth's surface. [. . .]

The Stars

The difference between the appearance of the planets and the fixed stars seems also deserving of notice. The planets present their discs perfectly round, just as if described with a pair of compasses, and appear as so many little moons, completely illuminated and of a globular shape; but the fixed stars do not look to the naked eye bounded by a circular circumference, but rather like blazes of light, shooting out beams on all sides and very sparkling, and with a telescope they appear of the same shape as when they are viewed by simply looking at them, but so much larger that a star of the fifth or sixth magnitude seems to equal Sirius, the largest of all the fixed stars.

But beyond the stars of the sixth magnitude you will behold through the telescope a host of other stars, which escape the unassisted sight, so numerous as to be almost beyond belief, for you may see more than six other differences of magnitude, and the largest of these, which I may call stars of the seventh magnitude, or of the first magnitude of invisible stars, appear with the aid of the telescope larger and brighter than stars of the second magnitude seen with the unassisted sight. [. . .]

Asterism of the belt and sword of Orion

Jupiter

There remains the matter, which seems to me to deserve to be considered the most important in this work, namely, that I should disclose and publish to the world the occasion of discovering and observing four PLANETS, never seen from the very beginning of the world up to our own times, their positions, and the observations made during the last two months about their movements and their changes of magnitude; and I summon all astronomers to apply themselves to examine and determine their periodic times, which it has not been permitted me to achieve up to this day, owing to the restriction of my time. I give them warning however again, so that they may not approach such an inquiry to no purpose, that they will want a very accurate telescope, and such as I have described in the beginning of this account.

On the 7th day of January in the present year, 1610, in the first hour of the following night, when I was viewing the constellations of the heavens through a telescope, the planet Jupiter presented itself to my view, and as I had prepared for myself a very excellent instrument, I noticed a circumstance which I had never been able to notice before, owing to want of power in my other telescope, namely, that three little stars, small but very bright, were near the planet; and although I believed them to belong to the number of the fixed stars, yet they made me somewhat wonder, because they seemed to be arranged exactly in a straight line, parallel to the ecliptic, and to be brighter than the rest of the stars, equal to them in magnitude. The position of them with reference to one another and to Jupiter was as follows.

On the east side there were two stars, and a single one towards the west. The star which was furthest towards the east, and the western star, appeared rather larger than the third. I scarcely troubled at all about the distance between them and Jupiter, for, as I have already said, at first I believed them to be fixed stars; but when on January 8th, led by some fatality, I turned again to look at the same part of the heavens, I found a very different state of things, for there were three little stars all west of Jupiter, and nearer together than on the previous night, and they were separated from one another by equal intervals, as the accompanying illustration shows. At this point, although I had not turned my thoughts at all upon the approximation of the stars to one another, yet my surprise began to be excited, how Jupiter could one day be found to the east of all the aforesaid fixed

stars when the day before it had been west of two of them; and forthwith I became afraid lest the planet might have moved differently
from the calculation of astronomers, and so had passed those stars by
its own proper motion. I therefore waited for the next night with the
most intense longing, but I was disappointed of my hope, for the sky
was covered with clouds in every direction.

But on January 10th the stars appeared in the following position
with regard to Jupiter; there were two only, and both on the east side
of Jupiter, the third, as I thought, being hidden by the planet. They
were situated just as before, exactly in the same straight line with
Jupiter, and along the Zodiac.

When I had seen these phenomena, as I knew that corresponding
changes of position could not by any means belong to Jupiter, and as,
moreover, I perceived that the stars which I saw had been always the
same, for there were no others either in front or behind, within a
great distance, along the Zodiac,—at length, changing from doubt
into surprise, I discovered that the interchange of position which I
saw belonged not to Jupiter, but to the stars to which my attention
had been drawn, and I thought therefore that they ought to be
observed henceforward with more attention and precision.

Accordingly, on January 11th I saw an arrangement of the following kind, namely, only two stars to the east of Jupiter, the nearer of which was distant from Jupiter three times as far as from the star further to the east; and the star furthest to the east was nearly twice as large as the other one; whereas on the previous night they had appeared nearly of equal magnitude. I therefore concluded, and decided unhesitatingly, that there are three stars in the heavens moving about Jupiter, as Venus and Mercury round the Sun; which at length was established as clear as daylight by numerous other subsequent observations. These observations also established that there are not only three, but four, erratic sidereal bodies performing their revolutions round Jupiter.

LEONARDO DA VINCI

Seashells in the Mountains, from
Notebooks (ca. 1480, 1506–1509, ca. 1515)

Along with his painting, sculpture, architecture, and engineering, Leonardo da Vinci (1452–1519) regularly jotted down scientific inquiries in a series of notebooks. Once, while he was working in Milan, some peasants brought him a sack of shells and corals they had found on the mountains of Parma and Piacenza. Leonardo then worried over their meaning for 25 years. His notebooks make it clear that he went and looked at the scene. Furthermore, unlike many of the thinkers of his age, Leonardo seems never to have doubted that these were the remains of sea creatures and not just "sports of nature," stones that merely looked like animal remains. What Leonardo wanted to understand was how seashells got to those mountains. He had an answer: the area had once been at the edge of the sea. By the end of his picking at the question, Leonardo was doubting not only the idea that Noah's flood might have put the fossils there but also that a flood had ever taken place.

Leonardo's meditations show a fine scientific imagination striving in an unscientific age. There were as yet no catalogs of sea species to tell Leonardo whether the creatures he saw were extinct or thriving. There were no geological maps of ancient seas to help him understand what he saw. So Leonardo was unable to push his observations toward proof. His notebooks indicate that scientific imagining is a human characteristic that might appear in any age but that without the organized companionship of like-minded people, science does not advance.

Leonardo's notebooks are not linear diaries but hodgepodges. The arrangement given here uses a very general dating proposed by the Leonardo scholar Carlo Pedretti.

[Entry from about the year 1480]

From the two lines of shells we are forced to say that the earth indignantly submerged under the sea and so the first layer was made; and then the deluge made the second.

[Entries from the years 1506-1509]

In this work you have first to prove that the shells at a thousand
braccia[1] of elevation were not carried there by the deluge, because
they are seen to be all at one level, and many mountains are seen to
be above that level; and to inquire whether the deluge was caused by
rain or by the swelling of the sea; and then you must show how, neither by rain which makes the rivers swell, nor by the overflow of this
sea, could the shells—being heavy objects—be floated up the mountains by the sea, nor have been carried there by rivers against the
course of their waters. [. . .]

That in the drifts, among one and another, there are still to be
found the traces of the worms which crawled upon them when they
were not yet dry. And all marine clays still contain shells, and the
shells are petrified together with the clay. Of the silliness and stupidity
of those who will have it that these animals were carried up to places
remote from the sea by the deluge. Another sect of ignorant persons
declare that Nature or Heaven created them in these places by celestial
influences, as if in these places we did not also find the bones of fishes
which have taken a long time to grow; and as if we could not count, in
the shells of cockles and snails, the years and months of their life, as
we do in the horns of bulls and oxen, and in the branches of plants
that have never been cut in any part. And having proved by these signs
the length of their lives, it is evident, and it must be admitted, that
these animals could not live without moving to fetch their food; and
we find in them no instrument for penetrating the earth or the rock
where we find them enclosed. But how could we find in a large snail-
shell the fragments and portions of many other sorts of shells, of various sorts, if they had not been thrown in when dead, by the waves of
the sea like the other light objects which it throws on the earth? Why
do we find so many fragments and whole shells between layer and
layer of stone, if this had not formerly been covered on the shore by a
layer of earth thrown up by the sea, and which was afterwards petrified? And if the before-mentioned deluge had carried them to these
parts of the sea, you might find these shells at the boundary of one

---

[1] *A measure based on the length of an arm, or about 2 English feet.*

drift but not at the boundary between many drifts. We must also account for the winters of the years during which the sea multiplied the drifts of sand and mud brought down by the neighbouring rivers by washing down the shores; and if you choose to say that there were several deluges to produce these rifts and the shells among them, you would also have to affirm that such a deluge took place every year. Again, among the fragments of these shells, it must be presumed that in those places there were sea-coasts where all the shells were thrown up, broken, and divided, and never in pairs, since they are found alive in the sea, with two valves, each serving as a lid to the other; and in the drifts of rivers and on the shores of the sea they are found in fragments. And within the limits of the separate strata of rocks they are found, separated and in pairs like those which were left by the sea, buried alive in the mud, which subsequently dried up and, in time, was petrified. [. . .]

Why do we find the bones of great fishes and oysters and corals and various other shells and sea-snails on the high summits of mountains by the sea, just as we find them in low seas?

You now have to prove that the shells cannot have originated if not in salt water, almost all being of that sort; and that the shells in Lombardy are at four levels, and thus it is everywhere, having been made at various times. And they all occur in valleys that open towards the seas. [. . .]

If you were to say that the shells which are to be seen within the confines of Italy now, in our days, far from the sea and at such heights, had been brought there by the deluge which left them there, I should answer that if you believe that this deluge rose seven cubits above the highest mountains—as he who measured it has written— these shells, which always live near the seashore, should have been left on the mountains and not such a little way from the foot of the mountains; nor all at one level, nor in layers upon layers. And if you were to say that these shells are desirous of remaining near to the margin of the sea, and that, as it rose in height, the shells quitted their first home, and followed the increase of the waters up to their highest level; to this I answer that the cockle is an animal of not more rapid movement than the snail is out of water, or even somewhat slower, because it does not swim, on the contrary, it makes a furrow in the sand; by leaning against the sides of this furrow it will travel each day from three to four braccia; therefore this creature, with so slow a motion, could not have travelled from the Adriatic sea as far as Monferrato in Lombardy, a distance of 250 miles, in 40 days; which he has said who took account of the time. And if you say that the waves carried them there, by their gravity they could not move, excepting at the bottom. And if you will not grant me this, confess at least that they would have to stay at the summits of the highest mountains, and in the lakes enclosed among the mountains, like the lakes of Lario, and il Maggiore, and of Como, and of Fiesole, and of Perugia, and others.

And if you should say that the shells were carried by the waves, being empty and dead, I say that where the dead went they were not far removed from the living; for in these mountains living ones are found, which are recognizable by the shells being in pairs; and they are in a layer where there are no dead ones; and a little higher up they are found, where they were thrown by the waves, all the dead ones with their shells separated, near to where the rivers fell into the sea, to a great depth; like the Arno which fell from the Gonfolina near to Monte Lupo where it left a deposit of gravel which may still be seen, and which has agglomerated; and of stones of various districts, natures, and colours, and hardness, making one single conglomerate. And a little beyond the sandstone conglomerate a tufa has been formed, where it turned towards Castel Florentino; farther on the mud was deposited in which the shells lived, and which rose in layers according to the levels at which the turbid Arno flowed into that sea. And from time to time the bottom of the sea was raised, depositing these shells in layers, as may be seen in the cutting at Colle Gonzoli, laid open by the Arno which is wearing away the base of it; in which cutting the said layers of shells are very plainly to be seen in clay of a bluish colour, and various marine objects are found there. And if the earth of our hemisphere is indeed raised by so much higher than it used to be, it must have become by so much lighter by the waters which it lost through the rift between Gibraltar and Ceuta; and all the more the higher it rose, because the weight of the waters which were thus lost would be added to the earth in the other hemisphere.

And if the shells had been carried by the muddy deluge they would have been mixed up and separated from each other amidst the mud, and not in regular steps and layers—as we see them now in our time.

As to those who say that shells existed for a long time and were born at a distance from the sea, from the nature of the place and of the cycles, which can influence a place to produce such creatures—to them it may be answered: such an influence could not place the animals all on one line, except those of the same sort and age; and not the old with the young, nor some with an operculum[2] and others without their operculum, nor some broken and others whole, nor some filled with sea-sand and large and small fragments of other shells inside the whole shells which remained open; nor the claws of crabs without the rest of their bodies; nor the shells of other species stuck on to them like animals which have moved about on them; since the traces of their track still remain, on the outside, after the manner of worms in the wood which they consume. Nor would there be found among them the bones and teeth of fish which some call arrows and others serpents' tongues, nor would so many portions of various animals be found all together if they had not been thrown on the seashore. And the deluge cannot have carried them there, because things that are heavier than water do not float on the water. But these things could not be at so great a height if they had not been carried there by the water, such a thing being impossible from their weight. In places where the valleys have not been filled with salt sea-water shells are never to be seen; as is plainly visible in the great valley of the Arno above Gonfolina, a rock formerly united to Monte Albano, in the form of a very high bank which kept the river pent up, so that before it could flow into the sea, which was then at its foot, it formed two great lakes; of which the first was where we now see the flourishing city of Florence together with Prato and Pistoia. And Monte Albano followed the rest of its bank as far as where Serravalle now stands. From the Val d'Arno upwards, as far as Arezzo, another lake was formed, which discharged its waters into the former lake. It was

---

[2] *A gill slit.*

closed at about the spot where now we see Girone, and occupied the whole of that valley above for a distance of 40 miles in length. [. . .]

A great quantity of shells are to be seen when the rivers flow into the sea, because on such shores the waters are not so salt owing to the admixture of the fresh water which is poured into it. Evidence of this is to be seen where, of old, the Apennines poured their rivers into the Adriatic Sea; for there in most places great quantities of shells are to be found, among the mountains, together with bluish marine clay; and all the rocks which are quarried in such places are full of shells. The same may be observed to have been done by the Arno when it fell from the rock of Gonfolina into the sea, which was not so very far below; for at that time it was higher than the top of San Miniato al Tedesco, since at the highest summit of this the shores may be seen full of shells and oysters within its flanks. The shells did not extend towards Val di Nievole, because the fresh waters of the Arno did not extend so far. [. . .]

If the deluge had to carry shells three hundred and four hundred miles from the sea, it would have carried them mixed with various other natural objects heaped together; and we see at such distances oysters all together, and sea-snails and cuttlefish, and all the other shells which congregate together, all to be found together and dead; and the solitary shells are found wide apart from each other, as we may see them on seashores every day. And if we find oysters of very large shells joined together and among them very many which still have the covering attached, indicating that they were left here by the sea, and still living when the strait of Gibraltar was cut through; there are to be seen, in the mountains of Parma and Piacenza, a multitude of shells and corals, full of holes, and still sticking to the rocks there. When I was making the great horse for Milan, a large sack full was brought to me in my workshop by certain peasants; these were found in that place and among them were many preserved in their first freshness. [. . .]

And if you were to say that these shells were created, and were continually being created in such places by the nature of the spot and of the heavens which might have some influence there, such an opinion cannot exist in a brain of much reason; because here are the years of their growth, numbered on their shells, and there are large and small ones to be seen which could not have grown without food, and could not have fed without motion—and here they could not move.

[Entries from about 1515]

Here a doubt arises, and that is: whether the deluge which happened at the time of Noah was universal or not. And it would seem not, for the reasons now to be given: We have it in the Bible that this deluge lasted 40 days and 40 nights of incessant and universal rain, and that this rain rose to ten cubits above the highest mountain in the world. And if it had been that the rain was universal, it would have covered our globe which is spherical in shape. And this spherical surface is equally distant, in every part, from the centre of its sphere; hence the sphere of the waters being under the same conditions, it is impossible that the water upon it should move, because water does not move of its own accord unless to descend; therefore how could the waters of such a deluge depart, if it is proved that it has no motion? and if it departed, how could it move unless it went upwards? Here, then, natural reasons are wanting; hence to remove this doubt it is necessary to call in a miracle to aid us, or else to say that all this water was evaporated by the heat of the sun. [. . .]

Let them show you where are the shells on Monte Mario.

CHARLES DARWIN

Birds in the Galapagos, from
The Voyage of H.M.S. Beagle (1839)

Charles Darwin (1809–1882) was famous even before he published his account of the origin of species. Twenty years earlier
his *Journal of Researches into the Geology and Natural History of
the Various Countries Visited by H.M.S.* Beagle had been a best-
seller. Darwin had gone on the voyage somewhat in the spirit of
Herman Melville's Ishmael: He was young, his prospects were uncertain, and the opportunity for adventure and the use of his training
(more geological than biological at that time) seemed like a salvation.
The *Beagle* visited places like Tierra del Fuego and the Galapagos
Islands, remote lands that most Europeans had never heard of. When
Darwin described these sites, the readers of 1840 did not know that
his research would lead to the century's greatest scientific idea, but
they could certainly tell that a keen observer was asking important
questions. In the Galapagos chapter, Darwin could see that the island
species were related to those of South America, yet they were different. He felt that the phenomenon touched on a "mystery of mysteries." By that time, enough fossils had been cataloged and geological
changes had been charted for all naturalists to know that species varied over time. They understood that the "old" species died off. But
where did new species come from? It could not have been exactly like
the description in Genesis. One obvious answer was that perhaps the
species themselves changed over time. Darwin's grandfather had
urged that idea, although nobody could find a natural explanation
for such a change. Not knowing, Darwin was forced to take the first
step in scientific progress: precise observation.

The natural history of [the Galapagos] islands is eminently curious, and well deserves attention. Most of the organic productions are
aboriginal creations, found nowhere else; there is even a difference
between the inhabitants of the different islands; yet all show a
marked relationship with those of America, though separated from
that continent by an open space of ocean, between 500 and 600 miles

in width. The archipelago is a little world within itself, or rather a satellite attached to America, whence it has derived a few stray colonists, and has received the general character of its indigenous productions. Considering the small size of these islands, we feel the more astonished at the number of their aboriginal beings and at their confined range. Seeing every height crowned with its crater, and the boundaries of most of the lava-streams still distinct, we are led to believe that within a period, geologically recent, the unbroken ocean was here spread out. Hence, both in space and time, we seem to be brought somewhat near to that great fact—that mystery of mysteries—the first appearance of new beings on this earth. [. . .]

Of land-birds, I obtained twenty-six kinds, all peculiar to the group and found nowhere else, with the exception of one lark-like finch from North America *(Dolichonyx oryzivorus)*, which ranges on that continent as far north as 54°, and generally frequents marshes. The other twenty-five birds consist, firstly, of a hawk, curiously intermediate in structure between a Buzzard and the American group of carrion-feeding Polybori; and with these latter birds it agrees most closely in every habit and even tone of voice. Secondly, there are two owls, representing the short-eared and white barn-owls of Europe. Thirdly, a wren, three tyrant fly-catchers (two of them species of Pyrocephalus, one or both of which would be ranked by some ornithologists as only varieties), and a dove—all analogous to, but distinct from, American species. Fourthly, a swallow, which though differing from the Progne purpurea of both Americas, only in being rather duller coloured, smaller, and slenderer, is considered by Mr. Gould[3] as specifically distinct. Fifthly, there are three species of mocking-thrush—a form highly characteristic of America. The remaining land-birds form a most singular group of finches, related to each other in the structure of their beaks, short tails, form of body, and plumage: there are thirteen species, which Mr. Gould has divided into four sub-groups. All these species are peculiar to this archipelago; and so is the whole group, with the exception of one species of the sub-group

---

[3] *John Gould (1804–1881), a British ornithologist whose celebrated illustrations of bird life were popular in the nineteenth century. Many of the species he identified are still recognized by biologists.*

1. Geospiza magnirostris.   2. Geospiza fortis.
3. Geospiza parvula.   4. Certhidea olivacea.

FINCHES FROM GALAPAGOS ARCHIPELAGO

Cactornis, lately brought from Bow Island, in the Low Archipelago. Of Cactornis, the two species may be often seen climbing about the flowers of the great cactus-trees; but all the other species of this group of finches, mingled together in flocks, feed on the dry and sterile ground of the lower districts. The males of all, or certainly of the greater number, are jet black; and the females (with perhaps one or two exceptions) are brown. The most curious fact is the perfect gradation in the size of the beaks in the different species of Geospiza, from one as large as that of a hawfinch to that of a chaffinch, and (if Mr. Gould is right in including his sub-group, Certhidea, in the main group) even to that of a warbler. The largest beak in the genus Geospiza is shown in Fig. 1, and the smallest in Fig. 3; but instead of their being only one intermediate species, with a beak of the size shown in Fig. 2, there are no less than six species with insensibly graduated beaks. The beak of the sub-group Certhidea, is shown in Fig. 4. The beak of Cactornis is somewhat like that of a starling; and that of the fourth sub-group, Camarhynchus, is slightly parrot-shaped. Seeing this gradation and diversity of structure in one small, intimately related group of birds, one might really fancy that from an original paucity of birds in this archipelago, one species had been taken and modified for different ends. In a like manner, it might be fancied that a bird originally a buzzard, had been induced here to undertake the office of the carrion-feeding Polybori of the American continent.

Of waders and water-birds I was able to get only eleven kinds, and of these only three (including a rail confined to the damp summits of the islands) are new species. Considering the wandering habits of the gulls, I was surprised to find that the species inhabiting these islands is peculiar, but allied to one from the southern parts of South America. The far greater peculiarity of the land-birds, namely twenty-five out of twenty-six being new species or at least new races, compared with the waders and web-footed birds, is in accordance with the greater range which these latter orders have in all parts of the world. We shall hereafter see this law of aquatic forms, whether marine or fresh-water, being less peculiar at any given point of the earth's surface than the terrestrial forms of the same classes, strikingly illustrated in the shells, and in a lesser degree in the insects of this archipelago.

Two of the waders are rather smaller than the same species brought from other places: the swallow is also smaller, though it is doubtful whether or not it is distinct from its analogue. The two owls, the two tyrant fly-catchers (Pyrocephalus) and the dove, are also smaller than the analogous but distinct species, to which they are most nearly related; on the other hand, the gull is rather larger. The two owls, the swallow, all three species of mocking-thrush, the dove in its separate colours though not in its whole plumage, the Totanus, and the gull, are likewise duskier coloured than their analogous species; and in the case of the mocking-thrush and Totanus, than any other species of the two genera. With the exception of a wren with a fine yellow breast, and of a tyrant fly-catcher with a scarlet tuft and breast, none of the birds are brilliantly coloured, as might have been expected in an equatorial district. Hence it would appear probable, that the same causes which here make the immigrants of some species smaller, make most of the peculiar Galapageian species also smaller, as well as very generally more dusky coloured. All the plants have a wretched, weedy appearance, and I did not see one beautiful flower. The insects, again, are small sized and dull coloured, and, as Mr. Waterhouse informs me, there is nothing in their general appearance which would

have led him to imagine that they had come from under the equator.
The birds, plants, and insects have a desert character, and are not
more brilliantly coloured than those from southern Patagonia; we
may, therefore, conclude that the usual gaudy colouring of the in-
tertropical productions, is not related either to the heat or light of
those zones, but to some other cause, perhaps to the conditions of
existence being generally favourable to life.

GEORGE B. SCHALLER

Mating Seasons, from
Stones of Silence (1980)

George Schaller (1933–) is a biologist in the naturalist tradition. He travels, observes, and writes about what he sees. He went into the equatorial forest and spent years observing gorillas. He spent time on Africa's Serengeti Plain watching lions. He went to South America to observe an ecosystem and then on to Asia. His ambition is to look and report. Compared with the laboratory analysis of bio-chemicals or paleontological digging for fossils, Schaller's brand of research seems romantic and old-fashioned, but even so, this kind of work has transformed our knowledge of animals in the wild. When Schaller first went to observe gorillas, we knew very little about what free animals do with themselves. Today that ignorance has been exchanged for a mass of facts.

Frequently Schaller's travels result in two books-—the first, a technical volume written in academic prose that presents his data and, the second, a more readable account, again reporting what he observed, but this time giving readers a sense of the look and feel of what he saw. The selection here is an account of the mating season of the wild goats of the Himalayas. Schaller puts you there.

More than a hundred million years ago, during the early Cretaceous, that land which is now the Himalayan region was cov-ered by the Tethys Sea, which separated Eurasia from the southern continent, Gondwanaland. [. . .] The *Caprinae*, the subfamily to which sheep and goats belong, arose during the Miocene when mountains appeared and the Tethys Sea was in its final retreat. And today's sheep and goats made their appearance in the late Pliocene and Pleistocene when the earth once again buckled and climates cooled. Sheep and goats colonized many ranges of Eurasia together. They usually occupied simple habitats, such as deserts and terrain recently vacated by ice, where plant growth is sparse. [. . .] The goats—wild goat, markhor, ibex, depending on area—prefer cliffs and their immediate vicinity, whereas sheep occupy the plateaus above cliffs and the undulating terrain along their bases. [. . .]

At first glance sheep and goats would seem to be rather dissimilar animals. However, their bones are notoriously difficult to place into the correct genus, and the animals are genetically similar enough to be able to produce living hybrids on occasion. [. . .]

For several years I had been intrigued by the possible similarities and differences in the ecology and behavior of Eurasian sheep and goats, wondering, for instance, how life on cliffs affected goat society and how it differed from that of sheep. As one step in gaining such an understanding, I was now in the Karchat Hills to study wild goat. The Karchat Hills consist of a small masif, about thirteen miles long and four wide, at the southern end of the Kirthar Range bordering the western edge of the Indus Plain. [. . .] Herds of black-haired goats, thin, bony creatures, scour the terrain, leaving only thorny and ill-tasting plants in their wake. Had man not misused this land for thousands of years, I would be driving through woodland, with wild asses standing in the broad-crowned shade of acacias and cheetah stalking unsuspecting Indian gazelle through swards of golden grass. Perhaps down by the river a pride of lions would be resting after the night's hunt. The forests are gone now, the rivers dry except after a downpour, and the lion, cheetah, and asses are dead. Only a few gazelle remain. No wonder the land seems lonely as one drives toward the distant hills, trailing a funnel of red dust made incandescent by the sun. [. . .]

The bloody sun of morning finds me scanning the plateau for wild goat. The sky is yellow with sand, and sharp gusts of wind still hammer my body. Finally, far away I note a white spot moving up a cliff; and then, through the scope, a section of cliff moves as dozens of wild goat—large silver-colored males, gray-brown females, and youngsters—file out of a canyon to feed on the plateau. I count sixty-eight. Females resemble young males at a distance, and I am too far away to classify each animal according to age and sex. Shy from being hunted, the wild goat must be approached with care. I angle into a defile. The limestone along its bottom has been scoured smooth by rare torrents, and potholes have been carved into the rock. Brush grows here and stunted trees; a white-checked bulbul calls. I climb out of the canyon, and protected by a low ridge, follow the plateau until I judge the goats to be near. They are still ahead and below, about five hundred feet away. Lying with the scope propped on my rucksack, I watch them.

Most are foraging, cropping brittle tufts of grass that grow in cracks where soil has accumulated, and nibbling on the leafless twigs of *Leptodenia* and fleshy leaves of *Capparis*. Five adult males, sleek and glossy, have congregated in a bachelor club at the herd's edge where they walk around with stately steps as if to impress each other and the nearby females. It is now early September, and in preparation for the rut, they have molted from drab summer pelage to shining nuptial coats. Their necks and backs are a shining silver-gray, but their faces and chests are black, as are the collars encircling their shoulders and the dark crests of hair following along their spines. Sweeping up and back like scimitars, are thinly elegant horns over forty inches long. Seeing such magnificent beasts, it is difficult to remember that they are the probable ancestors of our domestic goats. Adorned with conspicuous horns and a striking pelage, wild goat males are designed by evolution to impress others, they are perambulatory status symbols. Any herd member knows at a glance that here is a prime animal, powerful, adult both physically and psychologically, a male which by his mere survival has shown himself to be of superior stock. Few dare challenge such an individual even in such minor matters as the right-of-way along a ledge. There is a dominance hierarchy among adult males based on size, especially the size of horns. Growing a bit more each year, horns reflect the age and by implication the strength of the bearer. Since any male usually defers to another with larger horns, wasteful interactions are avoided. Of course, wild goats do fight on occasion. Like markhor, they sometimes spar, pushing horn against horn, or one may rear up and then plunge down to bash the horns of his opponent. But not once this morning, in over an hour of observation, do I see a fight. Strife is most likely during the rut, which I suspect has not yet begun.

As the sun grows hot on my back and bleaches sky and hills, the goats become lethargic. Some recline, others drift toward a canyon. One by one they move out of sight to spend the day beneath trees, in shallow caves, or in other fragments of shade. Not until late afternoon will they reappear. I amble circuitously back toward camp, along the rims of canyons and the bottoms of ravines; I explore secret clefts known only to porcupines, and I search for traces of leopard. Probably no more than two leopards hunt this part of the Karchat Hills. Wild and domestic goats are their main food, supplemented with whatever else they can catch—porcupine, hare, gazelle, striped hyena.

Although I occasionally find leopard scrapes and chalky white drop-
pings on the stony trails, I never see the animals themselves. Life is
sparse and cautious in these desert mountains. I peer into *Euphorbia*
thickets; beneath the spiny arms of these plants many night creatures
seek refuge, among them Indian mongoose and long-eared hedgehog.
During their first fragile years acacia seedlings find shelter there too,
but later repay this hospitality by killing their hosts with shade. [. . .]

My favorite guide on trips to new areas is Bachhal, a short, bow-
legged Sindhi from a nearby village. His face is pock-marked; he has
a black beard, and an untidy plaid turban swathes his head. Dressed
in a loose shirt, baggy trousers, and frayed sandals, he looks unprepossessing, with an ugliness that is somehow decorative. Not long
after our arrival, I decided to check the cliffs on the other side of the
plateau. Bachhal leads the way, and another local man, a game guard
employed by the Forest Department, brings up the rear. I only
remember his face, not his name. He is lean and taciturn, his eyes
stare like loaded pistols, his golden earring flashes. As usual we cross
the gravel plain toward the escarpment. In the shadows of Kira,
beside some boulders, is a saw-scaled viper, pale like the desert rocks;
it is often there before the first flares of sunrise penetrate the canyon
and I always look for it—acquaintances are difficult to come by here.
The guides and I walk to the crest of the plateau. As always the wind
is with us, sometimes subdued, sometimes wildly panting as it comes
across the maze of canyons and ridges. Bachhal borrows my binocu-
lars and scans the terrain for wild goats, his crouched little body like
a promontory of rock. Then I look, too. There is no sign of life
except for the forlorn cooing of a collared dove in a nearby ravine.
We walk on. The heat makes us indolent and we soon halt, retreating
into a shallow cavern. I eat an orange; within a few minutes its peel-
ings are shriveled and hard. Bachhal has a goatskin with water,
smoky in taste and fetid in odor. Small bees are attracted by the moisture and swarm into our refuge, clustering on the goatskin and clinging to our faces and arms, but the sun is too hostile to tempt us back
into the open. Instead we withdraw into ourselves, each of us lying in
a quiet torpor, ignoring thirst, heat, bees.

At 4:00 P.M. we continue our exploration. In the depths of a
ravine is a pool, the water slimy green with dead wasps floating in it.
According to Bachhal there are only two such pools and a tiny spring
in the whole southern half of the Karchat Hills at this season. The

pools will soon be dry. Desert wildlife must be able to subsist without drinking, and indeed, wild goat live for months each year on whatever moisture they can obtain from their forage. Like many animals of arid lands they conserve body fluids. They concentrate their urine, and waste little water in defecation, eliminating dry, crumbly pellets. Like some antelopes they may also have a flexible temperature-regulating mechanism; instead of eliminating excess heat through perspiration, they permit the body temperature to rise several degrees, thereby conserving water until at night their temperature drops back to normal. It is almost dusk when we descend to the plains on the other side of the range. [. . .]

In the morning we continue our search for wild goat, arcing northward over the plateau and then home. Based on this trip and others, I estimate a population of four to five hundred goats in the Karchat Hills. Most of them are in small herds with fewer than twenty individuals but occasionally as many as one hundred may briefly band together.

The 1972 rut began slowly. [. . .] As with markhor, the male goat often walks around with his tail folded up over his rump, presumably wafting enticing aromas from his tail glands, and occasionally one squirts urine on his face and chest. Restlessly he wanders, stopping to sniff or nuzzle the rump of a female, checking whether by chance she is approaching estrus. More often than not she trots aside, avoiding the crass overture. Persistently, the male may then tempt her with a display: holding neck low and muzzle forward, he approaches the females in a slight crouch, perhaps rotating his head while twisting his horns away from her when close. And sometimes he raises a foreleg as if to kick. All the while his tongue flicks and occasionally he grunts. Such displays probably arouse her interest or at least overcome her aversion. Sometimes she halts to urinate, but while he sniffs the urine, she hurries away. Indeed, she gives the impression of having urinated to divert his attention.

By recording how often males of various ages display, [. . .] I become aware that the rut's tempo increases dramatically on September 18 and 19. Instead of giving a cursory head twist or a casual kick, males now display again and again, insistently demanding attention. The adults are most active. Not that young males lack interest, but whenever one tarries near a female he is displaced, the mere approach of a powerful adult usually being enough to cause retreat. If not, a horn threat or lunge forces him to flee. Animals can assert themselves aggressively in two ways. In one they do so directly, threatening an opponent overtly or actually attacking him. In the other, they do so indirectly, attempting to achieve dominance not by a test of strength but by means of intimidation through the use of rank symbols. Wild goat, we note, waste little time on indirect methods: they usually attack. Although competition may seem intense, actual combat is rare because males do have a rank order to which they usually adhere. Competition follows certain rules. For example, once a medium-sized four-and-a-half-year-old male twisted his head and then kicked behind a female. Another male of the same age, but somewhat larger, hurried up, cut in front of the first, and appropriated the female without being challenged. An adult male then arrived and lunged at the second male so suddenly that the latter barely had time to turn around to catch the blow harmlessly with his horns. Another clash of horns convinced the smaller of the two to retreat. But the newcomer had barely claimed the female when a still larger adult appeared. With a self-assured jerk of his head he bid the rival leave and then followed the female. Thus, with a minimum of strife, the largest male appropriates the estrous female. If, as sometimes happens, two or three females are in heat simultaneously, the two or three largest males in the herd each claim one.

When a male finds a female in heat, he tends her closely. With his muzzle almost touching her rump, he follows her, moving and halting whenever she does. If she flees, he pursues, and the two then race in and out of ravines, along ledges, and down precipices. She may halt abruptly and butt her ardent suitor in the neck but he does not retaliate. Finally she accepts his overtures. As she walks slowly he kicks his foreleg behind her and twists his head sideways. Gently he licks her neck and just as gently he mounts. She may then rub her face against his. He mounts once more, and then again. Young males sometimes trail behind the courting pair, keeping a safe distance.

The wild goat rut reached its peak in early October. With a gestation period of about 165 days, most young would be born in March. And on March 1, the following spring, I was back at Karchat. It is somewhat cooler than in autumn and even drier. The grass is a hard stubble and most trees have shed their leaves. There is a Pashtu proverb that describes the desert seasons: spring is teeming, summer sweltering, autumn sickly, and winter needy. In this gaunt land every season looks needy to my undiscerning eye. Only after a rare summer rain are the hills briefly verdant. Yet a few trees select the harsh spring sea-son in which to bloom; there are large orange blossoms of *Tecoma,* strangely flamboyant amidst all this restraint, and pink-flowering pen-dants of *Capparis* which attract iridescent purple sunbirds.

The wild goat are around the same cliffs as during the previous autumn, many males and females still together with some of the latter heavily pregnant. Some of the females are alone among inaccessible precipices, so reluctant to leave that it is obvious their newborns must be concealed in a nearby retreat. Finally, after several days of search-ing, I see a young goat following its mother. It is like a tiny gray hare, and I am amazed that its shaky legs can carry it safely over such jagged terrain. However, young travel little for the first three to four days after birth, remaining hidden until their gait has steadied. Judg-ing by their slack abdomens and enlarged udders, most females have given birth by mid-March, but no more than a third of them are ac-companied by young. I wait. Surely more females have young cached somewhere. But reluctantly I come to a grim conclusion: about half of the young died at or within a few days after birth. In the following months half of the survivors will die too. By autumn only one female out of five still has a young at heel. What is wrong?

Accidents, predation, disease, starvation, so many factors con-spire against a young between birth and weaning that one can only marvel that any survive at all. However, nutrition of both mother and young is the most critical factor determining success or failure of a breeding season. Studies of domestic sheep show that poor nutrition decreases ovulation rates and causes loss of ova. This may explain why some wild goat females apparently failed to produce young, and why few if any had twins. Their nutrition during the rut certainly had been poor. The Karchat Hills had been gripped by drought for several years, and the summer of 1972 was especially severe. Monsoon clouds gathered, then dissipated, releasing only three brief sprinkles. By contrast, the summer of 1973 brought more rain, several heavy showers greening the land, and the next spring twice as many young appeared, though still few if any twins. Another study of domestic sheep shows that if ewes are put on a poor diet during the second half of pregnancy, many young will die within the first four days after birth. Poorly fed ewes not only lack interest in their lambs, but also give less milk. However, if nutrition is ample, the fetus deposits fat, affecting its subsequent chances of survival, and the ewes give more milk. Valerius Geist has pointed out that the length of time a young suckles is an indication of the mother's milk supply. Wild goat females permitted young to drink for an average of only fourteen seconds before abruptly stepping aside; they apparently had little milk. It thus seemed that not only were the young born weak, but also they were deprived of milk afterward.

Not every year is one of hardship for the young. Conditions in 1975 were good, as stated in a letter to me from Tom Roberts:

> I hardly recognized the area because of the changed landscape following very good rains last summer and considerable rains this winter. There was green grass and vegetation everywhere and the place looked like savannah rather than desert. . . . We had nice views of 2 wild goats going to where they had concealed their babies and were pleased to note twins in both cases.

How tenuous is life in the desert. One good summer shower can determine whether a young will survive its birth the following spring.

Having observed one rut and one birth season, I thought that the annual reproductive cycle of wild goat at Karchat had been accurately delineated. Confidently I returned on March 20, 1974—a date on which in the previous year some females were still pregnant—and found the birth season well over. I came to watch the rut on October 9, 1974—a time when in 1972 mating was at its peak—and found it completely finished. Obviously the time of rut and subsequent birth season can vary from year to year. I should have suspected this for Tom Roberts had written a short report in which he noted that newborns are observed "from mid-January to early February." Therefore the rut may peak in mid-August in some years, at least one and a half months earlier than in 1972. Good food may advance estrus in domestic ewes by as much as twenty days. Wild goat live in an unpredictable environment where their forage supply depends on a few erratic showers. By keeping their mating season flexible, they can take advantage of a sudden rain and the subsequent nutritious forage. Being well fed, most females will conceive, and if conditions remain favorable, many will have twins and most young will survive. But during droughts most young will die.

"But What Are They?"

Robert Kennedy Duncan

HERODOTUS

The Creation of Egypt, from
The History (ca. 444 B.C.)

Herodotus (ca. 484 B.C.–ca. 425 B.C.) was the first great prose writer in the Western tradition, and his *History* includes the oldest science writing that has survived to our own day. His account of the silting of the Nile is based on personal observation, on curiosity about the natural processes that account for the observations, and on speculation about what these processes mean generally for the natural world. It is in his speculation that we see a scientific imagination take a step beyond straightforward observation toward making a generalization about a natural process. His conclusion that—in what we would call the geologically recent past—Egypt had been a basin like the Red Sea is false. Geologists now believe the Nile to be a very ancient river, although at one time it did flow east to the Red Sea rather than north to the Mediterranean. But despite the error, understanding grew. Readers come away from this passage appreciating the basic geological doctrine that the world they see is the result of natural changes and that future changes will take this world away.

My own observation bears out the statement made to me by the priests that the greater part of [Egypt] has been built up by silt from the Nile. I formed the opinion that the whole region above Memphis between the two ranges of hills was originally a gulf of the sea, and resembles (if I may compare small things with great) the country around Troy, Teuthrania,

Ephesus and the plain of the Maeander—not that any of the rivers which have caused the alluvial deposits in those neighbourhoods are comparable in size to any one of the five mouths of the Nile. There are other rivers too I could mention, much smaller than the Nile, which have effected important changes in the coastline: for instance, the Achelous which flows through Acarnania and has already joined to the mainland half the islands of the Echinades group.

In Arabia not far from Egypt there is a very long narrow gulf, [the Red Sea]. It is only half a day's voyage across in its narrowest part, but its length from its extreme limit to the open sea is a voyage of forty days for a vessel under oars. It is tidal. Now it is my belief that Egypt itself was originally some such arm of the sea—there were two gulfs, that is, one running from the Mediterranean southwards towards Ethiopia, and the other northwards from the Indian Ocean towards Syria, and the two almost met at their extreme ends, leaving only a small stretch of country between them. Suppose, now, that the Nile should change its course and flow into this gulf—the Red Sea—what is to prevent it from being silted up by the stream within, say, twenty thousand years? Personally I think even ten thousand would be enough. That being so, surely in the vast stretch of time which has passed before I was born, a much bigger gulf than this could have been turned into dry land by silt brought down by the Nile—for the Nile is a great river and does, in fact, work great changes. So I not only believe the people who gave me this account of Egypt, but my own conclusions strongly support what they said. I have observed for myself that Egypt at the Nile Delta projects into the sea beyond the coast on either side; I have seen shells on the hills and noticed how salt exudes from the soil to such an extent that it affects even the pyramids; I have noticed, too, that the only hill where there is sand is the hill above Memphis, and—a further point—that the soil of Egypt does not resemble that of the neighbouring country of Arabia, or of Libya, or even of Syria [. . .], but is black and friable as one would expect of an alluvial soil formed of the silt brought down by the river from Ethiopia. The soil of Libya is, as we know, reddish and sandy, while in Arabia and Syria it has a larger proportion of stone and clay. I had from the priests another striking piece of evidence about the origin of the country: namely that in the reign of Moeris the whole area below Memphis used to be flooded when the river rose only twelve feet—and when I got that information Moeris had been dead for less than nine hundred years. To-day, however, the river never floods unless it rises at least twenty-three and a half, or twenty-four, feet.[1] It seems to me therefore that, if the land continues to increase at the same rate in height and extent, the Egyptians who live below Lake Moeris in the Delta and thereabouts will, if the Nile fails to flood, suffer permanently the same fate as they say would some day overtake the Greeks; for when they learned that all Greece is watered by rain and not, as Egypt is, by the flooding of rivers, they remarked that the day would come when the Greeks would be sadly disappointed and starve—in other words, if God sees fit to send no rain but afflicts us with a drought, we shall all die from famine because we have no source of water other than the rain which God chooses to grant us. All this is only too true—but let me point out in answer how the case stands with the Egyptians themselves: if, as I said before, the land below Memphis (for this is the part which is always rising) continues to increase in height at the same rate as in the past, is it not obvious that when the river can no longer flood the fields—and there is no chance of rain either—the people who live there will have to go hungry? As things are at present these people get their harvests with less labour than anyone else in the world, the rest of the Egyptians included; they have no need to work with plough or hoe, or to use any other of the ordinary methods of cultivating their land; they merely wait for the river of its own accord to flood their fields; then, when the water has receded, each farmer sows his plot, turns pigs into it to tread in the seed, and then waits for the harvest. [. . .]

The Ionians maintain that Egypt proper is confined to the Nile Delta, a stretch of country running along the coast from what is known as Perseus' Watchtower to the Pelusian Salt-pans—a distance of forty *schoeni*[2]—and inland as far as Cercasorus, where the Nile

---

[1] *In paleolithic Egypt, there was an ancient lake, now called Lake Moeris, that shrank and silted exactly in the manner Herodotus imagined that the whole Nile Valley was silting up. About 1200 years before Herodotus's visit, a pharaoh of the Middle Kingdom, probably Ammenemes III (1842–1797 B.C.), had the lake's connection to the Nile widened and deepened. By the time Herodotus arrived, the story of this reclamation project had become a bit garbled.*

divides into the two branches which enter the sea at Pelusium and Canopus. The rest of what is usually called Egypt belongs, according to this view, either to Libya or Arabia. If, therefore, we accept it, we are forced to the conclusion that there was a time when the Egyptians had no country at all; for I am convinced—and the Egyptians themselves admit the fact—that the Delta is alluvial land and has only recently (if I may so put it) appeared above water. If, then, they once had no place to live in, why did they make such a business of the theory that they are the oldest race in the world? [. . .] But the fact is, I do not believe that the Egyptians came into being at the same period as the Delta (as the Ionians call it); on the contrary, they have existed ever since men appeared upon the earth, and as the Delta increased with the passage of time, many of them moved down into the new territory and many remained where they originally were.

▸┼◂▸─O─◂▸┼◂

---

[2] *Schoeni, a Persian measure of distance equal to 5 ± 2 miles; thus, 40 schoeni was between 120 and 280 miles. The Nile Delta today, as measured along the coastline, is about 200 miles long.*

HORACE BÉNÉDICT DE SAUSSURE

The Movement of Glaciers, from
Travels in the Alps (1796)

The Swiss physicist Horace Bénédict de Saussure (1740–1799) is credited with coining the word *geology*. His classic *Travels in the Alps* presents the results of more than thirty years of geological observations in the mountains around him. The selection given here reflects Saussure's physics training. He could see that traditional explanations for glacial movements and the changes they wrought could not be right because their physics did not make sense. Like Herodotus two thousand years before him, Saussure is trying to understand the ancient processes of change. Unlike Herodotus, Saussure lived in a time when many physical processes were already understood, so he was better able to investigate an entirely novel observation.

Of course, Saussure's ideas were incomplete, and later geologists would have to reconsider glaciers and their movements, especially in light of the ice-age theory that proposed ice caps of a power and force that Saussure dismissed as inconceivable. But we would miss his contribution if we considered only his mistakes. Before Saussure, science knew nothing of glaciers, their movement, the debris they carry with them, or the mounds they create.

The excerpt presented here also introduces the word *moraine,* a technical term for a mound created by a glacier. Saussure reports it as a term used by mountain peasants. Before Saussure, scientists were not aware of what these peasants knew.

The sliding of snow masses in the form of avalanches is a well-known phenomenon and one to which we shall have occasion to refer elsewhere. That of glaciers, taking place more slowly and ordinarily with less noise, is not so well known.

Nearly all the glaciers [. . .] repose on inclined floors, and all those of any considerable size have currents of water below them even in winter, which flow between the ice and the floor on which it rests.

One understands then that these frozen masses, drawn by the slope of the floor on which they rest, loosened by the waters from any binding they might have formed with this same floor, sometimes even raised by these waters, must slide little by little and descend along the slope of the valleys or hills that they cover.

It is this slow but continual sliding of the ice masses on their inclined bases that carries them as far as the low valleys and maintains the heaps of ice in lowlands warm enough to produce great trees and even abundant harvests. In the bottom of the valley of Chamonix, for example, no glacier forms. Even the snow disappears in the months of May and June. Nevertheless the glaciers of Buisson, Bois, and Argentière descend to the bottom of this valley. But the lower ice of these glaciers has not been formed in this place. It carries, so to speak, the evidence of the place of its birth, for it is full of debris from the rocks that border the highest extremity of the glacial valley, and these rocks differ from those found in the mountains that border the lower part of this same

All the great glaciers have, at their lower end and along their edges, great heaps of sand and debris, products of landslides from the mountains that tower above them. Often the glaciers are even encased for their entire length by a species of parapet or earthworks composed of the same debris, which the lateral ice of these glaciers has deposited on their ridges. In glaciers that were once larger than they are today, the parapets rise above the ice itself. On the other hand, in those which are greater than they ever have been before, these parapets are lower than the ice. Finally, some are seen where they are at the same level. The peasants of Chamonix call these piles of debris the moraine of the glacier.

The stones which heap up to form these parapets are mostly rounded, either because their angles have been dulled in rolling down from the mountain tops or because the ice has broken them by rubbing and holding them against its floor and edges. But those which have remained at the surface of the ice, without having undergone any great amount of rubbing, have kept their jagged corners intact and sharp. As for their composition, those found at the upper end of the glaciers are the same kinds of stone as the mountains that rise above them; but as the ice carries them down toward the valleys, they lie among mountains of a nature entirely different from theirs.

It seems somewhat more difficult to account for the heaps of rock and sand found piled in the middle of the *vallés de glace*, at such a great distance from the bordering mountains as to make it appear impossible that they came from them.

The stones are ordinarily arranged in lines parallel to the edge of the glacier, and one often sees many of these lines separated by bands of live, pure ice. On crossing the great glacier, two leagues above Montanvert, one has to pass over four or five earthworks of this kind, some of which are raised thirty or forty feet above the surface of the glacier. This is due in part to the quantity of stones gathered there, in part to the ice itself, which, sheltered from the sun and rain by the debris, stands higher than those parts that were originally higher but are bare and exposed to all the destructive action of the atmosphere.

I have known inhabitants of the Alps who, not knowing how to explain the origin of these ridges, said that the ice pushed up and thrust to the surface all foreign bodies found in its interior, and even the loose rocks and sand which were below it. But aside from the fact that such a force would be absolutely inconceivable, there is a still greater difficulty; namely, that the ice, as I have just said, is much higher under these benches of debris than in the rest of the glacier, so that the debris covers only the peaks of the ice, which are sometimes fifteen or twenty feet higher than the bare ice which separates them. It would be necessary then to suppose that the ice thrust itself up, and did that solely and precisely in the places where it is laden with the greatest weight. This is entirely absurd; all the more so because a perfect continuity is observed between this covered ice and that which is not covered—the same openings, the same fractures are seen continuing from one to the other so that it cannot be maintained that one originated at the bottom and the other at the surface. I believe the following is the true explanation of the phenomenon.

In the high Alps, as in the plains, mountains are found in such a state of decay that they continually detach fragments, either whole or crumbled to the form of earth and sand. And this takes place either because the mountains break naturally into fragments of different form or because the destructive action of the atmosphere wears them down and decomposes them. Especially in springtime—at the time of thaw, warm rain, and melting of snow—the particles of rock, sand, and earth that the frost has raised and moved fall on the ice in the high valleys. These stones, piled up on the edges of the glaciers, then follow the motion of the ice that carries them. But we have already seen that all this ice has a progressive movement, that it slides on its inclined floor, that it descends gradually to the low valleys, that there it is melted by the summer heat, and that what is thus destroyed is continually replaced by the progressive movement of the glacier. But the lower part of the glacial valleys is not the only place where ice melts. On fine summer days, especially when south winds prevail or warm rains fall, it melts throughout all the extent of the glaciers. The waters produced by this melting unite and form wide, deep ravines on the ice itself. The glaciers are divided by great crevasses, and as the valleys all have more or less the form of a cradle with their floor carved deeper than their sides, the ice compresses and contracts toward the middle of the valleys. The ice at the edge withdraws from the slopes, sliding toward the lowest point and carrying with it the earth and stones with which it is covered.

The proof of this truth is that toward the end of summer one sees in many localities, especially in the widest valleys, considerable spaces between the foot of the mountain and the edge of the glacier. These spaces arise, not only from the melting of the lateral ice but also because it is diverted from the margin in descending toward the center of the valley. During the next winter, these spaces are filled with snow. The snow absorbs water and is converted into ice. The edges of this new ice nearest to the mountain are covered anew with debris. These covered lines, in their turn, advance toward the center of the glacier. And thus are formed these parallel benches that move obliquely in a composite motion arising from the slope of the ground toward the center of the valley and from the slope of this same valley toward the foot of the mountain.

Finally, the proof of the origin of these benches is completed by the fact that none of them form in localities where glaciers are bordered by rocks of indestructible granite or when the slopes of the mountains are covered with snow or ice.

It seems, at first sight, that these parallel lines of sand and debris ought to mark the years and so serve to determine the age of different parts of the glaciers. But when the benches come from the two sides of a glacier, they mingle near the middle; also, the irregular slope of their bed often disturbs their order and parallelism.

There are localities where there are mountains which are breaking up on only one side of the glacier, and there this calculation could be made with less uncertainty.

>-+-◆>--O--◆+-◄

JAMES CLERK MAXWELL
Molecules (1873)

Although he lived for a relatively short time, James Clerk Maxwell (1831–1879) was one of nineteenth-century science's most productive members. He is best remembered today for his theory that electricity and magnetism are two sides of the same thing and that this thing is not material but a field of energy. But Maxwell studied much else besides, from the chemistry of gases to the nature of color vision and color blindness. In the same year he published his four equations supporting his field theory of electromagnetism, Maxwell gave a lecture on molecules to the British Association for the Advancement of Science. The lecture was later published in *Nature*, the association's journal. In this talk he presented his thinking about gases and their behavior, still another area in which he made many original contributions. His talk was based on observations and demonstrations. As he spoke, Maxwell opened jars of gases for the audience to observe, even though what he was talking about—molecules—was invisible. Everything Maxwell and his audience observed was, therefore, interpreted according to an idea, an idea also defensible through observation. These observations at second hand, as it were, seem different from those of Galileo or Darwin or Schaller—all of them being firsthand, in the field. But Maxwell's interpretive observation was inevitable if science was to progress from description toward understanding. And Maxwell knew, of course, exactly what he was doing and also that scientists could not run too far ahead of their observations. That sense of approaching a limit accounts for the—to modern tastes—astonishing close of this lecture.

We do not know much about the science organisation of Thrace twenty-two centuries ago, or of the machinery then employed for diffusing an interest in physical research. There were men, however, in those days, who devoted their lives to the pursuit of knowledge with an ardour worthy of the most distinguished members of the British Association, and the lectures in which Democritus explained the atomic theory to his fellow-citizens of Abdera realised, not in golden opinions only, but in golden talents, a sum hardly equalled even in America.

To another very eminent philosopher, Anaxagoras, best known to the world as the teacher of Socrates, we are indebted for the most important service to the atomic theory, which, after its statement by Democritus, remained to be done. Anaxagoras, in fact, stated a theory which so exactly contradicts the atomic theory of Democritus that the truth or falsehood of the one theory implies the falsehood or truth of the other. [. . .]

Take any portion of matter, say a drop of water, and observe its properties. Like every other portion of matter we have ever seen, it is divisible. Divide it in two, each portion appears to retain all the properties of the original drop, and among others that of being divisible. The parts are similar to the whole in every respect except in absolute size.

Now go on repeating the process of division till the separate portions of water are so small that we can no longer perceive or handle them. Still we have no doubt that the sub-division might be carried further, if our senses were more acute and our instruments more delicate. Thus far all are agreed, but now the question arises, Can this sub-division be repeated for ever?

According to Democritus and the atomic school, we must answer in the negative. After a certain number of sub-divisions, the drop would be divided into a number of parts each of which is incapable of further sub-division. We should thus, in imagination, arrive at the atom, which, as its name literally signifies, cannot be cut in two. This is the atomic doctrine of Democritus, Epicurus, and Lucretius, and, I may add, of your lecturer. [. . .]

But we must now go on to molecules. Molecule is a modern word. It does not occur in *Johnson's Dictionary*.[3] The ideas it embodies are those belonging to modern chemistry.

A drop of water, to return to our former example, may be divided into a certain number, and no more, of portions similar to each other. Each of these the modern chemist calls a molecule of water. But it is by

---

[3] Samuel Johnson's Dictionary of the English Language *was published in 1755 and immediately became the standard source for English vocabulary and spelling.*

no means an atom, for it contains two different substances, oxygen and hydrogen, and by a certain process the molecule may be actually divided into two parts, one consisting of oxygen and the other of hydrogen. According to the received doctrine, in each molecule of water there are two molecules of hydrogen and one of oxygen. Whether these are or are not ultimate atoms I shall not attempt to decide.

We now see what a molecule is, as distinguished from an atom.

A molecule of a substance is a small body such that if, on the one hand, a number of similar molecules were assembled together they would form a mass of that substance, while on the other hand, if any portion of this molecule were removed, it would no longer be able, along with an assemblage of other molecules similarly treated, to make up a mass of the original substance.

Every substance, simple or compound, has its own molecule. If this molecule be divided, its parts are molecules of a different substance or substances from that of which the whole is a molecule. An atom, if there is such a thing, must be a molecule of an elementary substance. Since, therefore, every molecule is not an atom, but every atom is a molecule, I shall use the word molecule as the more general term. [. . .]

Our business this evening is to describe some researches in molecular science, and in particular to place before you any definite information which has been obtained respecting the molecules themselves. The old atomic theory, as described by Lucretius and revived in modern times, asserts that the molecules of all bodies are in motion, even when the body itself appears to be at rest. These motions of molecules are in the case of solid bodies confined within so narrow a range that even with our best microscopes we cannot detect that they alter their places at all. In liquids and gases, however, the molecules are not confined within any definite limits, but work their way through the whole mass, even when that mass is not disturbed by any visible motion.

This process of diffusion, as it is called, which goes on in gases and liquids and even in some solids, can be subjected to experiment, and forms one of the most convincing proofs of the motion of molecules.

Now the recent progress of molecular science began with the study of the mechanical effect of the impact of these moving molecules when they strike against any solid body. Of course these flying molecules must beat against whatever is placed among them, and the constant succession of these strokes is, according to our theory, the sole cause of what is called the pressure of air and other gases.

This appears to have been first suspected by Daniel Bernoulli, but he had not the means which we now have of verifying the theory. The same theory was afterwards brought forward independently by Lesage, of Geneva, who, however, devoted most of his labour to the explanation of gravitation by the impact of atoms. [. . .]

If the velocity of the molecules is given, and the number varied, then since each molecule, on an average, strikes the side of the vessel the same number of times, and with an impulse of the same magnitude, each will contribute an equal share to the whole pressure. The pressure in a vessel of given size is therefore proportional to the number of molecules in it, that is to the quantity of gas in it.

This is the complete dynamical explanation of the fact discovered by Robert Boyle,[4] that the pressure of air is proportional to its density. [. . .]

We have now to conceive the molecules of the air in this hall flying about in all directions, at a rate of about seventeen miles in a minute.

If all these molecules were flying in the same direction, they would constitute a wind blowing at the rate of seventeen miles a minute, and the only wind which approaches this velocity is that which proceeds from the mouth of a cannon. How, then, are you and I able to stand here? Only because the molecules happen to be flying in different directions, so that those which strike against our backs enable us to support the storm which is beating against our faces. Indeed, if this molecular bombardment were to cease, even for an instant, our veins would swell, our breath would leave us, and we should, literally, expire. But it is not only against us or against the walls of the room that the molecules are striking. Consider the immense number of them, and the fact that they are flying in every possible direction, and you will see that they cannot avoid striking

---

[4] *Robert Boyle (1627–1691) is a contributor to this volume. His law appeared in 1662, when he published the second edition of his* New Experiments Physico-Mechanical, Touching the Spring of the Air.

each other. Every time that two molecules come into collision, the paths of both are changed, and they go off in new directions. Thus each molecule is continually getting its course altered, so that in spite of its great velocity it may be a long time before it reaches any great distance from the point at which it set out.

I have here a bottle containing ammonia. Ammonia is a gas which you can recognise by its smell. Its molecules have a velocity of six hundred metres per second, so that if their course had not been interrupted by striking against the molecules of air in the hall, every-one in the most distant gallery would have smelt ammonia before I was able to pronounce the name of the gas. But instead of this, each molecule of ammonia is so jostled about by the molecules of air, that it is sometimes going one way and sometimes another. It is like a hare which is always doubling,[5] and though it goes a great pace, it makes very little progress. Nevertheless, the smell of ammonia is now begin-ning to be perceptible at some distance from the bottle. The gas does diffuse itself through the air, though the process is a slow one, and if we could close up every opening of this hall so as to make it air-tight, and leave everything to itself for some weeks, the ammonia would become uniformly mixed through every part of the air in the hall. [. . .]

As most gases are invisible, I shall exhibit gaseous diffusion to you by means of two gases, ammonia and hydrochloric acid, which, when they meet, form a solid product. The ammonia, being the lighter gas, is placed above the hydrochloric acid, with a stratum of air between, but you will soon see that the gases can diffuse through this stratum of air, and produce a cloud of white smoke when they meet. During the whole of this process no currents or any other visible motion can be detected. Every part of the vessel appears as calm as a jar of undis-turbed air.

But, according to our theory, the same kind of motion is going on in calm air as in the inter-diffusing gases, the only difference being that we can trace the molecules from one place to another more easily when they are of a different nature from those through which they are diffusing.

---

[5] *Doubling, retracing its steps, doubling back.*

If we wish to form a mental representation of what is going on among the molecules in calm air, we cannot do better than observe a swarm of bees, when every individual bee is flying furiously, first in one direction, and then in another, while the swarm, as a whole, either remains at rest, or sails slowly through the air.

In certain seasons, swarms of bees are apt to fly off to a great distance, and the owners, in order to identify their property when they find them on other people's ground, sometimes throw handfulls of flour at the swarm. Now let us suppose that the flour thrown at the flying swarm has whitened those bees only which happened to be in the lower half of the swarm, leaving those in the upper half free from flour.

If the bees still go on flying hither and thither in an irregular manner, the floury bees will be found in continually increasing proportions in the upper part of the swarm, till they have become equally diffused through every part of it. But the reason of this diffusion is not because the bees were marked with flour, but because they are flying about. The only effect of the marking is to enable us to identity certain bees. [. . .]

The dynamical theory of liquids is not so well understood as that of gases, but the principal difference between a gas and a liquid seems to be that in a gas each molecule spends the greater part of its time in describing its free path, and is for a very small portion of its time engaged in encounters with other molecules, whereas in a liquid the molecule has hardly any free path, and is always in a state of close encounter with other molecules.

Hence in a liquid the diffusion of motion from one molecule to another takes place much more rapidly than the diffusion of the molecules themselves, for the same reason that it is more expeditious in a dense crowd to pass on a letter from hand to hand than to give it to a special messenger to work his way through the crowd. [. . .]

In solids the molecules are still in motion, but their motions are confined within very narrow limits. Hence the diffusion of matter does not take place in solid bodies, though that of motion and heat takes place very freely. Nevertheless, certain liquids can diffuse through colloid solids, such as jelly and gum, and hydrogen can make its way through iron and palladium. [. . .]

Thus far we have been considering molecular science as an inquiry into natural phenomena. But though the professed aim of all scientific work is to unravel the secrets of nature, it has another effect, not less valuable, on the mind of the worker. It leaves him in possession of methods which nothing but scientific work could have led him to invent, and it places him in a position from which many regions of nature, besides that which he has been studying, appear under a new aspect.

The study of molecules has developed a method of its own, and it has also opened up new views of nature. [. . .]

As long as we have to deal with only two molecules, and have all the data given us, we can calculate the result of their encounter, but when we have to deal with millions of molecules, each of which has millions of encounters in a second, the complexity of the problem seems to shut out all hope of a legitimate solution.

The modern atomists have therefore adopted a method which is I believe new in the department of mathematical physics, though it has long been in use in the Section of Statistics.[6] When the working members of Section F get hold of a Report of the Census, or any other document containing the numerical data of Economic and Social Science, they begin by distributing the whole population into groups, according to age, income-tax, education, religious belief, or criminal convictions. The number of individuals is far too great to allow of their tracing the history of each separately, so that, in order to reduce their labour within human limits, they concentrate their attention on a small number of artificial groups. The varying number of individuals in each group, and not the varying state of each individual, is the primary datum from which they work.

This, of course, is not the only method of studying human nature. We may observe the conduct of individual men and compare it with that conduct which their previous character and their present circumstances, according to the best existing theory, would lead us to expect. Those who practise this method endeavour to improve their knowledge of the elements of human nature, in much the same way as an astronomer corrects the elements of a planet by comparing its

---

[6] *Study section, known as Section F, in the British Association.*

actual position with that deduced from the received elements. The study of human nature by parents and schoolmasters, by historians and statesmen, is therefore to be distinguished from that carried on by registrars and tabulators, and by those statesmen who put their faith in figures. The one may be called the historical, and the other the statistical method.

The equations of dynamics completely express the laws of the historical method as applied to matter, but the application of these equations implies a perfect knowledge of all the data. But the smallest portion of matter which we can subject to experiment consists of millions of molecules, not one of which ever becomes individually sensible to us. We cannot, therefore, ascertain the actual motion of any one of these molecules, so that we are obliged to abandon the strict historical method, and to adopt the statistical method of dealing with large groups of molecules.

The data of the statistical method as applied to molecular science are the sums of large numbers of molecular quantities. In studying the relations between quantities of this kind, we meet with a new kind of regularity, the regularity of averages, which we can depend upon quite sufficiently for all practical purposes, but which can make no claim to that character of absolute precision which belongs to the laws of abstract dynamics.

Thus molecular science teaches us that our experiments can never give us anything more than statistical information, and that no law deduced from them can pretend to absolute precision. But when we pass from the contemplation of our experiments to that of the molecules themselves, we leave the world of chance and change, and enter a region where everything is certain and immutable.

The molecules are conformed to a constant type with a precision which is not to be found in the sensible properties of the bodies which they constitute. In the first place the mass of each individual molecule, and all its other properties, are absolutely unalterable. In the second place the properties of all molecules of the same kind are absolutely identical. [. . .]

The molecule, though indestructible, is not a hard rigid body, but is capable of internal movements, and when these are excited it emits rays, the wave-length of which is a measure of the time of vibration of the molecule.

By means of the spectroscope the wave-lengths of different kinds of light may be compared to within one ten-thousandth part. In this way it has been ascertained, not only that molecules taken from every specimen of hydrogen in our laboratories have the same set of periods of vibration, but that light, having the same set of periods of vibration, is emitted from the sun and from the fixed stars.

We are thus assured that molecules of the same nature as those of our hydrogen exist in those distant regions, or at least did exist when the light by which we see them was emitted.

From a comparison of the dimensions of the buildings of the Egyptians with those of the Greeks, it appears that they have a common measure. Hence, even if no ancient author had recorded the fact that the two nations employed the same cubit as a standard of length, we might prove it from the buildings themselves. We should also be justified in asserting that at some time or other a material standard of length must have been carried from one country to the other, or that both countries had obtained their standards from a common source.

But in the heavens we discover by their light, and by their light alone, stars so distant from each other that no material thing can ever have passed from one to another, and yet this light, which is to us the sole evidence of the existence of these distant worlds, tells us also that each of them is built up of molecules of the same kinds as those which we find on earth. A molecule of hydrogen, for example, whether in Sirius or in Arcturus, executes its vibrations in precisely the same time.

Each molecule, therefore, throughout the universe, bears impressed on it the stamp of a metric system as distinctly as does the metre of the Archives at Paris, or the double royal cubit of the Temple of Karnac.

No theory of evolution can be formed to account for the similarity of molecules, for evolution necessarily implies continuous change, and the molecule is incapable of growth or decay, of generation or destruction.

None of the processes of Nature, since the time when Nature began, have produced the slightest difference in the properties of any molecule. We are therefore unable to ascribe either the existence of the molecules or the identity of their properties to the operation of any of the causes which we call natural.

On the other hand, the exact quality of each molecule to all others of the same kind gives it, as Sir John Herschel has well said, the essential character of a manufactured article, and precludes the idea of its being eternal and self existent.

Thus we have been led, along a strictly scientific path, very near to the point at which Science must stop. Not that Science is debarred from studying the internal mechanism of a molecule which she cannot take to pieces, any more than from investigating an organism which she cannot put together. But in tracing back the history of matter Science is arrested when she assures herself, on the one hand, that the molecule has been made, and on the other that it has not been made by any of the processes we call natural.

Science is incompetent to reason upon the creation of matter itself out of nothing. We have reached the utmost limit of our thinking faculties when we have admitted that because matter cannot be eternal and self-existent it must have been created.

―――――●―――――

ROBERT KENNEDY DUNCAN
Radio-Activity: A New Property of Matter,
from *Harper's* (1902)

Robert Kennedy Duncan (1868–1914) was an American chemist and science writer who enjoyed great popularity at the start of the twentieth century. Although now forgotten, he was as popular in his day as Asimov or Gould is today. His essays, including this selection, frequently appeared in *Harper's* magazine as he chronicled the great changes in our understanding of matter. Like Asimov, Duncan was a chemist who understood science on a technical level, yet he was able to make it concrete and clear to the nonspecialist. Reading his essays today, I am struck by his tremendous faith in progress, a faith that has largely been lost in our own age, even though science has grown in a way that Duncan could only dream of. Particularly striking is how in the opening passage, he takes for granted an image later used by the philosopher Ludwig Wittgenstein: climbing a ladder to knowledge and then throwing away the ladder once perfection had been attained.

Duncan is as open as Galileo, reporting astonishment at the unexpected, but like Saussure, he draws from a great many observations to make his point. And from them he introduces new knowledge by categorizing what has been observed about the nature of radioactivity. The A and B types of radiation he cites would soon be dignified with Greek names, alpha and beta rays. A third type, the gamma ray, also was identified. (Note that his table lists a ray as traveling at 1.7 times the speed of light, a speed that a few years later, Einstein declared to be impossible.)

"In the beginning God created," and in the midst of His creation He set down man with a little spark of the Godhead in him to make him strive to know,—and in the striving, to grow, and to progress to some great, worthy, unknown end in this world. He gave him hands to do, a will to drive, and seven senses to apprehend,—just a working equipment; and so he has won his way, so far, out of the horrible conditions of pre-history.

To know, is to work and to do; and a new thing done is forever a rung in the ladder by which man climbs,—necessary, and good for all generations, until the summit is attained and the ladder can be cast aside.

The theme of this present article is of a new Thing Done—the discovery of a new property of matter. All we can do is simply to place it on its feet before you in a collection of experiments. It is hoped that outside of its extrinsic interest you will see deep within it the beauty and poetry of reasoned Action.

If you questioned the discoverer—the doer of the work—about himself, he would probably tell you that his work, possibly was something—he himself was nothing; and in a measure he is right; for in a few years he will pass, while his work will endure forever. Still, we wish to know him for his work's sake, and surely it will not be amiss to say something at least about him.

Let us say, then, that Henri Becquerel, Mèmbre de l'Institut, is the discoverer of Becquerel rays, the basis of the phenomena of radio-activity. He comes very honestly by his powers. His grandfather, Antoine-César (1788–1878), through sixty years of indefatigable labor, contributed more than five hundred memoirs, works of note on mineralogy and electricity. His father, Alexandre-Edmond (1820–1891) was the author of so many memoirs that they constitute practically a history of the relations of optics to electricity through the past fifty years. Henri Becquerel, the son, was subjected to the training and influence of these honored men, and it is little wonder, then, that through heredity and environment, he should bear the face of one who sends his soul into the invisible—for that, in good solid truth, is what every true experimenter literally does. In due time he succeeded to the Professorship of Physics, the chair of his fathers, and began his work in their laboratory in the quaint old home of Cuvier in the Jardin des Plantes,—"a laboratory to which I had gone," he says, "from the time I was able to walk." There he wrought nobly for the credit of his name, until Röntgen's discovery of the X-rays initiated an investigation which culminated in the discovery of the Becquerel rays and radio-activity. [. . .]

X-rays are in some way entangled with the phosphorescence in a Crookes tube.[7] Consequently the discovery of Professor Röntgen[8] set men wondering as to whether the power of emitting penetrating rays might not be a property of phosphorescent bodies in general. Thought always advances in waves; and there are always several men on the top of the same wave. In this instance there was Niewenglowski, who made the interesting discovery that [. . .] a certain compound of sulphur and calcium, calcium sulphide, which is the basis of luminous paint, shines in the dark after exposure to sunlight—that is, it is phosphorescent. [. . .] The question was asked of Nature, "Does this substance, this calcium sulphide, emit rays which will penetrate glass and metal to affect a photographic plate?" And Nature answered, in her legible signature, "This substance will."

But men are children, and one question fathers another. Are these rays light? The answer is [. . .] affirmative. [. . .] It should be remembered, however, that these penetrating light rays were not given off by the sulphide in its natural condition. It had previously to be exposed to sunlight whence it derived its energy.

But Becquerel was abreast of the same wave of investigative thought as Niewenglowski. He says, "For my part, from the day on which I first had knowledge of the discovery of Professor Röntgen, there came to me, too, the idea of seeing whether the property of emitting very penetrating rays was not intimately bound up with phosphorescence." His thought was soon represented concretely; for, taking fragments of various phosphorescent substances, he placed them one after another on a photographic plate enveloped in black paper, and thus gave them an opportunity of telling their secrets by penetrating the paper and affecting the plate beneath.

In this his work resembled that of Niewenglowski; but the importance of it is, and the luck of it was, that he experimented with different substances. Out of all the different substances tried, there was one, a substance containing the metal uranium, that had waited aeons for this one precious day. For one day of twenty-four hours this substance

---

[7] *A vacuum tube, named for Sir William Crookes (1832–1919).*

[8] *Wilhelm Röntgen had just received the Nobel Prize in physics when this article appeared and was a world celebrity, thanks to X rays.*

lay upon a photographic plate enveloped in black paper, and thus, after ages upon ages of waiting, found utterance. The plate *was* affected. [. . .] The plate is obscure, as would be the picture of the approach of dawn; and it is equally significant. It reveals nothing but the presence of penetrating rays.

"Here I am," said Nature. "Now tell me, am I Niewenglowski's rays?"[. . .]

"I thought, then," says Becquerel, "that it was necessary previously to expose the substance to light in order to provoke this penetrating emission, but a short time after *I recognized that the emission of the rays was produced spontaneously, even when the substance had been kept completely sheltered from any previous exposure to light.*"

This settles the question. Niewenglowski's rays were directly due to the action of the sun upon the substance which emitted them; Becquerel rays arise from a substance whose natural property it is not only to emit them, but, apparently, to manufacture them. It may be stated here that since this discovery the rays emitted by this particular fragment of uranium compound have shown no signs whatever of diminution. They are apparently a permanent property of this form of matter. Furthermore, it was soon seen to be a matter of indifference what uranium substance was employed. Any substance containing uranium gave off the rays. Metallic uranium itself [. . .] gave out more rays than any of its compounds. More than that, the emission of the rays turned out to be altogether independent even of phosphorescence. Uranium bodies, whether phosphorescent or not, emitted rays. Here, then, was no stored-up, transformed sunlight, such as Niewenglowski's rays, but penetrating, continuous emissions from a substance having no relation to light. The emission of rays capable of passing straight through copper from a chemical substance in its normal condition constitutes to us a new property of matter,—a new thing in nature! So, as Becquerel stood in his laboratory that night, with this thought in his mind and the plate in his hand, he appears sharply silhouetted against the background of the ages; he is comparable with that Theophrastus who, two thousand years ago, rubbed a piece of amber on his coat sleeve and noticed that it attracted bits of paper, unknowing that his bit of amber was equal to the lamp of Aladdin,—or to that paleolithic savage who, the first of all men, noticed the attractive powers of loadstone. New properties of matter are not so common that their significance can be exaggerated. This new property of matter was called *radio-activity*, and as such it takes its place beside magnetism, electricity, light, and heat.

Radio-activity, a new property of matter, had been discovered, but whence its source? "The metal uranium itself," you say, "since *it* gives off the rays." Yes; but still a doubt—a little, tiny doubt—remained.

Was it not possible that the power of emitting rays, the radio-activity, was due to some small impurity present in the uranium? That doubt was the key which unlocked the door to a roomful of other discoveries.

It arose in the minds of two investigators who had been interested observers of Becquerel's work. M. Pierre Curie, Professor of Physics in the School of Physics and Industrial Chemistry at Paris, and Madame Sklodowska Curie, his wife. They resolved to investigate the ray-emitting powers of pitchblende, the parent substance from which all uranium is extracted. To their gratification they discovered that selected specimens of pitchblende possessed a radio-activity four times greater than metallic uranium itself.

Nature never insults us by caprice, and consequently we find the Curies saying: "It becomes then, very probable that if pitchblende has so strong an activity it is because the mineral contains, in small quantities, a substance wonderfully radio-active, different from uranium or any of the simple bodies actually known. We proposed to ourselves to extract this substance from pitchblende, and we have, in fact, been able to prove that it is possible, by the methods of ordinary chemical analysis, to extract from pitchblende *substances of which the radio-activity is in the neighborhood of 100,000 times greater than that of metallic uranium.*"

In this simple manner did the Curies announce their discovery of the new elements with transcendent ray-emitting powers—radium, polonium, actinium. Of these three strangers, radium has been selected for purposes of research into the character of Becquerel rays because it was most easily obtained. Its discovery, with its ray-emitting power 100,000 times greater than uranium, placed in the hands of Becquerel a mighty engine of research for determining the properties of his rays.

Radium has never been isolated. As a free element it has never been seen, never been touched, never been handled, as gold and iron may be, but it is manifest in the properties of its compounds. It has been studied only in combination with other elements. We know that it exists as an element different from every other body in nature solely and completely through the fact that every element has its own sign-manual, or spectrum, but means of which it signifies its existence, whether it is found in the sun, the stars, or the laboratory. [. . .] The spectrum of radium [. . .] [is] caused by no other known element on the earth or in the heavens. Therefore radium is a new element. The amount of radium in pitchblende is less than one ten-millionth per cent, and the quantities of the much rarer polonium and actinium are literally infinitesimal.

Considering only the cost of the pitchblende from which it is extracted, the value of the radium would be at least $10,000 a gram. As a matter of fact, less than a gram exists. M. Curie possesses about two to three hundredths of a gram of chemically pure radium chloride, which was utilized by M. Demarcay in obtaining [a] spectrum, and about three-tenths of a gram of a comparably pure product containing barium chloride.

With the sample of impure radium chloride generously provided by M. and Madame Curie, Becquerel proceeded with the study of the properties of his rays. Their surpassing power of penetrating matter generally considered opaque led to their discovery, as we have seen, and it was therefore one of the first properties to be investigated. It soon became evident that this power was quite independent of the kind of matter through which they passed. It was influenced only by the density of the substance interposed. Aluminum, for example, being light in weight, is to Becquerel rays what glass is to light—comparatively transparent. Lead, on the contrary, being heavy, is comparatively opaque.

In the power to take radiographs, Becquerel rays resemble X-rays. Many substances when they are exposed to Becquerel rays shine in the dark—that is, they phosphoresce. The diamond and the ruby shine out vividly on being help up in the invisible rays emitted by a pinch of chloride of radium. So do fluor-spar, calcium sulphide, barium platino-cyanide, and many others. So powerful is the phosphorescence caused by Becquerel rays that if a tube of radium chloride be held to the forehead, and the experimenter close his eyes, he will still see light. The retina itself becomes phosphorescent. They even react upon the radium substance itself, so that it too becomes luminous, and shines vividly with a light which, since the discovery of radium, has shown no shadow of variableness. Becquerel rays will photograph the substance that emits them. [. . .] They are strange things, then, these Becquerel rays. The light which took the picture shone when the morning of creation broke, and will shine with the dawn of the last day of reckoning; for Becquerel rays are a property of the atom of the substance, and are therefore indestructible. It is a matter of indifference what physical stress is brought to bear, or what chemical transformation is effected. The light will shine, undiminished and undiminishable,—in the gram, a soft radiance; in the pound, if we could get it, a new sun.

The physiological effect of Becquerel rays is most intense—almost incredible. A pinch of radium salt, contained in a sealed glass tube, was placed in a card-board box, which was then tied to the sleeve of Professor Curie for a hour and a half. An intense inflammation resulted, followed by a suppurating sore which took more than three months to heal. Professor Becquerel, as he went about his work one day, chanced to carry a sealed glass tube of radium salt in his pocket, placed there for convenience. He was sorry; for the sore was painful and most tedious in healing.

Photographic plates and electrified bodies are widely different. Yet Becquerel discovered, at about the same time, that they were both affected by his rays. A photographic plate was blackened; an electrified body was discharged: either was a detector of radio-activity. With the discovery of radium, the discharging effect became, of course, exceedingly apparent. [. . .] An electroscope [had] its little gold leaves spread apart by electrification. On the approach of a glass tube containing a tiny amount of radium chloride, the leaves at once collapse through the discharge of their electrification. The approach of the radium and the discharge of the leaves are simultaneous. Investigation showed that the effect was due to the fact that the rays emitted by the radium *spontaneously rendered the air a conductor of electricity,* and naturally the electrification of the leaves flew away with as much ease as if they had been touched by a copper wire. As a matter of fact, an electrified body is a more sensitive detector of radio-activity than a photographic plate.

Any substance placed near radium becomes itself a *false* radium. "We have found," say the Curies, "that any substance placed in the neighborhood of radium acquires a radio-activity which persists for many hours, and even many days, after the removal of the radium. This induced radio-activity increases with the time during which it is exposed to the action of the radium up to a certain limit. After the radium is removed, it decreases rapidly and tends to disappear. The kind of substance exposed to the action of the radium is almost a matter of indifference. They all acquire a radio-acitivity of their own." This fact has been verified over and over again by every experimenter in the field. The zinc, iron, and lead fittings, the air of the laboratory, the water, the clothing of the workers, their very persons, in the presence of radium start into activity and give out rays comparable to radium in affecting a photographic plate and discharging electricity. This becomes very vexatious and disconcerting, and extreme care is necessary to prevent the radium giving out rays altogether misdirected. For days Professor Curie was unable to approach his electrometers, or even enter his laboratory, owning to his acquired radio-activity. These secondary radiations, in the case of zinc, were four times as intense as in ordinary uranium. Moreover, this acquired radio-activity cannot be removed by washing, and the Becquerel rays from radium will impart it even after passing through metal screens. It must be remembered, however, that the radio-activity is only temporary. It vanishes sooner or later upon the removal from the neighborhood of the potent radium. [. . .]

Becquerel rays cause chemical effects. Emitted from radium, they will discolor paper, cause glass to take a permanent violet tint, turn oxygen into ozone, yellow phosphorus into red phosphorus, mercury perchloride into calomel.

We have learned how Becquerel discovered his rays, we have studied their properties, and we are now face to face with the problem most important of all. What are they? Now Becquerel has discovered in some measure what they are; but before exposing his proofs and results it is necessary to describe one more strange property which he discovered them to possess, and which, as it turned out, appears to afford a master-key to their nature.

Becquerel rays are bent by a magnet. Let us follow him while he proves this by one of his characteristic, fecund, simple experiments. Taking a narrow photographic plate enveloped in black paper, he placed it horizontally between the poles of a powerful electro-magnet. On the back envelope of the plate he then placed a little lead trough containing a small amount of the radium compound. The rays could thus affect the plate only by bending over, for the lead trough is opaque. He energized the magnet; then, after a certain time, he reversed the polarity, thinking that, if the rays bent at all, they would then bend in the opposite direction. [. . .] The plate [showed] two broad bands, proving that the rays must have been curved down to meet it; that there are two bands instead of one proves that reversing the polarity causes the rays to bend in the opposite direction. Becquerel rays are deviable by a magnet.

But are they all equally deviable? Are they homogenous? Let us follow Becquerel. Upon another plate of the same kind he placed strips of platinum, aluminum, and paper, and at the end of the plate, as before, the little lead trough containing the radium compound. If they were all equally deviable, they would form a line where they bent to meet the plate; if not, they would form a band. After energizing the magnet and developing the plate, [. . .] the rays [. . .] form a broad band, a veritable spectrum of an infinity of radiations unequally deviable. The same plate shows as well that the rays penetrate the screens in this order: the platinum least, the aluminum next, and the paper most of all.

Are they *all* deviable? Is it not possible that some of them are totally unaffected by the magnet, and do not bend at all? To find out why not place the narrow photographic plate vertically instead of horizontally between the poles of the magnet? The idea was carried out, and the result is plain beyond all question. The Becquerel rays consist of two distinct kinds of radiation. One kind, A, is bent by the magnet; the other, B, is totally unaffected by it, and passes undeviatingly on. [. . .] Whatever they are, Becquerel rays are a mixture of deviable and undeviable radiations.

But what are they? Examine the tabular statement [. . .] of the properties of the different kinds of rays, and you will see this curious fact: The properties of the deviable Becquerel rays are identical with those of the cathode rays of a Crookes tube, and the properties of the undeviable rays are identical with those of the X-rays of Röntgen. Identity of property means identity of nature, and we are therefore forced to conclude that the Becquerel rays from radium are nothing more nor less than a mixture of cathode and X-rays, their progenitors in the history of discovery.

Tabular Statement of Properties of Different Kinds of Rays

Name of Ray	Cathode	Deviable Becquerel Rays	X-rays	Undeviable Becquerel Rays	Niewen-glowski's Rays	Ultra-violet Light	Red Light
Existence of interference, polarization, reflection, refraction	o	o	o	o	x	x	x
Photographic effect	x	x	x	x	x	x	o
Excite phosphorescence	x	x	x	x	?	x	o
Render air conductive	x	x	x	x	x	x	o
Penetrate opaque bodies	x	x	x	x	x	o	o
Undergo deflection by a magnet	x	x	o	o	?	o	o
Velocity relative to that of light	1.7	1.2	1	?	?	1	1

x, property exists; o, property does not exist; ?, property uncertain

"What an anti-climax!" says the reader. "We started out to study a new property of matter, and here we end up with an old one."

Not a bit of it. We called the new property of matter radio-activity—not Becquerel rays. "What is the difference?" All the difference between a natural intrinsic property and a property of condition. The light of an arc-lamp is a property of condition; suppose you found, deep in the earth, a substance blazing forever with a light as great, that would be a natural intrinsic property—and a very curious one—radio-activity. So with the cathode and X rays. They arise from a Crookes tube, a mechanism which is the complicated result of centuries of thought; they are a property of condition. The Becquerel rays from radium, on the contrary, arise from a substance dug out of the ground which emits them, apparently, forever and forever, as it has emitted them through the countless centuries of the past, without any extrinsic influence. It is their natural intrinsic property—a new property of matter—radio-activity.

The cathode rays are streams of material particles. These particles are projected from radium with a velocity anywhere from sixty millions to ninety millions of miles per second. They fly out laden with electricity and hence naturally enough discharge an electroscope. They are so small that the atoms of the chemist are giants in comparison. Since these particles flying off from radium are decomposed atoms, their properties are not the properties of iron, or gold, or copper, but the properties of matter in general. These particles, or corpuscles, as they are called, appear to be the primary atoms of some parent form of matter out of which the elements, as we know them, have been evolved. It is interesting, in this connection, to recall the words of Huxley, written long ago, before Becquerel rays had entered into the dreams of the wildest speculator. "It seems safe to assume," he wrote, "that the hypothesis of the evolution of the elements from a primitive form of matter will in the future play no less a part in the history of science than the atomic hypothesis, which, to begin with, had no greater, if as great, an empirical foundation." These words were written with the prescience of a master.

Possibly the most interesting thought in all the strange, eventful history of these interesting bodies is the question of their energy. Whence does it come? It is suggested by Madame Curie that the radium receives its energy from, and responds to, radiations which traverse all space, much as some article of bric-à-brac in a room will vibrate responsively to a certain tone of the piano. This may be. Heaven only *knows*. One thing we do know—space is all aquiver with waves of radiant energy, ranging in length from many feet to a size infinitesimally small. To only a few of these are our bodily senses fitted to correspond, or our mechanisms to detect. Waves of radiant energy constitute what has been called "the harp of life." We vibrate in sympathy with a few strings here and there—with the tiny X-rays, actinic waves,[9] light waves, heat waves, in the treble, and the huge electro-magnetic waves of Hertz and Marconi, and the grand air waves of sound, in the bass; but there are great spaces, numberless strings, an infinity of possible radiations, to which we are deaf—stone-deaf. Some day, a thousand years hence, we shall know the full sweep of this magnificent harmony, and with it we shall vibrate in accord with the Master Musician of it all.

---

[9] *Waves carrying the force by which chemical effects, such as changes in photographic plates, were thought to occur. In the last half of the nineteenth century, sunlight was said to contain an "actinic power" that could produce effects. The modern understanding of light as energy came a few years after this article appeared.*

I. P. PAVLOV

The Atoms of Activity, from
Conditioned Reflexes (1924)

Ivan Petrovich Pavlov (1849–1936) is most widely remembered through the adjective *Pavlovian*, meaning automatic, knee-jerk behavior. The term is a warning not to allow one's work to be reduced to one's own name, for Pavlov's work was much more subtle than a single name is able to convey. His most famous book, *Conditioned Reflexes*, was published in 1926 and is based on a series of lectures he gave in 1924, when he was 75 years old. The book summarizes decades of experimental observations aimed at understanding the physical workings of the brain. Pavlov's book was subtitled *An Investigation of the Physiological Activity of the Cerebral Cortex*. Nothing would have astounded him more than the idea that became popular in America, that his work showed that everything could be explained by the environment and did away with any need to understand the brain's role in behavior.

Like other compilers of observations, Pavlov was interested in organizing his observations into categories. He hoped to find the basic units of animal activity and then to show how they could be combined. It was a chemical approach—to identify the elements and then to show how they could be compounded—and not surprisingly, he began his work at a time when chemistry, especially Russian chemistry, was making great progress. As do the selections in this chapter by Duncan and Cannon, this essay shows that understanding often starts with categorizing.

Our starting point has been Descartes' idea of the nervous reflex. This is a genuine scientific conception, since it implies necessity. It may be summed up as follows: An external or internal stimulus falls on some one or other nervous receptor and gives rise to a nervous impulse; this nervous impulse is transmitted along nerve fibres to the central nervous system, and here, on account of existing nervous connections, it gives rise to a fresh impulse which passes along outgoing nerve fibres to the active organ, where it excites a special activity of the cellular structures. Thus a stimulus appears to be connected of necessity with a definite response, as cause with effect. [. . .]

The aggregate of reflexes constitutes the foundation of the nervous activities both of men and of animals. It is therefore of great importance to study in detail all the fundamental reflexes of the organism. Up to the present, unfortunately, this is far from being accomplished, especially, as I have mentioned before, in the case of those reflexes which have been known vaguely as "instincts." Our knowledge of these latter is very limited and fragmentary. Their classification under such headings as "alimentary," "defensive," "sexual," "parental" and "social" instincts, is thoroughly inadequate. Under each of these heads is assembled often a large number of individual reflexes. Some of these are quite unidentified; some are confused with others; and many are still only partially appreciated. I can demonstrate from my own experience to what extent the subject remains inchoate and full of gaps. In the course of the researches which I shall presently explain, we were completely at a loss on one occasion to find any cause for the peculiar behaviour of an animal. It was evidently a very tractable dog, which soon became very friendly with us. We started off with a very simple experiment. The dog was placed in a stand with loose loops round its legs, but so as to be quite comfortable and free to move a pace or two. Nothing more was done except to present the animal repeatedly with food at intervals of some minutes. It stood quietly enough at first, and ate quite readily, but as time went on it became excited and struggled to get out of the stand, scratching at the floor, gnawing the supports, and so on. This ceaseless muscular exertion was accompanied by breathlessness and continuous salivation, which persisted at every experiment during several weeks, the animal getting worse and worse until it was no longer fitted for our researches. For a long time we remained puzzled over the unusual behaviour of this animal. We tried out experimentally numerous possible interpretations, but though we had had long experience with a great number of dogs in our laboratories we could not work out a satisfactory solution of this strange behaviour, until it occurred to us at last that it might be the expression of a special *freedom reflex,* and that the dog simply could not remain quiet when it was constrained in the stand. This reflex was overcome by setting off another against it—the reflex for food. We began to give the dog the whole of its food in the stand. At first the animal ate but little, and lost considerably in weight, but gradually it got to eat more, until at last the whole ration was consumed. At the same time the animal grew quieter during the course of the experiments: the freedom reflex was being inhibited. It is clear that the freedom reflex is one of the most important reflexes, or, if we use a more general term, reactions, of living beings. This reflex has even yet to find its final recognition. In James's writings[10] it is not even enumerated among the special human "instincts." But it is clear that if the animal were not provided with a reflex of protest against boundaries set to its freedom, the smallest obstacle in its path would interfere with the proper fulfilment of its natural functions. Some animals as we all know have this freedom reflex to such a degree that when placed in captivity they refuse all food, sicken and die.

As another example of a reflex which is very much neglected we may refer to what may be called the *investigatory reflex*. I call it the "What-is-it?" reflex. It is this reflex which brings about the immediate response in man and animals to the slightest changes in the world around them, so that they immediately orientate their appropriate receptor organ in accordance with the perceptible quality in the agent bringing about the change, making full investigation of it. The biological significance of this reflex is obvious. If the animal were not provided with such a reflex its life would hang at every moment by a thread. In man this reflex has been greatly developed with far-reaching results, being represented in its highest form by inquisitiveness—the parent of that scientific method through which we may hope one day to come to a true orientation in knowledge of the world around us.

Still less has been done towards the elucidation of the class of negative or inhibitory reflexes (instincts) which are often evoked by any strong stimulus or even by weak stimuli, if unusual. Animal hypnotism, so-called, belongs to this category.

As the fundamental nervous reactions both of men and of animals are inborn in the form of definite reflexes, I must again

---

[10] *William James (1842–1910), a psychologist and the brother of novelist Henry James (1843–1916). He published* The Principles of Psychology *in 1890, and it became the standard psychology text for many years afterward.*

emphasize how important it is to compile a complete list comprising all these reflexes with their adequate classification. For, as will be shown later on, all the remaining nervous functions of the animal organism are based upon these reflexes. Now, although the possession of such reflexes as those just described constitutes the fundamental condition for the natural survival of the animal, they are not in themselves sufficient to ensure a prolonged, stable and normal existence. This can be shown in dogs in which the cerebral hemispheres have been removed. Leaving out of account the internal reflexes, such a dog still retains the fundamental external reflexes. It is attracted by food; it is repelled by nocuous stimuli; it exhibits the investigatory reflex, raising its head and pricking up its ears to sound. In addition it exhibits the freedom reflex, offering a powerful resistance to any restraint. Nevertheless it is wholly incapable of looking after itself, and if left to itself will very soon die. Evidently something important is missing in its present nervous make-up. What nervous activities can it have lost? It is easily seen that, in this dog, the number of stimuli evoking reflex reaction is considerably diminished; those remaining are of an elemental, generalized nature, and act at a very short range. Consequently the dynamic equilibrium between the inner forces of the animal system and the external forces in its environment has become elemental as compared with the exquisite adaptability of the normal animal, and the simpler balance is obviously inadequate to life.

Let us return now to the simplest reflex from which our investigations started. If food or some rejectable substance finds its way into the mouth, a secretion of saliva is produced. The purpose of this secretion is in the case of food to alter it chemically, in the case of a rejectable substance to dilute and wash it out of the mouth. This is an example of a reflex due to the physical and chemical properties of a substance when it comes into contact with the mucous membrane of the mouth and tongue. But, in addition to this, a similar reflex secretion is evoked when these substances are placed at a distance from the dog and the receptor organs affected are only those of smell and sight. Even the vessel from which the food has been given is sufficient to evoke an alimentary reflex complete in all its details; and, further, the secretion may be provoked even by the sight of the person who brought the vessel, or by the sound of his footsteps. All these innumerable stimuli falling upon the several finely discriminating distance receptors lose their power for ever as soon as the hemispheres are taken from the animal, and those only which have a direct effect on mouth and tongue still retain their power. The great advantage to the organism of a capacity to react to the former stimuli is evident, for it is in virtue of their action that food finding its way into the mouth immediately encounters plenty of moistening saliva, and rejectable substances, often nocuous to the mucous membrane, find a layer of protective saliva already in the mouth which rapidly dilutes and washes them out. Even greater is their importance when they evoke the motor component of the complex reflex of nutrition, *i.e.* when they act as stimuli to the reflex of seeking food.

Here is another example—the reflex of self-defence. The strong carnivorous animal preys on weaker animals, and these if they waited to defend themselves until the teeth of the foe were in their flesh would speedily be exterminated. The case takes on a different aspect when the defence reflex is called into play by the sights and sounds of the enemy's approach. Then the prey has a chance to save itself by hiding or by flight.

How can we describe, in general, this difference in the dynamic balance of life between the normal and the decorticated animal? What is the general mechanism and law of this distinction? It is pretty evident that under natural conditions the normal animal must respond not only to stimuli which themselves bring immediate benefit or harm, but also to other physical or chemical agencies—waves of sound, light, and the like—which in themselves only *signal* the approach of these stimuli; though it is not the sight and sound of the beast of prey which is in itself harmful to the smaller animal, but its teeth and claws.

Now although the *signalling stimuli* do play a part in those comparatively simple reflexes we have given as examples, yet this is not the most important point. The essential feature of the highest activity of the central nervous system, with which we are concerned and which in the higher animals most probably belongs entirely to the hemispheres, consists not in the fact that innumerable signalling stimuli do initiate reflex reactions in the animal, but in the fact that under different conditions these same stimuli may initiate quite different reflex reactions; and conversely the same reaction may be initiated by different stimuli.

In the above-mentioned example of the salivary reflex, the signal at one time is one particular vessel, at another time another; under certain conditions one man, under different conditions another—strictly depending upon which vessel had been used in feeding and which man had brought the vessel and given food to the dog. This evidently makes the machine-like responsive activities of the organism still more precise, and adds to it qualities of yet higher perfection. So infinitely complex, so continuously in flux, are the conditions in the world around, that that complex animal system which is itself in living flux, and that system only, has a chance to establish dynamic equilibrium with the environment. Thus we see that the fundamental and the most general function of the hemispheres is that of reacting to signals presented by innumerable stimuli of interchangeable signification.

୬ଈଈଈ୨

ANNIE J. CANNON

Classifying the Stars, from
The Universe of Stars (1926)

Annie Jump Cannon (1863–1941) was an assistant at the Harvard College Observatory for 14 years. In 1911 she was made the curator of Harvard's astronomical photographs, a title that belies the work's importance. It meant that Cannon was in charge of all observational data. She cataloged references to more than 200,000 stars, as well as the stars' spectra. Spectra are the colors found in the stars' light. When their twinkling light is observed through a prism, it reveals almost all the colors of the rainbow. But some colors are missing, which are revealed as black lines in the rainbow. Cannon prepared a catalog of the spectra of about 250,000 stars that remains the international standard. She also is remembered as the first great woman astronomer.

As seen in other writing about masses of observation, the first step toward understanding is categorizing. Although categorizing has no explanatory power, it sorts out things so that a person can say, oh this thing acts this way and that thing acts that way, and so the mix of varieties begins to look orderly. In Cannon's essay (originally a radio broadcast), she talks about classifying stars and what can be learned from the process. It is also charming to see that a scientist who spent decades cataloging stars still had a wit and sense of fun. Notice, too, that her cataloging is based on the interpretation of invisible processes that Maxwell described in his lecture fifty years earlier.

The very beginning of our knowledge of the nature of a star dates back to 1672, when Isaac Newton gave to the world the results of his experiments on passing sunlight through a prism. To describe the beautiful band of rainbow tints, produced when sunlight was dispersed by his three-cornered piece of glass, he took from the Latin the word *spectrum,* meaning an appearance. The rainbow is the spectrum of the Sun.

Hardly a more fascinating page in man's search for knowledge is to be found than this story of the analysis of light. May I rehearse just a few points of the story which are vital to the subject of classifying the stars? In 1814, more than a century after Newton, the spectrum of the Sun was obtained in such purity that an amazing detail was seen and studied by the German optician, Fraunhofer. He saw that the multiple spectral tints, ranging from delicate violet to deep red, were crossed by hundreds of fine dark lines. In other words, there were narrow gaps in the spectrum where certain shades were wholly blotted out.

For fifty years, many searched for a Rosetta-stone to solve the baffling mystery of these hieroglyphics traced by Nature's hand in the radiant sunbeams. The solution was actually found by studying the light of one of the best-known and commonest substances existing on our planet. It has sometimes happened that scientific men have been hardly able to earn their salt, but they have been able to prove that this omnipresent earthly substance, salt, or at least its sodium constituent, exists in the Sun and distant stars.

We must remember that the word spectrum is applied not only to sunlight, but also to the light of any glowing substance when its rays are sorted out by a prism or a grating. Each substance thus treated sends out its own vibrations of particular wave lengths, which may be likened to singing its own song. Now the spectrum of salt, called sodium chloride by chemists, is very simple and includes two bright yellow lines. In the spectrum of the Sun exactly the same shades of yellow are cut out by two black lines. Could there be any connection? Could the earthly yellow lines be made to change to black? Yes, it was found by experiment that they would do so instantly if a cooler vapor of salt were placed between the prism and a source of light that emits all wave lengths. Thus it was reasoned that some of the bright yellow light from the Sun's hot surface was absorbed by cooler sodium vapors in the Sun's atmosphere. Likewise two thousand black lines in the Sun's spectrum were traced to iron, and indeed all the common substances, so familiar to us here on the Earth, have been found to exist in the Sun by comparing its "absorption" spectrum with the bright line spectra given by these substances in laboratories.

It might have been expected that the Sun, our parent, would contain the familiar earthly elements, as we were certainly, in a distant age, bone of his bone and flesh of his flesh. But what about the Stars, so far away, apparently so faint? The Sun outshines even the brilliant Sirius, the Dog Star, ten billion times. But the light of a star may be magnified several thousand fold by a telescope. Then, with a Spectro-scope attached to the telescope, we may behold a radiant and beauti-ful sight, for the twinkling starlight becomes a band showing all the rainbow colors, also crossed by the telltale dark lines. The stars then are suns. Do the stars differ among themselves? If so, how may we learn about them? Although the human eye is such an admirable instrument, and, aided by the magnifying powers of a telescope, can penetrate far into space, it is not well adapted to observe the spectra of the stars. Fortunately, just at the time when astronomers were peering at starlight, chemists were as eagerly at work with com-pounds of bromides and silver, little dreaming that in their mixing bowl lay the means of solving the riddle of the centuries concerning the great inverted bowl above our heads. The photographic method that the chemists developed has now completely superseded visual observations of stellar spectra. The film is more successful in register-ing faint light than the eye, and is sensitive also to rays which are too short or too long for the small gamut of human vision.

The Harvard Observatory was the first to undertake the photog-raphy of stellar spectra on a large scale. With a prism placed over the lens of a telescope, the spectrum of every star of sufficient brightness in the field of view can be photographed. We lose the beauty of the colors in the process, for they affect the film only as a background of light on which the dark lines are engraved. But stars can be classified from the position and strength of these lines.

Some photographs, eight by ten inches in size, covering a portion of the sky about twice as large as the bowl of the Big Dipper, show the spectra of four thousand stars. When we examine the spectra, we notice at once that many look alike. We may then select a few of the brighter stars as typical of a class. A simple system of designating the various classes of stars by the best known of all symbols, the alphabet, was originated by the Harvard Observatory, and has been adopted by the whole astronomical world. Let us see how some of the familiar stars are thus lettered.

When an astronomer speaks of a class A star, he refers to White stars like Sirius and Vega, in whose spectra we see a very strong series of dark lines caused by hydrogen in the atmosphere. For blue-white stars like Rigel we use the letter B. The gas, helium, so precious and so scarce here on earth, is very abundant in the atmosphere of these stars. The letter G is used for our own Sun and other yellow stars; and for red stars like Betelgeuse, we use the letter M. Between A and G are the F stars; between G and M, the K stars. Other letters designate several rarer varieties. After a large number of stars had been classified, it was found that the letters B, A, F, G, K, M, stand for six divisions including the great majority of the stars. B must go before A in the astronomer's alphabet, because when it was too late to use the letters, the B stars were found to precede the A stars in history. Thus we have the so-called B A F classification, which is easy to remember, said an Irish astronomer, because B A F stands for Baffling.

Arranged in the given order, with intermediate divisions, the letters represent stars arranged in order of decreasing temperature and increasing redness. You have observed, I am sure, that stars differ in color. Do you not recall red Antares, the so-called Rival of Mars, and pure-white Spica in the summer sky? The color is a clue given us by Nature that stars differ in kind. But are all the stars growing colder and redder? No, the marvellous dark lines also tell the story of an ascending branch on a star's life tree. Stars pass first through the sequence from Class M to Class B, from red to blue; then later from B to M. Such is the life story of a star. In youth gigantic, rarefied, and red; in middle life very blue and hot and radiant; in old age shrunken, dense, and again red. This is not a fairy tale I am telling you, although the name of "giants" has been applied to stars in the youthful bloated stage, and the name of "dwarfs" to stars in the late condensed stage.

Twenty years ago it was assumed that stars differ in composition, that the A stars might have a monopoly of hydrogen, and the B stars, of helium, but the combined labors of chemists, physicists, and astronomers have pointed the way to the belief that the differences are mainly in temperature, and that the class of a star expresses the temperature which is required to produce the observed atmospheric conditions. The classifying of stars has moreover led to the belief that there is no new kind of matter in the universe, for the main features of the spectra of all stars can be accounted for in terms of substances known on Earth.

A quarter of a million stars have already been arranged in forty classes from a study of the Harvard photographs. We now have much material to study the architecture of the celestial mansions and the streaming of the celestial tribes. Do stars of the same class, like birds of a feather, flock together? Why, yes, they often seem to. Nearly all of the B stars live in the Milky Way. They are extremely hot and brilliant, being in the very apex of star life. They are distant, for their great brilliancy enables them to shine through vast spaces. The A stars also seem to prefer the Galaxy. Stars like our own Sun, of Class G, are scattered, however, over the whole sky, in the highways and byways of the universe. If perchance any of them has had past adventures similar to that of the Sun, and is now burdened with a family of planets, the inhabitants, if such there be, are blessed with a beneficent sunlight like our own. The very nearest star, Alpha Centauri, whose light takes but four and a third years to reach us, is of Class G.

The Great Nebula of Andromeda, which is faintly visible to the naked eye, looks something like a stationary comet. It is apparently not connected with our own universe of suns, but constitutes a separate system, to which the picturesque term, Island Universe, is sometimes applied. This system is estimated to be so far away that light travelling 186,000 miles a second requires nearly a million years to arrive; in other words, we see the nebula as it was a million years ago. Yet photographs exposed for many hours with large telescopes have revealed, in this Nebula of Andromeda, the existence of suns of Class G, like our own luminary.

Thus, peering into far-away spaces of the heavens, and looking back, as it were, into by-gone epochs of time, we find stars composed of the same elements necessary to us today, vibrating in the same rhythm, sending out waves of the same lengths.

Classifying the stars has helped materially in all studies of the structure of the universe, than which no greater problem is presented to the human mind. While teaching man his relatively small sphere in creation, it also encourages him by its lessons of the unity of nature, and shows him that his power of comprehension allies him with the great intelligence that encompasses all.

"The Demonstration That Cost So Much Effort"

Marie Curie

><>◦<><

GALILEO GALILEI

Where Is the Center of the Universe, from *The Two Chief World Systems* (1632)

This second selection by Galileo Galilei (1564–1643) is from a dialogue about whether the earth revolves around the sun. The participants are Simplicio, who defends the traditional wisdom of Aristotle; Salviati, who speaks for Galileo; and Sagredo, the intelligent layman whom the other two hope to convince. The Copernican theory challenged many ideas beside the astronomical notion of the sun's orbit. In physics, for example, things were said to fall to earth because matter had an inherent tendency to move toward the center of the universe, but if the earth orbited the sun, it could not occupy the center of the universe. So why did things fall to earth?

Questions like these were especially troubling to thoughtful scholars and were quite distinct from objections based on religious dogma. They had to be answered, and Galileo's *Dialogue Concerning the Two Chief World Systems—Ptolemaic & Copernican* addressed them directly. The selection here is the debate about the nature of gravity. Much that Newton later established is already suggested in this dialogue.

The tone of this selection is very different from that of Galileo's earlier piece. He had no need of dialogue in his *Sidereal Messenger*, as he was reporting observations. But in this dialogue it is not the observations that are in dispute. Rather, the question is how they are to be interpreted. As Mach remarked, science needs both "accurate observation and close, searching thought."

SIMPLICIO: Aristotle would not give assurance from his reasoning of more than was proper, despite his great genius. He held in his philosophizing that sensible experiments were to be preferred above any argument built by human ingenuity, and he said that those who would contradict the evidence of any sense deserved to be punished by the loss of that sense. Now who is there so blind as not to see that earthy and watery parts, as heavy things, move naturally downward—that is to say toward the center of the universe, assigned by nature itself as the end and terminus of straight motion *deorsum*?[1] Who does not likewise see fire and air move directly upward toward the arc of the moon's orbit, as the natural end of motion *sursum*? This being so obviously seen, and it being certain that *eadem est ratio totius et partium* [what applies to the whole applies to the part], why should he not call it a true and evident proposition that the natural motion of earth is straight motion *ad medium* [*to* the middle], and that of fire, straight *a medio* [*from* the middle]?

SALVIATI: The most that ought to be conceded to you by virtue of this argument of yours is that just as parts of the earth, removed from the whole (that is, from the place where they naturally rest) and, in short, reduced to a bad and disordered arrangement, return to their places naturally and spontaneously in a straight motion, so it may be inferred (granted that *eadem sit ratio totius et partium* [what applies to the whole applies to the part]) that if the terrestrial globe were forcibly removed from the place assigned to it by nature, it would return by a straight line. This, as I said, is the most that can be granted to you, even after giving you every sort of consideration. Anyone who wants to review these matters rigorously will deny at the outset that the parts of the earth, when returning to its whole, do move in a straight line and not in a circular or mixed one. You would surely have plenty of trouble demonstrating the contrary, as you will clearly see from the answers to the reasons and the particular experiments adopted by Ptolemy and Aristotle.

---

[1] Deorsum *is Latin for downward;* sursum, *upward. Aristotle saw these distinctions as the primary characteristics of motion, but Newton did not consider them important, at least in regard to distinguishing types of motion. One way to distinguish which direction is* deorsum *and which is* sursum *is to remember that* downward *begins with a d, as does* deorsum.

Secondly, if it should be said that the parts of the earth do not move so as to go toward the center of the universe, but so as to unite with the whole earth (and that consequently they have a natural tendency toward the center of the terrestrial globe, by which tendency they cooperate to form and preserve it), then what other "whole" and what other "center" would you find for the universe, to which the entire terrestrial globe would seek to return if removed therefrom, so that the rationale of the whole might still be like that of its parts?

I might add that neither Aristotle nor you can ever prove that the earth is *de facto* [in fact] the center of the universe; if any center may be assigned to the universe, we shall rather find the sun to be placed there, as you will understand in due course.

Now just as all the parts of the earth mutually cooperate to form its whole, from which it follows that they have equal tendencies to come together in order to unite in the best possible way and adapt themselves by taking a spherical shape, why may we not believe that the sun, moon, and other world bodies are also round in shape merely by a concordant instinct and natural tendency of all their component parts? If at any time one of these parts were forcibly separated from the whole, is it not reasonable to believe that it would return spontaneously and by natural tendency? And in this manner we should conclude that straight motion is equally suitable to all world bodies.

SIMPLICIO: There is no doubt whatever that since you wish to deny not only the principles of the sciences, but palpable experience and the very senses themselves, you can never be convinced, nor relieved from any preconceived opinion. Therefore I shall hold my peace because *contra negantes principia non est disputandum* [there is no arguing against the denial of axioms], and not because I am persuaded by your reasoning.

Concerning the things you have just said, questioning even whether the motion of heavy bodies is straight or not, how can you ever reasonably deny that parts of the earth—that is to say, heavy bodies—descend straight toward the center? For if you let a rock fall from a very high tower whose walls are straight and plumb, it will go down grazing the tower to the earth, and strike in the same place where a plummet would come to rest if hung on a cord fastened above, exactly where the rock was let drop. Isn't this only too obvious an argument that such motion is straight and toward the center?

In the second place you question whether parts of the earth move so as to go toward the center of the universe, as Aristotle affirms. As if he had not proved this conclusively by the doctrine of contrary motions, arguing as follows. The motion of heavy bodies is contrary to that of light ones. But the motion of light ones is seen to be directly upward; that is, toward the circumference of the universe. Therefore the motion of heavy bodies is directly toward the center of the universe, and it happens *per accidens* [by chance] that this is toward the center of the earth, because the latter coincides with the former and is united to it. [. . .]

SALVIATI: As I said before, you will learn how reasonable it is for me to doubt whether heavy bodies move by a straight and perpendicular line when I examine that particular argument. As to the second point, I am surprised that you should need to have Aristotle's fallacy revealed, it being so obvious, and I wonder at your failure to perceive that Aristotle assumes what is in question. For observe that . . .

SIMPLICIO: Please, Salviati, speak more respectfully of Aristotle. He having been the first, only, and admirable expounder of the syllogistic forms, of proofs, of disproofs, of the manner of discovering sophisms and fallacies—in short, of all logic—how can you ever convince anyone that he would subsequently equivocate so seriously as to take for granted that which is in question? Gentlemen, it would be better first to understand him perfectly, and then see whether you want to refute him.

SALVIATI: Simplicio, we are engaging in friendly discussion among ourselves in order to investigate certain truths. I shall never take it ill that you expose my errors; when I have not followed the thought of Aristotle, rebuke me freely, and I shall take it in good part. Only let me expound my doubts and reply somewhat to your last remarks. Logic, as it is generally understood, is the organ with which we philosophize. But just as it may be possible for a craftsman to excel in making organs and yet not know how to play them, so one might be a great logician and still be inexpert in making use of logic. Thus we have many people who theoretically understand the whole art of poetry and yet are inept at composing mere quatrains; others enjoy all the precepts of da Vinci and yet do not know how to paint a stool. Playing the organ is taught not by those who make organs, but by those who know how to play them; poetry is learned by continual reading of the poets; painting is acquired by continual painting and designing; the art of proof, by the reading of books filled with demonstrations—and these are exclusively mathematical works, not logical ones.

Now, returning to our purpose, I say all that Aristotle sees of the motion of light bodies is that fire leaves any part of the surface of the terrestrial globe and goes directly away from it, rising upward; this indeed is to move toward a circumference greater than that of the earth. Aristotle has it move to the arc of the moon's path. But he cannot affirm that this is the circumference of the universe, or is concentric with that, so that to move toward it is to move toward the circumference of the universe. To do so he must suppose that the center of the earth, from which we see these ascending light bodies depart, is the same as the center of the universe; which is as much as to say that the terrestrial globe is located in the center of the universe. Now that is just what we were questioning, and what Aristotle intended to prove. You say that this is not an obvious fallacy?

SAGREDO: This argument of Aristotle's appeared to me defective and inconclusive also in another respect, even if one concedes to him that the circumference toward which fire moves in straight lines is that which encloses the universe. For leaving not only from the center, but from any other point in a circle, every body moving in a straight line toward any point whatever will doubtless go toward the circumference and, continuing its motion, will arrive there. Thus we may say truly that this moves toward the circumference, but it will not always be true that anything moving by the same line in the opposite direction would go toward the center. This will be true only if the point taken is itself the center, or if the motion is made along that single line which, produced from the given point, passes through the center. So that to say "Fire moving in a straight line goes toward the circumference of the universe; therefore particles of earth, which move with a contrary motion along the same lines, go toward the center of the universe" is valid only when it is assumed that such lines of fire, produced, pass through the center of the universe. And though we know for certain that they pass through the center of the terrestrial globe (being perpendicular to its surface, not inclined), to draw any conclusion we must suppose that the center of the earth is the same as the center of the universe, or else that particles of fire and earth ascend and descend only by one particular line, which passes through the center of the universe. Now this is false and repugnant to experience, which shows us that particles of fire ascend always by lines perpendicular to the surface of the terrestrial globe, and not by any one line alone, but by infinitely various lines extending from the center of the earth toward every part of the universe.

SALVIATI: You most ingeniously lead Aristotle into the same difficulty, Sagredo, showing the obvious mistake, and adding to it yet another inconsistency. We observe the earth to be spherical, and therefore we are certain that it has a center, toward which we see that all its parts move. We are compelled to speak in this way, since their motions are all perpendicular to the surface of the earth, and we understand that as they move toward the center of the earth, they move to their whole, their universal mother. Now let us have the grace to abandon the argument that their natural instinct is to go not toward the center of the earth, but toward the center of the universe; for we do not know where that may be, or whether it exists at all. Even if it exists, it is but an imaginary point; a nothing, without any quality. [. . .]

SIMPLICIO: This way of philosophizing tends to subvert all natural philosophy, and to disorder and set in confusion heaven and earth and the whole universe. However, I believe the fundamental principles of the Peripatetics to be such that there is no danger of new sciences being erected upon their ruins.

SALVIATI: Do not worry yourself about heaven and earth, nor fear either their subversion or the ruin of philosophy. As to heaven, it is in vain that you fear for that which you yourself hold to be inalterable and invariant. As for the earth, we seek rather to ennoble and perfect it when we strive to make it like the celestial bodies, and, as it were, place it in heaven, from which your

philosophers have banished it. Philosophy itself cannot but benefit from our disputes, for if our conceptions prove true, new
achievements will be made; if false, their rebuttal will further
confirm the original doctrines. No, save your concern for certain
philosophers; come to their aid and defend them. As to science
itself, it can only improve. [. . .]

Our main goal being to bring forth and consider everything
that has been adopted for and against the two systems, Ptolemaic
and Copernican, it would not be good to pass by anything written on this subject.

SIMPLICIO: Then I shall begin with the objections contained in the
booklet of theses and later take up the others. First, the author
cleverly calculates how many miles per hour a point on the earth's
surface travels at the equator, and how many at other points, in
other latitudes. Not content with investigating such movements in
hourly times, he finds them also in minutes, and still unsatisfied
with minutes, he pursues them down to a single second. Moreover, he goes on to show precisely how many miles would be traveled in such a time by a cannon ball placed in the moon's orbit,
assuming this orbit to be as large as figured by Copernicus himself, so as to take away every subterfuge from his adversary. These
very ingenious and elegant reckonings made, he shows that a
heavy body falling from there would consume rather more than
six days to get to the center of the earth, toward which heavy
bodies tend naturally.

Now if by Divine power, or by means of some angel, a very
large cannon ball were miraculously transported there and placed
vertically over us and released, it is indeed a most incredible thing
(in his view and mine) that during its descent it should keep itself
always in our vertical line, continuing to turn with the earth
about its center for so many days, describing at the equator a spiral line in the plane of the great circle, and at all other latitudes
spiral lines about cones, and falling at the poles in a simple
straight line.

The great improbability of this he then establishes and con-
firms by advancing, through his method of interrogation, many
difficulties which it is impossible for the followers of Copernicus
to remove; these are, if I remember correctly . . .

SALVIATI: Just a moment, please, Simplicio. You do not want to lose
me with so many new things at one stretch; I have a poor memory, so I have to go step by step. And since I remember having
already calculated how long it would take such a heavy body
falling from the moon's orbit to arrive at the center of the earth,
and seem to recall that it would not take this long. [. . .]

SIMPLICIO: [It] seems to me a remarkable thing in any case that in
coming from the moon's orbit, distant by such a huge interval,
the ball should have a natural tendency to keep itself always over
the same point of the earth which it stood over at its departure,
rather than to fall behind in such a very long way.

SALVIATI: The effect might be remarkable or it might be not at all
remarkable, but natural and ordinary, depending upon what had
gone on before. If, in agreement with the supposition made by the
author, the ball had possessed the twenty-four-hour circular
motion while it remained in the moon's orbit, together with the
earth and everything else contained within that orbit, then that
same force which made it go around before it descended would
continue to make it do so during its descent too.[2] [. . .]

But if the ball had no rotation in the orbit, it would not in
descending be obliged to remain perpendicularly over that point
of the earth which was beneath it when the descent began. Nor
does Copernicus or any of his adherents say it would.

SIMPLICIO: But the author will object, as you see, asking upon what
principle this circular motion of heavy and light bodies depends—
whether upon an internal or an external principle.[3]

SALVIATI: Keeping to the problem in hand, I say that the principle
which would make the ball revolve while in the lunar orbit is the
same one which would maintain this revolving also during the
descent. I shall leave it to the author to make this be internal or
external, at his pleasure.

---

[2] *This argument that forces maintain motion was easily made after Newton presented his
laws of motion. Galileo anticipated them but could not have expected many people to
follow his thinking.*

[3] *Talk of internal versus external principles refers to the kinds of abstract causes that
Bacon sought to remove from science. Galileo played with this issue but could not quite
dismiss it.*

SIMPLICIO: The author will prove that it cannot be either internal or external.

SALVIATI: And I shall reply that the ball was not moving in the orbit, and thus be freed from any responsibility of explaining why, in descending, it remains vertically over the same point, since it will not remain so.

SIMPLICIO: Very well, but as heavy and light bodies can have neither an internal nor an external principle of moving circularly, then neither does the earth move circularly. And thus we have his meaning.

SALVIATI: I did not say that the earth has neither an external nor an internal principle of moving circularly; I say that I do not know which of the two it has. My not knowing this does not have the power to remove it.

But if this author knows by which principle other world bodies are moved in rotation, as they certainly are moved, then I say that that which makes the earth move is a thing similar to whatever moves Mars and Jupiter, and which he believes also moves the stellar sphere. If he will advise me as to the motive power of one of these movable bodies, I promise I shall be able to tell him what makes the earth move. Moreover, I shall do the same if he can teach me what it is that moves earthly things downward.[4]

SIMPLICIO: The cause of this effect is well known; everybody is aware that it is gravity.

SALVIATI: You are wrong, Simplicio; what you ought to say is that everyone knows that it is called "gravity." What I am asking you for is not the name of the thing, but its essence, of which essence you know not a bit more than you know about the essence of whatever moves the stars around. I except the name which has been attached to it and which has been made a familiar household word by the continual experience that we have of it daily. But we do not really understand what principle or what force it is that moves stones downward, any more than we understand what moves them upward after they leave the thrower's hand, or what

---

[4] *Newton is usually credited with linking the motion of the planets with the falling of things (like apples) on earth. Quite plainly, he was not first.*

moves the moon around. We have merely, as I said, assigned to the first the more specific and definite name "gravity", whereas to the second we assign the more general term "impressed force" (*virtù impressa*), and to the last-named we give "spirits" (*intelligenza*), either "assisting" (*assistente*) or "abiding" (*informante*); and as the cause of infinite other motions we give "Nature."

➤┼◆➤─○─◄➤┼◄

ROBERT BOYLE

Doubting the Four Elements, from
The Skeptical Chymist (1661)

Robert Boyle (1627–1691) holds a recognized place in literary as
well as scientific history. His dialogue *The Skeptical Chymist*
was widely read by educated Englishmen long after its scientific facts
and problems were obsolete. Rather, it was enjoyed for its energetic
insistence on the evidence of one's senses and also for the fun Boyle
has in portraying "the generality of alchymists," as he labeled his
opponents in his book's subtitle. Boyle's Aristotelian, Themistius, is
much more pompous than Simplicio is in Galileo's dialogues. Boyle's
structural problems in organizing the dialogue were also more complicated than Galileo's. Galileo was quarreling with only one idea—
that the sun moved—but Boyle was doubting both Aristotle's classic
theory of four elements and a late medieval doctrine of three elements. In this dialogue Philoponus speaks for the three-element doctrine. Boyle's representative, named Carneades, has two opponents.
A fourth figure in the dialogue is Eleutherius; like Galileo's Sagredo
figure, he is the educated layman whom all sides wish to convince.

The selected passage concerns the heart of the ancient doctrines,
that all things are composed of four (or maybe three) elements. To
modern readers, the three- (or four-) element doctrine seems absurd,
but—if we do not take the idea literally—we can still see that the
ancients had discovered something. The four elements neatly summa-
rize the forms of things known to normal experience: solids (earth),
gas (air), liquid (water), and energy (fire). Alchemists, however, did
not know they were speaking metaphorically, and in this selection
Boyle/Carneades points out the ambiguity of their traditional proofs.

[Themistius said,] "If I were allowed the freedom in pleading for
the four elements to employ the arguments suggested to me by reason
to demonstrate them, I should almost as little doubt of making you a
proselyte to those unsevered teachers, Truth and Aristotle, as I do of
your candor and your judgment. And I hope you will however consider, that that great favorite and interpreter of nature, Aristotle, who
was (as his *Organum* witnesses) the greatest master of logic that ever

lived, disclaimed the course taken by other petty philosophers (ancient and modern) who not attending the coherence and consequences of their opinions, are more solicitous to make each particular opinion plausible independently upon the rest, than to frame them all so, as not only to be consistent together, but to support each other. For that great man in his vast and comprehensive intellect so framed each of his notions that being curiously adapted into one system, they need not each of them any other defense than that which their mutual coherence gives them: as 'tis in an arch, where each single stone, which if severed from the rest would be perhaps defenseless, is sufficiently secured by the solidity and entireness of the whole fabric of which it is a part.

"How justly this may be applied to the present case, I could easily show you if I were permitted to declare to you how harmonious Aristotle's doctrine of the elements is with his other principles of philosophy; and how rationally he has deduced their number from that of the combinations of the four first qualities from the kinds of simple motion belonging to simple bodies, and from I know not how many other principles and phenomena of nature, which so conspire with his doctrine of the elements that they mutually strengthen and support each other. But since 'tis forbidden me to insist on reflections of this kind, I must proceed to tell you, that though the assertors of the four elements value reason so highly, and are furnished with arguments enough drawn from thence to be satisfied that there must be four elements, though no man had ever yet made any sensible trial to discover their number, yet they are not destitute of experience to satisfy others that are wont to be more swayed by their senses than their reason. And I shall proceed to consider the testimony of experience, when I shall have first advertised you, that if men were as perfectly rational as 'tis to be wished they were, this sensible way of probation[5] would be as needless as 'tis wont to be imperfect. For it is much more high and philosophical to discover things *a priori* than *a posteriori*. and therefore the peripatetics have not been very solicitous to gather experiments to prove their doctrines, contenting themselves with a few only, to satisfy those that are not capable of a nobler conviction.

---

[5] Probation *used to mean "putting to the proof."*

And indeed they employ experiments rather to illustrate than to demonstrate their doctrines, as astronomers use spheres of pasteboard to descend to the capacities of such as must be taught by their senses, for want of being arrived to a clear apprehension of purely mathematical notions and truths.

"I speak thus, Eleutherius, (adds Themistius) only to do right to reason, and not out of diffidence of the experimental proof I am to allege. For though I shall name but one, yet it is such a one as will make all other appear as needless as itself will be found satisfactory. For if you but consider a piece of green wood burning in a chimney, you will readily discern in the disbanded parts of it the four elements, of which we teach it and other mixed bodies to be composed. The fire discovers itself in the flame by its own light; the smoke by ascending to the top of the chimney, and there readily vanishing into air, like a river losing itself in the sea, sufficiently manifests to what element it belongs and gladly returns. The water in its own form boiling and hissing at the ends of the burning wood betrays itself to more than one of our senses; and the ashes by their weight, their fieriness, and their dryness, put it past doubt that they belong to the element of earth.

"If I spoke (continues Themistius) to less knowing persons, I would perhaps make some excuse for building upon such an obvious and easy analysis, but 'twould be, I fear, injurious not to think such an apology needless to you, who are too judicious either to think it necessary that experiments to prove obvious truths should be far fetched, or to wonder that among so many mixed bodies that are compounded of the four elements, some of them should upon a slight analysis manifestly exhibit the ingredients they consist of. Especially since it is very agreeable to the goodness of nature, to disclose, even in some of the most obvious experiments that men make, a truth so important, and so requisite to be taken notice of by them. Besides that our analysis by how much the more obvious we make it, by so much the more suitable it will be to the nature of that doctrine which 'tis alleged to prove, which being as clear and intelligible to the understanding as obvious to the sense, 'tis no marvel the learned part of mankind should so long and so generally embrace it.

"For this doctrine is very different from the whimsies of the chemists and other modern innovators, of whose hypotheses we may

observe, as naturalists do of less perfect animals, that as they are
hastily formed, so they are commonly short lived. For so these, as
they are often framed in one week, are perhaps thought fit to be
laughed at the next; and being built perchance but upon two or three
experiments are destroyed by a third or fourth, whereas the doctrine
of the four elements was framed by Aristotle after he had leisurely
considered those theories of former philosophers, which are now
with great applause revived, as discovered by these latter ages; and
had so judiciously detected and supplied the errors and defects of former hypotheses concerning the elements, that his doctrine of them
has been ever since deservedly embraced by the lettered part of mankind: All the philosophers that preceded him having in their several
ages contributed to the completeness of this doctrine, as those of succeeding times have acquiesced in it. Nor has an hypothesis so deliberately and maturely established been called in question till in the last
century Paracelsus[6] and some few other sooty empirics, rather than
(as they are fain to call themselves) philosophers, having their eyes
darkened, and their brains troubled with the smoke of their own furnaces, began to rail at the peripatetic doctrine, which they were too
illiterate to understand, and to tell the credulous world, that they
could see but three ingredients in mixed bodies; which to gain themselves the repute of inventors, they endeavored to disguise by calling
them instead of earth, and fire, and vapor, salt, sulfur, and mercury;
to which they gave the canting title of hypostatical principles: but
when they came to describe them, they showed how little they understood what they meant by them, by disagreeing as much from one
another, as from the truth they agreed in opposing: For they deliver
their hypotheses as darkly as their processes; and 'tis almost as impossible for any sober man to find their meaning, as 'tis for them to
find their elixir.[7] And indeed nothing has spread their philosophy, but
their great brags and undertakings; notwithstanding all which, (says

---

[6] *Paracelsus, the name taken by Philippius Aureolis Theophrastus Bombast von
Honnenheim (1493–1541), brought chemistry into medicine. Although he was a
notorious braggart, his name was* not *the source of the word* bombast.

[7] *In alchemy, an elixir was the long-sought preparation whose use would change metals
into gold.*

Themistius smiling) I scarce know any thing they have performed worth wondering at, save that they have been able to draw Philoponus to their party and to engage him to the defense of an unintelligible hypothesis, who knows so well as he does, that principles ought be like diamonds, as well very clear, as perfectly solid."

Themistius having after these last words declared by his silence that he had finished his discourse, Carneades addressed himself, as his adversary had done, to Eleutherius, returned this answer to it, "I hoped for demonstration, but I perceive Themistius hopes to put me off with a harangue, wherein he cannot have given me a greater opinion of his parts than he has given me distrust for his hypothesis, since for it even a man of such learning can bring no better arguments. The rhetorical parts of his discourse, though it make not the least part of it, I shall say nothing to, designing to examine only the argumentative part, and leaving it to Philoponus to answer those passages wherein either Paracelsus or chemists are concerned: I shall observe to you, that in what he has said besides, he makes it his business to do these two things. The one to propose and make out an experiment to demonstrate the common opinion about the four elements; and the other, to insinuate divers things which he thinks may repair the weakness of his argument, from experience, and upon other accounts bring some credit to the otherwise defenseless doctrine he maintains.

"To begin then with his experiment of the burning wood, it seems to me to be obnoxious to not a few considerable exceptions.

"And first, if I would now deal rigidly with my adversary, I might here make a great question of the very way of probation which he and others employ, without the least scruple, to evince, that the bodies commonly called mixed are made up of earth, air, water, and fire, which they are pleased also to call elements; namely that upon the supposed analysis made by the fire, of the former sort of concretes,[8] there are wont to emerge bodies resembling those which they take for the elements. For not to anticipate here what I foresee I shall have occasion to insist on, when I come to discourse with Philoponus concerning the right that fire has to pass for the proper and universal instrument of analyzing mixed bodies, not to anticipate that, I say, if I

---

[8] Today we would say "sort of compound substances."

were disposed to wrangle, I might allege that by Themistius his experiment it would appear rather that those he calls elements are made
of those he calls mixed bodies, than mixed bodies of the elements.
For in Themistius's analyzed wood, and in other bodies dissipated and
altered by the fire, it appears, and he confesses, that which he takes for
elementary fire and water, are made out of the concrete; but it appears
not that the concrete was made up of fire and water. Nor has either
he, or any man, for ought I know, of his persuasion, yet proved that
nothing can be obtained from a body by the fire that was not preexistent in it."

At this unexpected objection, not only Themistius, but the rest
of the company appeared not a little surprised; but after a while
Philoponus conceiving his opinion, as well as that of Aristotle, concerned in that objection, "You cannot sure (says he to Carneades)
propose this difficulty, not to call it cavil, otherwise than as an exercise of wit, and not as laying any weight upon it. For how can that be
separated from a thing that was not existent in it. When, for instance,
a refiner mingles gold and lead, and exposing this mixture upon a
cupel[9] to the violence of the fire, thereby separates it into pure and
refulgent[10] gold and lead (which driven off together with the dross of
the gold is then called *Lithargyrium Auri*) can any man doubt that
sees these two so differing substances separated from the mass, that
they were existent in it before it was committed to the fire."

"I should (replies Carneades) allow your argument to prove
something, if, as men see the refiners commonly take beforehand both
lead and gold to make the mass you speak of, so we did see nature
pull down a parcel of the element of fire, that is fancied to be placed I
know not how so many thousand leagues off, contiguous to the orb
of the moon, and to blend it with a quantity of each of the three other
elements, to compose every mixed body, upon whose resolution the
fire presents us with fire, and earth, and the rest. And let me add,
Philoponius, that to make your reasoning cogent, it must be first
proved, that the fire does only take the elementary ingredients asun-
der without otherwise altering them. For else 'tis obvious, that bodies
may afford substances which were not preexistent in them; as flesh

---

[9] *A cupel was a container used in assaying gold.*
[10] *Gleaming.*

too long kept produces maggots and old cheese mites, which I suppose you will not affirm to be ingredients of those bodies. Now that fire does not always barely separate the elementary parts, but sometimes at least alter also the ingredients of bodies, if I did not expect ere long a better occasion to prove it, I might make probable out of your very instance, wherein there is nothing elementary separated by the great violence of the refiner's fire: the gold and lead which are the two ingredients separated upon the analysis being confessedly yet perfectly mixed bodies, and the litharge[11] being lead indeed; but such lead as is differing in consistence and other qualities from what it was before. To which I must add that I have sometimes seen, and so questionless[12] have you much oftener, some parcels of glass adhering to the test or cupel, and this glass though emergent as well as the gold or litherge upon your analysis, you will not I hope allow to have been a third ingredient of the mass out of which the fire produced it."

⊷•⊶•O•⊷•⊶

---

[11] *A litharge was a compound of lead and oxygen produced in the course of an alchemical investigation.*

[12] *Without question.*

ISAAC NEWTON AND ROBERT HOOKE

Dispute on the Nature of Light, from
Letters (1672)

Isaac Newton (1643–1727) is generally admired as the ideal scientist who made a major contribution to everything he touched. Besides his theory of gravity, Newton developed a theory of light and color based on his belief that light was composed of particles moving in straight lines. He built a reflecting telescope that much improved on existing ones and brought him some celebrity. In a letter to Henry Oldenberg, secretary for correspondence at the Royal Society, Newton explained how he had built the instrument. By itself, this letter would have been a great expression of an active, observant mind. But it prompted a reply.

Robert Hooke (1635–1703) was another scientific wonder who contributed to many fields. He had built a reflecting telescope in 1664, probably three or four years before Newton built his first one, and Hooke had performed experiments showing that light could bend around corners. He concluded that light was a wave and disputed Newton's particle theory.

Neither man could have imagined the irony of this dispute—that modern physics claims that light acts as both a wave and a particle—as both men assumed that at least one of them had to be wrong. Newton and Hooke had many disputes. Hooke, for example, wrote an early version of Newton's gravitational equation and felt that Newton did not give him enough credit for his contribution.

Light Moves in Straight Lines. Newton to Henry Oldenberg, Feb. 6, 1672

Sir,

To perform my late promise to you, I shall without further ceremony acquaint, you that in the beginning of the Year 1666 (at which time I applied myself to the grinding of optic glasses of other figures than spherical) I procured me a triangular glass prism to try therewith the celebrated phenomena of colors.[13] And in order thereto hav-

---

[13] *The phenomenon he is referring to has become so celebrated that every schoolchild learns it and is bored by it. It is the simple fact that when sunlight passes through a prism, it forms a spectrum of colors.*

ing darkened my chamber and made a small hole in my window-shuts[14] to let in a convenient quantity of the sun's light, I placed my prism at its entrance, that it might be thereby refracted to the opposite wall. It was at first a very pleasing divertissement, to view the vivid and intense colors produced thereby; but after a while applying myself to consider them more circumspectly, I became surprised to see them in an oblong form; which, according to the received laws of refraction, I expected should have been circular.

They were terminated at the sides with straight lines, but at the ends, the decay of light was so gradual, that it was difficult to determine justly, what was their figure; yet they seemed semicircular.

Comparing the length of this colored spectrum with its breadth, I found it about five times greater; a disproportion so extravagant, that it excited me to a more than ordinary curiosity of examining, from whence it might proceed. I could scarce think, that the various thickness of the glass, or the termination with shadow or darkness, could have any influence on light to produce such an effect; yet I thought it not amiss to examine first these circumstances, and so tried, what would happen by transmitting light through parts of the glass of divers thicknesses, or through holes in the window of divers bignesses, or by setting the prism without, so that the light might pass through it, and be refracted before it was terminated by the hole: But I found none of those circumstances material. The fashion of the colors was in all these cases the same.

Then I suspected, whether by any unevenness in the glass, or other contingent irregularity, these colors might be thus dilated. And to try this, I took another prism like the former, and so placed it, that the light, passing through them both, might be refracted contrary ways, and so by the latter returned into that course, from which the former had diverted it. For, by this means I thought, the regular effects of the first prism would be destroyed by the second prism, but the irregular ones more augmented, by the multiplicity of refractions. The event was, that the light, which by the first prism was diffused into an oblong form, was by the second reduced into an orbicular one with as much regularity, as when it did not at all pass through them. So that, whatever was the cause of that length, 'twas not any contingent irregularity.

---

[14] *Shutters.*

I then proceeded to examine more critically, what might be effected by the difference of the incidence of rays coming from divers parts of the sun; and to that end, measured the several lines and angles, belonging to the image. [. . .] The angle at the hole, which that breadth subtended, was about 31′, answerable to the sun's diameter; but the angle, which its length subtended, was more than five such diameters, namely 2 deg. 49′.

Having made these observations, I first computed from them the refractive power of that glass, and found [. . .] that the emergent rays should have comprehended an angle of about 31′, as they did, before they were incident.[15]

But because this computation was founded on the hypothesis of the proportionality of the sines of incidence, and refraction,[16] which though by my own and others experience I could not imagine to be so erroneous, as to make that angle but 31′, which in reality was 2 deg. 49′; yet my curiosity caused me again to take my prism. And having placed it at my window, as before, I observed, that by turning it a little about its axis to and fro, so as to vary its obliquity to the light, more than by an angle of 4 or 5 degrees, the colors were not thereby sensibly translated from their place on the wall, and consequently by that variation of incidence, the quantity of refraction was not sensibly varied.[17] By this experiment therefore, as well as by the former computation, it was evident, that the difference of the incidence of rays, flowing from divers parts of the sun, could not make them after decussation[18] diverge at a sensibly greater angle, than that at which they before converged;[19] which being, at most, but about 31 or 32 minutes, there still remained some other cause to be found out, from whence it could be 2 deg. 49′.

Then I began to suspect, whether the rays, after their trajection through the prism, did not move in curve lines, and according to their

---

[15] *Before they were incident, or before they struck the glass.*

[16] Incidence *refers to the light falling on the prism, and* refraction *is the light being bent by the prism. Newton says he assumed the angles of these two circumstances maintained a proportional relationship.*

[17] *But here Newton finds that changing the angle at which light falls on the prism does not change by much the angle by which the light is deflected. In short, light does not behave like a billiard ball.*

[18] *Intersecting.*

[19] *Today's English would make it "than that at which they converged before."*

more or less curvity tend to divers parts of the wall. And it increased my suspicion, when I remembered that I had often seen a tennis ball, struck with an oblique racket, describe such a curve line. For, a circular as well as a progressive motion being communicated to it by that stroke, its parts on that side, where the motions conspire, must press and beat the contiguous air more violently than on the other, and there excite a reluctancy and reaction of the air proportionally greater. And for the same reason, if the rays of light should possibly be globular bodies, and by their oblique passage out of one medium into another acquire a circulating motion, they ought to feel the greater resistance from the ambient ether, on that side, where the motions conspire, and thence be continually bowed to the other. But notwithstanding this plausible ground of suspicion, when I came to examine it, I could observe no such curvity in them. And besides (which was enough for my purpose) I observed, that the difference betwixt the length of the image, and diameter of the hole, through which the light was transmitted, was proportionable to their distance.

The gradual removal of these suspicions at length led me to the *Experimentum Crucis,* which was this: I took two boards, and placed one of them close behind the prism at the window, so that the light might pass through a small hole, made in it for that purpose, and fall on the other board, which I placed at about 12 foot distance, having first made a small hole in it also, for some of that incident light to pass through. Then I placed another prism behind this second board, so that the light, trajected through both the boards, might pass through that also, and be again refracted before it arrived at the wall. This done, I took the first prism in my hand and turned it to and fro slowly about its axis so much as to make the several parts of the image, cast on the second board, successively pass through the hole in it, that I might observe to what places on the wall the second prism would refract them. And I saw by the variation of those places that the light, tending to that end of the image, towards which the refraction of the first prism was made, did in the second prism suffer a refraction considerably greater then the light tending to the other end. And so the true cause of the length of that image was detected to be no other, than that light consists of rays differently refrangible,[20] which, without any respect to a difference in their incidence, were,

---

[20] *Rays differently refrangible, or rays with different capacities for deflection.*

according to their degrees of refrangibility, transmitted towards divers parts of the wall.

When I understood this, I left off my aforesaid glassworks; for I saw, that the perfection of telescopes was hitherto limited, not so much for want of glasses truly figured according to the prescriptions of optic authors (which all men have hitherto imagined) as because that light itself is a heterogeneous mixture of differently refrangible rays. So that, were a glass so exactly figured, as to collect any one sort of rays into one point, it could not collect those also into the same point, which having the same incidence upon the same medium are apt to suffer a different refraction. Nay, I wondered, that seeing the difference of refrangibility was so great, as I found it, telescopes should arrive to that perfection they are now at. For, measuring the refractions in one of my prisms, I found, that supposing the common sine of incidence upon one of its planes was 44 parts, the sine of refraction of the utmost rays on the red end of the colors, made out of the glass into the air, would be 68 parts, and the sine of refraction of the utmost rays on the other end, 69 parts: So that the difference is about a 24th or 25th part of the whole refraction. And consequently, the object-glass of any telescope cannot collect all the rays, which come from one point of an object so as to make them convene at its focus in less room than in a circular space, whose diameter is the 50th part of the diameter of its aperture; which is an irregularity, some hundreds of times greater, then a circularly figured lens, of so small a section as the object glasses of long telescopes are, would cause by the unfitness of its figure, were light uniform.

This made me take reflections into consideration, and finding them regular, so that the angle of reflection of all sorts of rays was equal to their angle of incidence; I understood, that by their mediation optic instruments might be brought to any degree of perfection imaginable, provided a reflecting substance could be found, which would polish as finely as glass, and reflect as much light, as glass transmits, and the art of communicating to it a parabolic figure be also attained. But these seemed very great difficulties, and I almost thought them insuperable, when I further considered, that every irregularity in a reflecting superficies[21] makes the rays stray 5 or 6

---

[21] *Surface.*

times more out of their due course, than the like irregularities in a refracting one: So that a much greater curiosity would be here requisite, than in figuring glasses for refraction.

Amidst these thoughts I was forced from Cambridge by the intervening plague, and it was more than two years, before I proceeded further. But then having thought on a tender way of polishing, proper for metal, whereby, as I imagined, the figure also would be corrected to the last; I began to try, what might be effected in this kind, and by degrees so far perfected an instrument (in the essential parts of it like that I sent to London) by which I could discern Jupiter's 4 concomitants, and showed them divers times to two others of my acquaintance. I could also discern the moonlike phase of Venus, but not very distinctly, nor without some niceness in disposing the instrument.

From that time I was interrupted till this last autumn, when I made the other. And as that was sensibly better then the first (especially for day-objects) so I doubt not, but they will be still brought to a much greater perfection by their endeavors, who, as you inform me, are taking care about it at London.

Light Moves as a Wave. Hooke to Oldenburg, Feb. 15, 1672.

I have perused the excellent discourse of Mr. Newton about colors and refractions, and I was not a little pleased with the niceness and curiosity of his observations. But though I wholly agree with him as to the truth of those he hath alleged, as having by many hundreds of trials found them so, yet as to his hypothesis of solving the phenomena of colors thereby I confess I cannot yet see any undeniable argument to convince me of the certainty thereof. For all the experiments and observations I have hitherto made, nay and even those very experiments which he alleged, do seem to me to prove that light is nothing but a pulse or motion propagated through an homogeneous, uniform and transparent medium: And that color is nothing but the disturbance of that light by the communication of that pulse to other transparent mediums, that is by the refraction thereof: that whiteness and blackness are nothing but the plenty or scarcity of the undisturbed rays of light; and that the two colors (than which there are no more uncompounded in nature) are nothing but the effects of a compounded pulse or disturbed propagation of motion caused by refraction. But how certain soever I think myself of my hypothesis, which I did not take up without first trying some hundreds of experiments; yet I should be very glad to meet with one *Experimentum Crucis* from Mr. Newton, that should divorce me from it. But it is not that, which he so calls, will do the turn; for the same phenomenon will be solved by my hypothesis as well as by his without any manner of difficulty or straining: nay I will undertake to show another hypothesis differing from both his and mine, that shall do the same thing. That the ray of light is as twere split or rarefied by refraction, is most certain, and that thereby a differing pulse is propagated both on those sides, and in all the middle parts of the ray, is easy to be conceived; and also that differing pulses or compound motions should make differing impressions on the eye, brain, or sense, is also easy to be conceived; and that whatever refracting medium does again reduce it to its primitive simple motion by destroying the adventitious, does likewise restore it to its primitive whiteness and simplicity. But why there is a necessity, that all these motions, or whatever else it be that makes colors, should be originally in the simple rays of light I do not yet understand the necessity; no more than that all those sounds must be in the air of the bellows which are afterwards heard to issue from the organ-pipes, or in the string which are afterwards by differing stoppings and strikings produced: which string (by the way) is a pretty representation of the shape of a refracted ray to the eye; and the manner of it may be somewhat imagined by the similitude there; for the ray is like the string, strained between the luminous object and the eye, and the stop or finger is like the refracting surface, on the one side of which the string has no motion; on the other, a vibrating one. Now we may say indeed or imagine that the rest or straightness of the string is caused by the cessation of motions or coalition of all vibrations, and that all the vibrations are dormant in it; but yet it seems more natural to me, to imagine it the other way. And I am a little troubled that this supposition should make Mr. Newton wholly lay aside the thoughts of improving telescopes and microscopes by refractions, since it is not improbable, but that he that hath made so very good an improvement of telescopes by his own trials upon reflection, would, if he had prosecuted it, have done more by refraction.[22] [. . .]

---

[22] *The difference between reflection and refraction is the difference between a mirror and a lens. A mirror reflects light, halting its forward passage, whereas a lens bends light but allows it to keep moving forward.*

But It Can't Be a Wave. Newton to Oldenburg, June 11, 1672.

[. . .] Mr. Hooke's hypothesis as to the fundamental part of it is not against me. The fundamental supposition is, that the parts of bodies when briskly agitated, do excite vibrations in the ether which are propagated every way from those bodies in straight lines, and cause a sensation of light by beating and dashing against the bottom of the eye, something after the manner that vibrations in the air cause a sensation of sound by beating against the organs of hearing. Now the most free and natural application of this hypothesis to the solution of phenomena I take to be this: That the agitated parts of bodies according to their several sizes, figures, and motions, excite vibrations in the ether of various depths or bignesses, which being promiscuously propagated through that medium to our eyes, effect in us a sensation of light of a white color; but if by any means those of unequal bignesses be separated from one another, the largest beget a sensation of a red color, the least or shortest of a deep violet, and the intermediate ones of intermediate colors: Much after the manner that bodies according to their several sizes, shapes, and motions, excite vibrations in the air of various bignesses, which according to those bignesses make several tones in sound. That the largest vibrations are best able to overcome the resistance of a refracting superficies, and so break through it with least refraction: Whence the vibrations of several bignesses, that is, the rays of several colors, which are blended together in light, must be parted from one another by refraction, and so cause the phenomena of prisms and other refracting substances. And that it depends on the thickness of a thin transparent plate or bubble, whether a vibration shall be reflected at its further superficies or transmitted; so that according to the number of vibrations interceding the two superficies they may be reflected or transmitted for many successive thicknesses. And since the vibrations which make blue and violet are supposed shorter than those which make red and yellow, they must be reflected at a less thickness of the plate: which is sufficient to explicate all the ordinary phenomena of those plates or bubbles, and also of all natural bodies whose parts are like so many fragments of such plates.

These seem to be the most plain genuine and necessary conditions of this hypothesis. And they agree so justly with my theories, that if Mr. Hooke think fit to apply them, he need not on that account fear a divorce from it. But yet how he will defend it from other difficulties I know not: For to me the fundamental supposition itself seems impossible; namely that the waves or vibrations of any fluid can like the rays of light be propagated in straight lines, without a continual and very extravagant spreading and bending every way into the quiescent medium where they are terminated by it. I am mistaken if there be not both experiment and demonstration to the contrary. And as to the other two or three hypotheses, which he mentions, I had rather believe them subject to the like difficulties, than suspect that Mr. Hooke should select the worst for his own.

What I have said of this, may be easily applied to all other mechanical hypotheses in which light is supposed to be caused by any pression or motion whatsoever excited in the ether by the agitated parts of luminous bodies. For it seems impossible that any of those motions or pressions can be propagated in straight lines without the like spreading every way into the shadowed medium on which they border. But yet if any man can think it possible, he must at least allow that those motions or endeavors to motion caused in the ether by the several parts of any lucid body which differ in size, figure, and agitation, must necessarily be unequal. Which is enough to denominate light an aggregate of difform rays according to any of those hypotheses. And if those original inequalities may suffice to difference the rays in color and refrangibility, I see no reason why they that adhere to any of those hypotheses, should seek for other causes of these effects, unless (to use Mr. Hooke's argument) they will multiply entities without necessity.

❦

MARIE CURIE

Obtaining Radium, from
Pierre Curie (1923)

Sixty years after her death, Marie Sklowdowska Curie (1867–1934) remains a celebrated figure in the history of science. In 1896 she began studying radioactivity, and in 1903 she became the first woman ever to share a Nobel Prize. In 1911 she became the first person of either sex to win a second Nobel Prize. The essay by Robert Kennedy Duncan puts her work into the context of its time, but here we find Marie Curie's own account of her effort to obtain radium. The American edition of her biography of her husband, Pierre, contains additional autobiographical notes, and this selection includes material from both parts of her book.

The effort reported here, an attempt to obtain and study radium, differs from the observations reported in previous chapters. Unlike Galileo looking through his telescope for whatever he could find, the Curies had a definite expectation. They were seeking a new element with stronger radioactive properties than uranium. Although success gave them something new to study, it also encouraged them with the thought that they knew where they were going.

My experiments proved that the radiation of uranium compounds can be measured with precision under determined conditions, and that this radiation is an atomic property of the element of uranium. Its intensity is proportional to the quantity of uranium contained in the compound, and depends neither on conditions of chemical combination, nor on external circumstances, such as light or temperature.

I undertook next to discover if there were other elements possessing the same property, and with this aim I examined all the elements then known, either in their pure state or in compounds. I found that among these bodies, thorium compounds are the only ones which emit rays similar to those of uranium. The radiation of thorium has an intensity of the same order as that of uranium, and is, as in the case of uranium, an atomic property of the element.

It was necessary at this point to find a new term to define this new property of matter manifested by the elements of uranium and thorium. I proposed the word radioactivity which has since become generally adopted; the radioactive elements have been called radio elements.

During the course of my research, I had had occasion to examine not only simple compounds, salts and oxides, but also a great number of minerals. Certain ones proved radioactive; these were those containing uranium and thorium; but their radioactivity seemed abnormal, for it was much greater than the amount I had found in uranium and thorium had led me to expect.

This abnormality greatly surprised us. When I had assured myself that it was not due to an error in the experiment, it became necessary to find an explanation. I then made the hypothesis that the ores uranium and thorium contain in small quantity a substance much more strongly radioactive than either uranium or thorium. This substance could not be one of the known elements, because these had already been examined; it must, therefore, be a new chemical element.

I had a passionate desire to verify this hypothesis as rapidly as possible. And Pierre Curie, keenly interested in the question, abandoned his work on crystals (provisionally, he thought) to join me in the search for this unknown substance.

We chose, for our work, the ore pitchblende, a uranium ore, which in its pure state is about four times more active than oxide of uranium.

Since the composition of this ore was known through very careful chemical analysis, we could expect to find, at a maximum, 1 per cent of new substance. The result of our experiment proved that there were in reality new radioactive elements in pitchblende, but that their proportion did not reach even a millionth per cent!

The method we employed is a *new method in chemical research based on radioactivity*. It consists in inducing separation by the ordinary means of chemical analysis, and of measuring, under suitable conditions, the radioactivity of all the separate products. By this means one can note the chemical character of the radioactive element sought for, for it will become concentrated in those products which will become more and more radioactive as the separation progresses. We soon recognized that the radioactivity was concentrated principally in two different chemical fractions, and we became able to recognize in pitchblende the presence of at least two new radioactive elements: polonium and radium. We announced the existence of polonium in July, 1898, and of radium in December of the same year.

In spite of this relatively rapid progress, our work was far from finished. In our opinion, there could be no doubt of the existence of these new elements, but to make chemists admit their existence, it was necessary to isolate them. Now, in our most strongly radioactive products (several hundred times more active than uranium), the polonium and radium were present only as traces. The polonium occurred associated with bismuth extracted from pitchblende, and radium accompanied the barium extracted from the same mineral. We already knew by what methods we might hope to separate polonium from bismuth and radium from barium; but to accomplish such a separation we had to have at our disposition much larger quantities of the primary ore than we had. [. . .]

We were very poorly equipped with facilities for this purpose. It was necessary to subject large quantities of ore to careful chemical treatment. We had no money, no suitable laboratory, no personal help for our great and difficult undertaking. It was like creating something out of nothing, and if my earlier studying years had once been called by my brother-in-law the heroic period of my life, I can say without exaggeration that the period on which my husband and I now entered was truly the heroic one of our common life.

We knew by our experiments that in the treatment of pitchblende at the uranium plant of St. Joachimsthal, radium must have been left in the residues, and, with the permission of the Austrian government, which owned the plant, we succeeded in securing a certain quantity of these residues, then quite valueless,—and used them for extraction of radium. How glad I was when the sacks arrived, with the brown dust mixed with pine needles, and when the activity proved even greater than that of the primitive ore! It was a stroke of luck that the residues had not been thrown far away or disposed of in some way, but left in a heap in the pine wood near the plant. Some time later, the Austrian government, on the proposition of the Academy of Science of Vienna, let us have several tons of similar residues at a low price. With this material was prepared all the radium I had in my laboratory up to the date when I received the precious gift from the American women.

The School of Physics could give us no suitable premises, but for lack of anything better, the Director permitted us to use an abandoned shed which had been in service as a dissecting room of the School of Medicine. Its glass roof did not afford complete shelter against rain; the heat was suffocating in summer, and the bitter cold of winter was only a little lessened by the iron stove, except in its immediate vicinity. There was no question of obtaining the needed proper apparatus in common use by chemists. We simply had some old pine-wood tables with furnaces and gas burners. We had to use the adjoining yard for those of our chemical operations that involved producing irritating gases; even then the gas often filled our shed. With this equipment we entered on our exhausting work.

Yet it was in this miserable old shed that we passed the best and happiest years of our life, devoting our entire days to our work. Often I had to prepare our lunch in the shed, so as not to interrupt some particularly important operation. Sometimes I had to spend a whole day mixing a boiling mass with a heavy iron rod nearly as large as myself. I would be broken with fatigue at the day's end. Other days, on the contrary, the work would be a most minute and delicate fractional crystallization, in the effort to concentrate the radium. I was then annoyed by the floating dust of iron and coal from which I could not protect my precious products. But I shall never be able to express the joy of the untroubled quietness of this atmosphere of research and the excitement of actual progress with the confident hope of still better results. The feeling of discouragement that sometimes came after some unsuccessful toil did not last long and gave way to renewed activity. We had happy moments devoted to a quiet discussion of our work, walking around our shed.

One of our joys was to go into our workroom at night; we then perceived on all sides the feebly luminous silhouettes of the bottles or capsules containing our products. It was really a lovely sight and one always new to us. The glowing tubes looked like faint, fairy lights.

Thus the months passed, and our efforts, hardly interrupted by short vacations, brought forth more and more complete evidence. Our faith grew ever stronger, and our work being more and more known, we found means to get new quantities of raw material and to carry on some of our crude processes in a factory, allowing me to give more time to the delicate finishing treatment.

At this stage I devoted myself especially to the purification of the radium, my husband being absorbed by the study of the physical properties of the rays emitted by the new substances. It was only after treating one ton of pitchblende residues that I could get definite results. Indeed we know to-day that even in the best minerals there are not more than a few decigrammes of radium in a ton of raw material.

At last the time came when the isolated substances showed all the characters of a pure chemical body. This body, the radium, gives a characteristic spectrum, and I was able to determine for it an atomic weight much higher than that of the barium. This was achieved in 1902. I then possessed one decigramme of very pure radium chloride. It had taken me almost four years to produce the kind of evidence which chemical science demands, that radium is truly a new element. One year would probably have been enough for the same purpose, if reasonable means had been at my disposal. The demonstration that cost so much effort was the basis of the new science of radioactivity.

ALFRED WEGENER

Jigsaw Continents, from
The Origin of Continents and Oceans (1929)

Alfred Wegener (1880–1930) proposed a theory of continental drift that was greeted originally as nonsense, and even today the shadow of the crackpot label hangs over him. When I entered college, thirty years after Wegener's death and just a few years before the general theory was confirmed, I was given to understand that the notion of moving continents was crazy and that nobody who took it seriously could be taken seriously. So when I hunted up his book for possible inclusion in this volume, I expected it would be shallow and cranky. But to my surprise, I found something else. Wegener did much more than point out that the African and South American coasts look like matching sides in a jigsaw puzzle. He showed how, up and down the Atlantic, the pieces fit like real parts of a puzzle—that is, not only do the shape of the sides match, but the markings and lines on the pieces also fit together. And the mountains and rivers match. Today such links are explained by the theory of plate tectonics. According to this idea, the earth's continents sit on massive platforms, or "plates," that crack open and are pushed apart by a spreading sea floor.

Wegener was a meteorologist and an adventurer rather than a professional geologist, but he took full benefit of his membership in a scientific world in which much had already been cataloged. He sought out the reported observations of others, to see whether the geology of the land matched as well as the shapes did. He was criticized as simply looking for supportive data, but the pieces in this collection by Curie, Rutherford, Smoot, and Sullivan all show that scientists often have an idea and then look to see whether the evidence supports it.

The first concept of continental drift first came to me as far back as 1910, when considering the map of the world, under the direct impression produced by the congruence of the coastlines on either side of the Atlantic. At first I did not pay attention to the idea because I regarded it as improbable. In the fall of 1911, I came quite accidentally upon a synoptic report in which I learned for the first time of paleontological evidence for a former land bridge between Brazil and Africa. As a result I undertook a cursory examination of relevant research in the fields of geology and palaeontology, and this provided immediately such weighty corroboration that a conviction of the fundamental soundness of the idea took root in my mind. [. . .]

By comparing the geological structure of both sides of the Atlantic, we can provide a very clear-cut test of our theory that this ocean region is an enormously widened rift whose edges were once directly connected, or so nearly as makes no difference. This is because one would expect that many folds and other formations that arose before the split occurred would conform on both sides, and in fact their terminal sections on either side of the ocean must have been so situated that they appear as direct continuations of each other in a reconstruction of the original state of affairs. Since the reconstruction itself is necessarily unambiguous because of the well-marked outlines of the continental margins and allows no scope for juggling, we have here a totally independent criterion of the highest importance for assessing the correctness of drift theory.

The Atlantic rift is widest in the south, where it first started. The width here is 6220 km. Between Cape São Roque[23] and the Cameroons there is a gap of only 4880 km, between the Newfoundland Bank and the shelf of Great Britain only 2410 km, only 1300 km between Scoresby Sound[24] and Hammerfest[25] and probably only 200–300 km between the shelf margins of northeastern Greenland and Spitsbergen, where the rift appears to have occurred only in relatively recent times.

Let us begin by comparing the southern margins. In the southernmost part of Africa there is a Permian folded range[26] striking east-west, the Swartberg.[27] In the reconstruction, the westerly extension of this range meets the area south of Buenos Aires which, according to the map, does not seem to be marked by any special feature. It is very interesting to note that Keidel found in the local sierras, particularly the more strongly folded southern portion, ancient folds which

---

[23] *Cape São Roque (5°28'S, 38°53'W) is the point where Brazil's coast turns south.*

[24] *Scoresby Sound (70°30'N, 20°0'W), an eastern point on Greenland.*

[25] *Hammerfest (70°40'N, 23°44'E), in the north of Norway.*

[26] *The Permian era was 280 million to 230 million years ago.*

[27] *The Swartberg mountains (34°11'S, 19°29'E) are at the Cape of Good Hope.*

completely resembled in structure, rock series and fossil content not only the pre-cordilleras[28] of San Juan and Mendoza Provinces to the northwest, which abut the Andean folds, but what is more, the South African Cape mountains. He states: "In the sierras of Buenos Aires Province, particularly in the southern range, we find a succession of beds very like that of the Cape mountains of South Africa. There appears to be strong conformity among at least three cases: the lower sandstone of the Lower Devonian transgression,[29] the fossil-bearing schists[30] which mark the culmination of this transgression and a more recent and very characteristic structure, the glacial conglomerate of the Upper Paleozoic.[31] . . . Both the sedimentary rocks of the Devonian transgression and the glacial conglomerate are strongly folded just as in the Cape mountains; and here, as there, the direction of the folding movement is mainly a northerly one." All this is an indication that we have here an elongated, ancient fold that traverses the southern tip of Africa, then is continued across South America south of Buenos Aires and finally turns north to join the Andes. Today the fragments of this fold are separated from each other by an ocean more than 6000 km wide.[32] In our reconstruction, which here in particular permits of no manipulation, the fragments are brought into direct contact; their distances from Cape São Roque in one case and the Cameroons in the other are equal. This evidence for the correctness of our synthesis is very remarkable, and one is reminded of the torn visiting card used as a means of recognition. [. . .]

However, this corroboration of our viewpoint provided by the Cape mountains and their continuation in the sierras of Buenos Aires is by no means the only one; we can find many other items of evidence along the Atlantic coastlines. Even in its broad outlines the vast gneiss[33] plateau of Africa, last folded a long time ago, shows a striking similarity to that of Brazil, and this similarity is not confined only to generalities, as is revealed by the conformity between the igneous

---

[28] *The Andean foothills.*

[29] *The Lower Devonian transgression (boundary) took place about 400 million years ago.*

[30] *Schist is a coarse-grained rock.*

[31] *The Upper Paleozoic was about 250 million years ago.*

[32] *Although far apart, Buenos Aires (34°40'S) and Capetown (33°50'S) lie on almost precisely the same latitude.*

[33] *Gneiss is a layered rock similar to granite.*

rocks and between the sedimentary deposits of each area, and by the conformity between the original fold directions.

H. A. Brouwer has made a comparison of the igneous rocks. He finds no less than five parallels: (1) the older granite, (2) the younger granite, (3) alkali-rich rocks, (4) volcanic Jurassic rock and intrusive dolerite and (5) kimberlite, alnoite, etc.

In Brazil, the older granite is found in the so-called "Brazilian complex"; in Africa, in the "fundamental complex" of the southwest, and also in the "Malmesbury system" of southern Cape Colony and in the "Swaziland system" of the Transvaal and Rhodesia. Brouwer says: "Both the east coast of Brazil in the Serra do Mar and the opposite west coast of southern and central Africa consist mainly of these rocks, and in many ways they give the landscape of both continents a similar topography." [. . .]

Finally, the last of the rock groups (kimberlite, alnoite, etc.) is the best-known, since both in Brazil and South Africa the beds yield the famous diamond finds. In both these regions the peculiar type of stratification known as "pipes" occurs. There are white diamonds in Brazil in Minas Geraes State and in South Africa north of the Orange River only. However, the correspondence between the two regions is shown more clearly by the extent of the kimberlitic parent rock than by these rare diamond sites. The same thing has been established in the gangues of Rio de Janeiro State: "As in the case of the kimberlite rocks near the west coast of South Africa, the well-known Brazilian rocks almost all belong to the low-mica basaltic varieties."

Brouwer stresses, however, that even the sedimentary rocks correspond closely on both sides: "The similarity between some groups of sedimentary rocks on both sides of the Atlantic Ocean is also very striking. We mention only the South African Karroo system and the Brazilian Santa Catharina system. The Orleans conglomerate in Santa Catharina and Rio Grande do Sul matches the Dwyka conglomerate of South Africa, and in both continents the uppermost sections are formed by the thick volcanic rock series already mentioned, like those of the Drakenberg in Cape Colony and of the Serra Geral in Rio Grande do Sul." [. . .]

We find additional conformities in the directions of the ancient folds which extend throughout these large gneiss plateaux. [. . .] In the gneiss massif of the African continent there are two main strike directions (trend lines) of somewhat different age. In the Sudan the predominant one is the older, northeasterly strike, which is at once evident in the straight upper course of the Niger, running in a similar direction and visible as far as the Cameroons. It cuts the coastline at 45°. However, south of the Cameroons [. . .] the other, younger strike direction takes precedence, running roughly north-south and parallel to the curves of the coast.

In Brazil we find the same phenomenon. E. Suess wrote: "The map of eastern Guiana . . . shows the more or less east-west strikes of the old types of rock that constitute this area. The included Paleozoic strata which form the northern part of the Amazon basin also follow this direction, and the run of the coast from Cayenne towards the mouth of the Amazon is therefore across the strike direction. . . . So far as the geology of Brazil is known today, it forces one to assume that up to Cape São Roque the contour of the mainland crosses the strike direction of the mountains, but that from these foothills on, all the way down to Uruguay, the lie of the coast is marked by the mountains." Here also the courses of the rivers (the Amazon on one side, the Rio São Francisco and the Parana on the other) generally follow the strike direction. [. . .] Taking into account the large angle through which South America must be turned in our reconstruction, the direction of the Amazon becomes exactly parallel to that of the upper course of the Niger, so that the two strike directions coincide with the African ones. We may see in this a further confirmation for the direct connection that once existed between the two continents.

The similar structure of Brazil and southern Africa has been emphasised more and more strongly of late. Maack states: "Anyone who knows southern Africa will find the geology of this (the Brazilian) landscape startling. At every step I was reminded of the formations of Namaland and the Transvaal. The Brazilian strata correspond perfectly in every detail to the strata series of the southern African shield." On this journey Maack found five kimberlite pipes at Patos (ca. $18\frac{1}{2}°$ S, $46\frac{1}{2}°$ W). He concludes: "It is obvious that in view of the distance which today separates the corresponding formations, one must reject the idea of sunken land bridges which extended across the Atlantic. A displacement of continents in A. Wegener's sense is what comes to mind, a concept which finds support in the observation that, since the very oldest geological times, apart from the Permo-Carboniferous, a dry climate has predominated in western

South Africa, and that the Triassic sedimentary deposits in Minas are in accordance with a dry inland climate."

Particularly thorough comparative studies have been carried out by the well-known South African geologist du Toit, who made a journey of exploration in South America for this purpose. The results of this investigation, which includes a very complete survey of the literature, were published in 1927. [. . .] The whole work is a unique geological demonstration of the correctness of drift theory so far as these parts of the globe are concerned. If we wanted to cite every detail in the book which favours the theory, we would have to translate it from start to finish. There are many statements like the following: "Indeed, viewed even at short range, I had great difficulty in realising that this was another continent and not some portion of one of the southern districts in the Cape." [. . .]

If we continue our comparison of opposite Atlantic coastlines farther north, we find that the Atlas range, which is situated on the northern border of the African continent, and whose folding took place chiefly during the Oligocene, but had already begun in the Cretaceous, has no counterpart on the American side. This agrees with our assumption put forward in the reconstruction that the Atlantic rift had been open for a long period already in this area. In fact, it is possible that here also the rift had been nonexistent at one time, but the start of the split-off must have occurred before the Carboniferous. Further, the great depth of the ocean in the western part of the North Atlantic perhaps means that the sea floor is older here. One should also note the contrast between the Iberian peninsula and the opposite American coastal region, a contrast which makes it most unlikely that the coastlines were formerly in direct connection. In any case, however, drift theory would not imply such a view, because between Spain and America there lies the broad submarine massif of the Azores. As I attempted to prove from the earliest transatlantic echo-sounding profile, this massif probably represents a layer of detritus composed of continental material whose original extent can be estimated as possibly 1000 km or more. [. . .]

Farther north still, we find in direct sequence three ancient fold zones which carry over from one side of the Atlantic to the other, and once again provide very remarkable confirmation of the idea of direct connection at one time.

What most strikes the eye is the Carboniferous folds which E. Suess called the Armorican mountains and which make the North American coal fields seem to be the direct continuation of the European. These mountains, now much levelled-off, come from the continental inland region of Europe and extend first towards the west-northwest in an arcuate formation, then west to form a wild, irregular (so-called "ria") coastline in southwest Ireland and Brittany. The southernmost folded ranges of this system, which cross France, appear to turn completely south in the offshore continental shelf and to continue on the Iberian peninsula on the other side of the deep-sea rift of the Bay of Biscay, which is shaped like an opened book. Suess called this offshoot the "Asturian swirl." The main mountain chains, however, obviously continue through the northern part of the shelf in a westerly direction, though the tops have been eroded down by the breakers, and they point out to the Atlantic here, as though insisting on a continuation westwards.

This continuation on the American side is formed, as Bertrand was the first to discover in 1887, by the offshoots of the Appalachians in Nova Scotia and southeastern Newfoundland. Here also a Carboniferous range of fold mountains ends, folded in a northerly direction as in Europe; this produces a ria coastline, and the range probably crosses the shelf of the Newfoundland Bank. Its direction, elsewhere northeasterly, turns directly eastwards near the rift area. According to the ideas held to date, it was already assumed that a single large fold system was involved, described by E. Suess as the "transatlantic Altaides." Drift theory simplifies matters considerably: in the reconstruction based on the theory, the two components are brought into virtual contact, whereas up till now a sunken middle section had to be assumed, longer than the terminal sections as known to us—a difficulty that Penck had already experienced. On the junction line of the rift lie some sporadic elevations of the sea floor, regarded hitherto as peaks of the sunken chain. Our theory sees them as fragments of the edges of the separating blocks, whose detachment is easily understandable in just such a region of tectonic disturbance. [. . .]

There are also the terminal moraines of the great Pleistocene inland icecaps of North America and Europe. They were deposited at a time when Newfoundland had already been split off from Europe, while in the north near Greenland the blocks were still joined. In any case, North America must at that time have been much closer to Europe than today. If one considers the moraines in our reconstruction, which holds for the period before separation, they join up without gaps or breaks [. . .]; this would be highly improbable if the coastlines at the times of their deposition had already been separated by their present-day distance of 2500 km especially since the American end today lies $4\frac{1}{2}°$ degrees of latitude farther south than the European.

ERNEST RUTHERFORD

From *The Transmutation of the Atom* (1933)

E rnest Rutherford (1871–1937), later Baron Rutherford, was one of the original researchers in nuclear physics, even before anyone knew the atom had a nucleus. In fact, it was Rutherford who first proposed the existence of the atomic nucleus. The essay selected here is a lecture he gave on British radio (the BBC) in 1933 that recounts a scientific observation in which the observer knows what he is looking for. Comments like "A further attack on this problem had to await a better understanding," and "Actuated by these general ideas, I made in 1919 some experiments" indicate that Rutherford's progress came only when he realized what he was doing.

This lecture is a lay version of one Rutherford had given a few weeks earlier to the British Association. In that talk he stated that the hope of atomic energy was moonshine. A Czech physicist, Leo Szilard, was so irritated by this attitude that he had immediately figured out a way to achieve nuclear fission through a chain reaction. The egg seemed to be on Rutherford's face. He repeats his denial in this lecture and seems even to dismiss the possibility of a chain reaction. But in the last paragraph, you can see why he did not want to pursue this issue: he valued understanding more than practical results, making him very old-fashioned, indeed.

The possibility of the transmutation of one kind of matter into another has always exerted a strong fascination on the human mind since the early days of science. During the past thirty years, I have been actively engaged in investigations on various aspects of this question. [. . .]

The first successful experiments in transmutation are comparatively recent, dating back to the year 1919, but in a sense the problem is a very old one and has been the subject of much thought and investigation from the time when science was in its infancy. You have all heard of the alchemists, who were in fact the first chemical investigators, and the search for a substance called the philosopher's stone whereby it was hoped one element could be changed into another.

Looking back from the standpoint of our knowledge today, we see that there was no hope of success in this quest with the limited laboratory appliances and methods available in those days, and that the experimental evidence brought forward in its support was of a very doubtful and meagre character. However, the persistence of this idea through the centuries was mainly due to a philosophical conception of the nature of matter, based on the writings of Aristotle, which had a very great influence on the outlook of the intellectual world in the Middle Ages. According to Aristotle, all matter consisted of a fundamental substance, primordial matter, together with a mixture of the four elements called earth, air, fire and water. One substance differed from another only in the relative combination of these hypothetical elements. On these views, it appeared almost self-evident that one substance could be changed into another if only a suitable method could be found to alter the amount of one or more of these constituents. Naturally, the hope of changing the base metals into the noble element gold assumed a prominent place, and from time to time men arose who claimed that they had discovered the secret of changing copper, lead or other metals into gold. [. . .] The old ideas persisted in the general mind, so that even today impostors or deluded men occasionally appear who claim to have a recipe for making gold, but the only gold they make is extracted from the pockets of their credulous supporters. The poverty of imagination of this class of charlatan is shown by the fact that they have not at once claimed to produce instead of gold the more rare and costly elements like platinum and radium.

The great work of the chemists of the nineteenth century had shown that [. . .] the atoms of the elements are not unrelated but must possess in some respects very similar structures. This idea received strong support from the fundamental discovery of the electron in 1897, mainly due to the work of Sir J. J. Thomson. The electron, which carries a negative charge of electricity, was found to have a minute mass—only about 1/1840 of the mass of the lightest atom hydrogen—and to be a constituent of all atoms. [. . .]

The discovery of the radio-activity of uranium by Becquerel in 1896 was another landmark in the history of [transmutation]. The experiments of Rutherford and Soddy in 1903 showed clearly that radio-activity was a direct manifestation of atomic instability. It was shown that occasionally an atom broke up with explosive violence, hurling out with great swiftness either a massive particle called the α-particle, or a swift electron of light mass called the β-particle. As a consequence of this explosion, the residual atom had entirely different physical and chemical properties from the parent atom. It was found that the successive transformations of the two elements uranium and thorium gave rise to thirty or more new elements which had radio-active properties. Radium is the best-known example of these many elements which originate from the successive transformations of the element uranium. The then rare gas helium was found by Ramsay and Soddy to be generated by the transformation of the radium atoms, and we now know that this helium arises from the α-particles which are charged helium atoms expelled from the exploding atom. The study of the swift α- and β-particles and the penetrating X-rays, brought out clearly the extraordinary intensity of these atomic explosions, in which energy was emitted of the order of one million times that generated by a combination of two atoms in the most violent explosive known.

This transformation of the radio-active atoms is spontaneous and uncontrollable; neither extremes of heat or cold have any effect on the rate of transformation or on the energy of the expelled particles. At the moment, no definite explanation can be given why the atoms of uranium break up. We have to regard their transformation as a natural process which is governed by the laws of chance. [. . .]

A further attack on this problem had to await a better understanding of the essential structures of all atoms. It is now clear that all the atoms of the elements are in a sense electrical structures built on the same general pattern. At the centre of each atom is a minute but massive nucleus which has a resultant charge of so many units of positive electricity varying from 1 to 92 for the various atoms. At a distance from the nucleus is a distribution of negative electrons in whirling motion, whose number is equal to the number of positive charges carried by the nucleus. It must be borne in mind that the radius of the nucleus, supposed spherical, is extremely minute and in general less than one ten-thousandth part of the radius of the atom as a whole. Yet this diminutive nucleus not only contains nearly the whole mass of the atom, but through its nuclear charge controls the number and motions of the external or planetary electrons, as they have been termed from analogy with our solar system. [. . .]

The broad features of the constitution of all atoms are now well established. As a result of the splendid work of Bohr and those who have followed him, we are able to understand the arrangement and motions of the planetary electrons and the way in which light or X-rays are emitted when the atom is disturbed. Unfortunately we have much less information about the constitution of the minute central nucleus. We know the value of the nuclear charge and the mass of each atom, but we have no precise information on the nature and arrangement of the particles composing it. Until a year or so ago, it was generally supposed that the nucleus of an atom was ultimately composed of two electrical units, the negative electrons of small mass and the positively charged protons of mass 1. At the same time it became clear that secondary units were also present and that the helium nucleus of mass 4—the α-particle—played a prominent part. Recently, however, we have had to extend our views, for undoubted evidence has been obtained of the existence of a new type of particle called the neutron which has a mass about 1, but no electrical charge. At the same time, the discovery this year of what is believed to be the positive electron of light mass—the counterpart of the negative electron—has complicated the problem. However, we may, I think, assume with some confidence that the nucleus of a heavy atom is in general composed of a large number of particles, some charged like the α-particle and proton, and others, like the neutron electrically neutral. These are held together by powerful forces in an extraordinarily minute volume and from a very stable structure. [. . .]

In order to transmute one atom into another, it appears essential to alter the charge on the nucleus. This can be done in imagination by adding another charged particle, say a proton or α-particle, to the nucleus, or removing a charged particle from it. We must, however, bear in mind that the nucleus is a strongly guarded structure held firmly together by strong attractive forces. In order to disrupt a nucleus, it thus seemed likely that very intense forces must be brought to bear directly upon it. One method of accomplishing this is to bombard the nucleus with very swift particles. Now the α-particle, which is spontaneously ejected from radium, is one of the most energetic particles known to science. It was recognised that if a stream of swift α-particles fell on matter, there was a small chance that one out of a great number might have an almost head-on collision with a nucleus. Under these conditions it must approach very near to it before it was turned back by the strong repulsive forces due to the electric charges on the two particles. It must be emphasized that the close collision of an α-particle with a nucleus involves the setting up of gigantic forces between the two nuclei concerned. In the case of light atoms where the nuclear charge is small, calculation indicated that the colliding α-particle, if it did not enter the struck nucleus, must at least approach sufficiently near to distort greatly its electrical structure. Under such disturbing forces, the nucleus might be expected to become unstable and then break up into other nuclei. Actuated by these general ideas, I made in 1919 some experiments to test whether any evidence of transformation could be obtained when α-particles were used to bombard matter. The experiments were of a simple type; a preparation of radium served as a source of α-rays and the scintillation method was used to detect the presence of any new types of particles. It is well known that each α-particle falling on a preparation of zinc sulphide gives a flash of light, a scintillation, which is easily seen in a darkened room, and it was to be expected that any fast charged particle liberated from the bombarded matter would indicate its presence by a scintillation. When α-particles were used to bombard the gas oxygen, no new effect was observed. When, however, nitrogen gas was substituted, a number of scintillations were observed far beyond the distance of travel of the α-particles. Special experiments showed that these scintillations were produced by charged hydrogen atoms which we now call protons. The appearance of fast protons in these experiments could only be explained by supposing that they arose from the transformation of some of the nitrogen nuclei as a result of α-particle bombardment. This was the first time that definite evidence was obtained that an atom could be transformed by artificial methods. In the light of later experiments by Blackett, the general mechanism of this transformation became clear. It was found that the α-particle must actually penetrate into the nitrogen nucleus and be captured by it. As a consequence of this profound disturbance, a proton was ejected with high speed from the new nucleus. [. . .] In subsequent investigations with Dr. Chadwick, the methods of observation were much improved and it became clear that at least twelve light elements could be transformed by α-particle bombardment, and in every case protons were ejected, the number and speed varying from element to element. It seems probable that the process of transformation is similar in all cases to that found for nitrogen. The α-particle is captured and the new atom that is formed has a mass 3 units greater and a charge of 1 unit higher than the original atom. In other words, we have succeeded with each of these twelve elements in turning one of their atoms into the atom of the element next higher in the normal order of the elements.

It must be borne in mind that the amount of new matter formed in this way is exceedingly minute—far too small to examine by ordinary chemical methods. Success has only been obtained by the development of delicate methods of counting single atoms of matter. On account of the minute target area of a nucleus, the chance of a direct hit by an α-particle is very small. In the case of nitrogen, only about one α-particle in 100,000 is effective in causing a transformation, and for the element aluminium the chance is still smaller—about one in a million. Of course, the α-particles in these experiments are not aimed at the nuclei, but are emitted at random in all directions, and occasionally one of them by chance happens to approach closely enough to a nucleus to be captured. [. . .]

The transformation of beryllium is [. . .] of a different kind from that of most of the other light elements. As before, the α-particle is captured, but a high speed neutron—not a proton—is expelled. By this process the nucleus of beryllium of mass 9 is changed into an atom of carbon accompanied by the ejection of a neutron.

This strange type of projectile has remarkable properties. On account of the absence of charge, the neutron is able to pass freely through atoms of matter with little if any loss of energy. It only makes its presence manifest when it collides with the nucleus of another atom. [. . .]

So far we have dealt with the transformations produced by fast particles which are themselves derived from the spontaneous disintegration of radio-active elements. It soon became clear, however, that in order to extend our knowledge of this subject, far more abundant streams of fast projectiles of different kinds were essential. It was well known that the passage of an electric discharge through a gas at low pressure gives rise to a multitude of charged atoms and molecules of various kinds. If the charged atoms obtained in this way were accelerated in a vacuum by passing through a strong electric field, we might hope to obtain a copious stream of fast projectiles of different kinds for bombardment purposes. For example, when a discharge is passed through hydrogen it is easy to produce a stream of as many projectiles as are emitted by 100,000 grams of radium in the same time. By the use of high voltages, of the order of one million volts, it seemed likely that we could hope to obtain sufficiently fast particles to effect atomic transformations. [. . .]

Cockcroft and Walton found that the number of α-particles observed increased rapidly with voltage, and was very large at 500,000 volts, while Oliphant has shown that the transformation of lithium can readily be detected for an accelerating voltage of only 30,000 volts. This result would have been difficult to understand on the old ideas, but is quite in accord with the new wave theory of matter. Gamow has shown that there is a chance, even if a very small one, for a comparatively slow particle to penetrate the strong electric barrier round the nucleus.

Cockcroft and Walton found that not only lithium, but also boron and fluorine, emitted α-particles under proton bombardment. The exact nature of these transformations has not yet been settled, but it may be that the boron nucleus of mass 11 captures a proton and then breaks up into three α-particles each of mass 4. In general, it appears that proton bombardment leads to the emission of α-particles and the formation of elements of smaller mass, while α-particle bombardment causes a rise in mass of the resulting element. [. . .]

It is interesting to consider the energy changes involved in these transformations. We have seen that a proton of energy corresponding to 30,000 volts can effect the transformation of lithium into two fast α-particles, which together have an energy equivalent of more than 16 million volts. Considering the individual process, the output of energy in the transmutation is more than 500 times greater than the energy carried by the proton. There is thus a great gain of energy in the single transmutation, but we must not forget that on an average more than 1000 million protons of equal energy must be fired into the lithium before one happens to hit and enter the lithium nucleus. It is clear in this case that on the whole the energy derived from transmutation of the atom is small compared with the energy of the bombarding particles. There thus seems to be little prospect that we can hope to obtain a new source of power by these processes. It has sometimes been suggested, from analogy with ordinary explosives, that the transmutation of one atom might cause the transmutation of a neighbouring nucleus, so that the explosion would spread throughout all the material. If this were true, we should long ago have had a gigantic explosion in our laboratories with no one remaining to tell the tale. The absence of these accidents indicates, as we should expect, that the explosion is confined to the individual nucleus and does not spread to the neighbouring nuclei, which may be regarded as relatively far removed from the centre of the explosion. [. . .]

My listeners may quite naturally ask why these experiments on transmutation should excite such interest in the scientific world. It is not that the experimenter is searching for a new source of power or the production of rare and costly elements by new methods. The real reason lies deeper, and is bound up with the urge and fascination of a search into one of the deepest secrets of Nature. Until a few years ago, we had to be content with the knowledge that the whole of matter in the universe, including our own bodies, was made up of ninety or more distinct chemical elements, but we had little definite knowledge of the inner structure of their atoms or of the processes by which one element could be converted into another. Now, for the first time, we are able to investigate these problems by direct experiments in the laboratory, and we are hopeful we shall soon add widely to our knowledge. The information so gained cannot but widen our outlook on the nature of matter, but must also have a direct bearing on many problems of cosmical physics. For example, in the furnace of the sun and other hot stars, the electrons, protons, neutrons and atoms present must be endowed with high average velocities owing to thermal agitation. It is thus to be expected that the processes both of disintegration and aggregation of nuclei, such as are observed in the laboratory, should be operative on a vast scale for all nuclei, and that a kind of equilibrium should be set up between those two opposing agencies of dissociation and association for each type of atomic nucleus. It is well known that the abundance of the elements in our earth's crust varies very widely. Some elements like iron, nickel and oxygen are abundant, whilst others like lithium, platinum and gold are relatively rare. The information to be gained in our laboratories on the efficiency of various types of agencies in transforming atoms may help us to throw light on the reason for the relative abundance of different elements in our earth, and thus in the sun from which our earth is believed to be derived.

JAMES WATSON

The Double Bases, from
The Double Helix (1968)

Although conversations and e-mail led to many recommendations for contributors to this volume, James Watson (1928–) is unique in that people urged me not to include him. His classic account of how he and Francis Crick worked out the DNA molecule's double helix structure is filled with gossip and boasting. Pride and ambition, as much as a thirst for knowledge, contributed to the process that Watson describes. Although it made for an entertaining, readable book, the science part of the story tended to get second billing. But science is part of the story, and in this selection Watson tells how he and Crick—especially Crick—came to understand that somebody else's observation might be just the clue they needed.

Watson and Crick's solution to the structure of DNA was one of science's greatest events because it explained one of nature's most widely recognized mysteries. Living things reproduce themselves. How? The answer turned out to be built into the DNA molecule. Its structure contains two filaments twisting through space like a Renaissance staircase and allows one thing to become two when the filaments come apart, turning one complicated strand into two simple threads. Then, thanks to the peculiarities of chemical bonding, each thread automatically replaces its missing half. Suddenly one molecule has turned itself into two identical molecules. In this passage, Watson describes the period when he and Crick began to appreciate the importance of evidence that pointed to just this kind of doubling.

The moment was thus appropriate to think seriously about some curious regularities in DNA chemistry first observed at Columbia by the Austrian-born biochemist Erwin Chargaff. Since the war, Chargaff and his students had been painstakingly analyzing various DNA samples for the relative proportions of their purine and pyrimidine[34] bases. In all their DNA preparations the number of adenine (A) mole-

---

[34] *Watson is talking about the chemical organization of the DNA molecule. The "purine" base puts carbon and nitrogen into a two-ringed structure. Although pyrimidine also uses the same carbon and nitrogen atoms, it organizes them into a one-ring structure.*

cules was very similar to the number of thymine (T) molecules, while the number of guanine (G) molecules was very close to the number of cytosine (C) molecules.[35] Moreover, the proportion of adenine and thymine groups varied with their biological origin. The DNA of some organisms had an excess of A and T, while in other forms of life there was an excess of G and C. No explanation for his striking results was offered by Chargaff, though he obviously thought they were significant. When I first reported them to Francis [Crick] they did not ring a bell, and he went on thinking about other matters.

Soon afterwards, however, the suspicion that the regularities were important clicked inside his head as the result of several conversations with the young theoretical chemist John Griffith. One occurred while they were drinking beer after an evening talk by the astronomer Tommy Gold on "the perfect cosmological principle." Tommy's facility for making a far-out idea seem plausible set Francis to wondering whether an argument could be made for a "perfect biological principle." Knowing that Griffith was interested in theoretical schemes for gene replication, he popped out with the idea that the perfect biological principle was the self-replication of the gene—that is, the ability of a gene to be exactly copied when the chromosome number doubles during cell division. Griffith, however, did not go along, since for some months he had preferred a scheme where gene copying was based upon the alternative formation of complementary surfaces.

This was not an original hypothesis. It had been floating about for almost thirty years in the circle of theoretically inclined geneticists intrigued by gene duplication. The argument went that gene duplication required the formation of a complementary (negative) image where shape was related to the original (positive) surface like a lock to a key. The complementary negative image would then function as the mold (template) for the synthesis of a new positive image. A smaller number of geneticists, however, balked at complementary replication. Prominent among them was H. J. Muller, who was impressed that several well-known theoretical physicists, especially Pascual Jordan, thought forces existed by which like attracted like. But [Linus] Pauling abhorred this direct mechanism and was especially irritated by the

---

[35] *The DNA molecule was known to be composed of smaller molecules, symbolized by the letters A, T, G, and C. The problem that Watson and Crick were trying to solve was how the A, T, G, and C was organized so that heredity and reproduction would be possible.*

suggestion that it was supported by quantum mechanics. Just before the war, he asked [Max] Delbrück (who had drawn his attention to Jordan's papers) to coauthor a note to *Science* firmly stating that quantum mechanics favored a gene-duplicating mechanism involving the synthesis of complementary replicas.

Neither Francis nor Griffith was long satisfied that evening by restatements of well-worn hypotheses. Both knew that the important task was now to pinpoint the attractive forces. Here Francis forcefully argued that specific hydrogen bonds were not the answer. They could not provide the necessary exact specificity, since our chemist friends repeatedly told us that the hydrogen atoms in the purine and pyrimidine bases did not have fixed locations but randomly moved from one spot to another. Instead, Francis had the feeling that DNA replication involved specific attractive forces between the flat surfaces of the bases.

Luckily, this was the sort of force that Griffith might just be able to calculate. If the complementary scheme was right, he might find attractive forces between bases with different structures. On the other hand, if direct copying existed, his calculations might reveal attraction between identical bases. Thus, at the closing hour they parted with the understanding that Griffith would see if the calculations were feasible. Several days later, when they bumped into each other in the Cavendish tea queue,[36] Francis learned that a semirigorous argument hinted that adenine and thymine should stick to each other by their flat surfaces. A similar argument could be put forward for attractive forces between guanine and cytosine.

Francis immediately jumped at the answer. If his memory served him right, these were the pairs of bases that Chargaff had shown to occur in equal amounts. Excitedly he told Griffith that I had recently muttered to him some odd results of Chargaff's. At the moment, though, he wasn't sure that the same base pairs were involved. But as soon as the data were checked he would drop by Griffith's rooms to set him straight.

At lunch I confirmed that Francis had got Chargaff's results right. But by then he was only routinely enthusiastic as he went over Griffith's quantum-mechanical arguments. For one thing, Griffith, when

---

[36] *A Cavendish tea queue is a line for fetching tea at the Cavendish lab in Cambridge, England.*

pressed, did not want to defend his exact reasoning too strongly. Too many variables had been ignored to make the calculations possible in a reasonable time. Moreover, though each base has two flat sides, no explanation existed for why only one side would be chosen. And there was no reason for ruling out the idea that Chargaff's regularities had their origin in the genetic code. In some way specific groups of nucleotides must code for specific amino acids. Conceivably, adenine equaled thymine because of a yet undiscovered role in the ordering of the bases. There was in addition Roy Markham's assurance that, if Chargaff said that guanine equaled cytosine, he was equally certain it did not. In Markham's eyes, Chargaff's experimental methods inevitably underestimated the true amount of cytosine.

Nonetheless, Francis was not yet ready to dump Griffith's scheme when, early in July, John Kendrew walked into our newly acquired office to tell us that Chargaff himself would soon be in Cambridge for an evening. John had arranged for him to have dinner at Peterhouse, and Francis and I were invited to join them later for drinks in John's room. At High Table John kept the conversation away from serious matters, letting loose only the possibility that Francis and I were going to solve the DNA structure by model building. Chargaff, as one of the world's experts on DNA, was at first not amused by dark horses trying to win the race. Only when John reassured him by mentioning that I was not a typical American did he realize that he was about to listen to a nut. Seeing me quickly reinforced his intuition. Immediately he derided my hair and accent, for since I came from Chicago I had no right to act otherwise. Blandly telling him that I kept my hair long to avoid confusion with American Air Force personnel proved my mental instability.

The high point in Chargaff's scorn came when he led Francis into admitting that he did not remember the chemical differences among the four bases. The faux pas slipped out when Francis mentioned Griffith's calculations. Not remembering which of the bases had amino groups, he could not qualitatively describe the quantum-mechanical argument until he asked Chargaff to write out their formulas. Francis' subsequent retort that he could always look them up got nowhere in persuading Chargaff that we knew where we were going or how to get there.

CARL SAGAN

Has the Earth Already Been Visited?, from
The Cosmic Connection (1973)

Carl Sagan (1934–1996) was an astronomer and general enthusiast for science. He wrote many popular books and hosted a television series about modern astronomy. Among his many interests was the question of life on other planets. Although there are logical grounds for believing that life exists on many planets beyond our own solar system, people who claim to know something about that life, or to have been approached by ambassadors from other planets, tend to be viewed as crackpots. The essay in this book by Loren Eiseley sums up the general scientific attitude toward that last proposition.

In this selection, Sagan makes the case for doubting that aliens do regularly visit earth, but he never denies that life may nonetheless exist elsewhere. His essay is an unusually entertaining example of building an argument through mathematical reasoning. But of course, no one can observe a negative, and this essay cannot really be taken as proof that tomorrow night some alien Columbus will not land his spaceship, with alien Cortéses and Pizarros soon to follow. Instead, the essay helps us understand why—even if there are billions of populated planets besides our own—we can be skeptical about claims that their residents have already passed this way.

By far the cheapest way of communicating with the Earth, if you're a representative of an advanced extraterrestrial civilization, is by radio. A single bit of radio information, sent winging across space to the Earth, would cost far less than a penny. A radio search for extraterrestrial intelligence seems, therefore, a very reasonable place for us to begin. But should we not examine other possibilities closer to home? Wouldn't we look silly if we expended a major effort listening for radio messages or searching for life on Mars if, all the while, there was here on Earth evidence of extraterrestrial life?

There are two hypotheses of this sort that have gained a following in the popular literature. The first postulates that the Earth is today being visited by spacecraft from other worlds—this is the extraterrestrial flying saucer or unidentified flying object (UFO) hypothesis. The second also postulates that the Earth has been visited by such spacecraft, but in the past, before written history.

The extraterrestrial hypothesis of UFO origins is a complex subject, powerfully dependent on the reliability of witnesses. A comprehensive discussion of this problem has recently been published in *UFO's: A Scientific Debate* (Carl Sagan and Thornton Page, editors, Ithaca, N.Y., Cornell University Press, 1972), in which all sides of the subject have been aired. My own view is that there are no cases that are simultaneously very reliable (reported independently by a large number of witnesses) and very exotic (not explicable in terms of reasonably postulated phenomena—as a strange moving light could be a searchlight from a weather airplane or a military aerial refueling operation). There are no reliably reported cases of strange machines landing and taking off, for example.

There is another approach to the extraterrestrial hypothesis of UFO origins. This assessment depends on a large number of factors about which we know little, and a few about which we know literally nothing. I want to make some crude numerical estimate of the probability that we are frequently visited by extraterrestrial beings.

Now, there is a range of hypotheses that can be examined in such a way. Let me give a simple example: Consider the Santa Claus hypothesis, which maintains that, in a period of eight hours or so on December 24–25 of each year, an outsized elf visits one hundred million homes in the United States. This is an interesting and widely discussed hypothesis. Some strong emotions ride on it, and it is argued that at least it does no harm.

We can do some calculations. Suppose that the elf in question spends one second per house. This isn't quite the usual picture—"Ho, Ho, Ho," and so on—but imagine that he is terribly efficient and very speedy. That would explain why nobody ever sees him very much—only one second per house, after all. With a hundred million houses he has to spend three years just filling stockings. I have assumed he spends no time at all in going from house to house. Even with relativistic reindeer, the time spent in a hundred million houses is three years and not eight hours. This is an example of hypothesis-testing independent of reindeer propulsion mechanisms or debates on the origins of elves. We examine the hypothesis itself, making very straightforward assumptions, and derive a result inconsistent with the hypothesis by many orders of magnitude. We would then suggest that the hypothesis is untenable.

We can make a similar examination, but with greater uncertainty, of the extraterrestrial hypothesis that holds that a wide range of UFOs viewed on the planet Earth are space vehicles from planets of other stars. The report rates, at least in recent years, have been several per day, at the very least. I will not make that assumption. I will make the much more conservative assumption that one such report per year corresponds to a true interstellar visitation. Let's see what this implies.

We have to have some feeling for the number, N, of extant technical civilizations in the Galaxy—that is, civilizations vastly in advance of our own, civilizations that are able, by whatever means, to perform interstellar space flight. (While the means are difficult, they don't enter into this discussion, just as reindeer propulsion mechanisms don't affect our discussion of the Santa Claus hypothesis.)

An attempt has been made to specify explicitly the factors that enter a determination of the number of such technical civilizations in the Galaxy. I will not here run through what numbers have been assigned to the various quantities involved—it's a multiplication of many probabilities, and the likelihood that we can make a good judgment decreases as we proceed down the list. N depends first on the mean rate at which stars are formed in the Galaxy, a number that is known reasonably well. It depends on the number of stars that have planets, which is less well known, but there are some data on that. It depends on the fraction of such planets that are so suitably located with respect to their star that the environment is a feasible one for the origin of life. It depends on the fraction of such otherwise feasible planets on which the origin of life, in fact, occurs. It depends on the fraction of *those* planets on which the origin of life occurs in which, after life has arisen, an intelligent form comes into being. It depends on the fraction of *those* planets in which intelligent forms have arisen that evolve a technical civilization substantially in advance of our own. And it depends on the average lifetime of such a technical civilization.

It is clear that we are rapidly running out of examples as we go farther and farther along. We have many stars, but only one instance of the origin of life, and only a very limited number—some would say only one—of instances of the evolution of intelligent beings and technical civilizations on this planet. And we have no cases whatever to make a judgment on the mean lifetime of a technical civilization. Nevertheless, there is an entertainment that some of us have been engaged in, making our best estimates about these numbers and coming out with a value of N. The result that emerges is that N roughly equals one tenth the average lifetime of a technical civilization in years.

If we put in a number like ten million (10^7) years for the average lifetime of advanced technical civilizations, we come out with a number for such technical civilizations in the Galaxy of about a million (10^6)—that is, a million other stars with planets on which today there are advanced civilizations. This is quite a difficult calculation to do accurately. The choice of ten million years for the average lifetime of a technical civilization is rather optimistic. But let's take these optimistic numbers and see where they lead us.

Let's assume that each of these million technical civilizations launches Q interstellar space vehicles a year, so that 10^6Q interstellar space vehicles are launched per year. Let's assume that there's only one contact made per journey. In the steady-state situation, there are something like 10^6Q arrivals somewhere or other per year. Now, there surely are something like 10^{10} interesting places in the Galaxy to go visit (we have several times 10^{11} stars) and, therefore, an average of $1/10^4 = 10^{-4}$ arrivals at a given interesting place (let's say a planet) per year. So if only one UFO is to visit the Earth each year, we can calculate what mean launch rate is required at each of these million worlds. The number turns out to be ten thousand launches per year per civilization, and ten billion launches in the Galaxy per year. This seems excessive. Even if we imagine a civilization much more advanced than ours, to launch ten thousand such vehicles for only one to appear here is probably asking too much. And if we were more pessimistic on the lifetime of advanced civilizations, we would require a proportionately larger launch rate. But as the lifetime decreases, the probability that a civilization would develop interstellar flight very likely decreases as well.

There is a related point made by the American physicist Hong-Yee Chiu; he takes more than one UFO arriving at Earth per year, but his argument follows along the same lines as the one I have just presented. He calculates the total mass of metals involved in all of these space vehicles during this history of the Galaxy. The vehicle has to be of some size—it should be bigger than the Apollo capsule, let's say—and we can calculate how much metal is required. It turns out that the total mass of half a million stars has to be processed and all their metals extracted. Or if we extend the argument and assume that only the outer few hundred miles or so of stars like the Sun can be mined by advanced technologies (farther in, it's too hot), we find that two billion such stars must be processed, or about 1 percent of the stars in the Galaxy. This also sounds unlikely.

Now you may say, "Well, that's a very parochial approach; maybe they have plastic spaceships." Yes, I suppose that's possible. But the plastic has to come from somewhere, and plastics vs. metals changes the conclusions very little. This calculation gives some feeling for the magnitude of the task when we are asked to believe that there are routine and frequent interstellar visits to our planet.

What about possible counterarguments? For example, it might be argued that we are the object of special attention—we have just developed all sorts of signs of civilization and high intelligence like nuclear weapons, and maybe, therefore, we are of particular interest to interstellar anthropologists. Perhaps. But we have only signaled the presence of our technical civilization in the past few decades. The news can be only some tens of light-years from us. Also, all the anthropologists in the world do not converge on the Andaman Islands because the fish net has just been invented there. There are a few fish net specialists and a few Andaman specialists; and these guys say, "Well, there's something terrific going on in the Andaman Islands. I've got to spend a year there right away because if I don't go now, I'll miss out." But the pottery experts and the specialists in Australian aborigines don't pack up their bags and leave for the Indian Ocean.

To imagine that there is something absolutely fascinating about what is happening right here is precisely contrary to the idea that there are lots of civilizations around. Because if the latter is true, the development of our sort of civilization must be pretty common. And if we are not pretty common, then there are not going to be many civilizations advanced enough to send visitors.

Even so, is it not possible that the second UFO hypothesis is true—that in historical or recent prehistoric times an extraterrestrial space vehicle made landfall on Earth? There is surely no way in which we can exclude such a contingency. How could we prove it?

A number of popular books have recently been written that allege to demonstrate such a visitation. The arguments are of two sorts, legend and artifact. I broached this subject in the book *Intelligent Life in the Universe*, written with the Soviet astrophysicist I. S. Shklovskii and published in 1966. I examined a typical legend suggestive of contact between our ancestors and an apparent representative of a superior society. The legend, taken from the earliest Sumerian mythology, is important because the Sumerians are the direct cultural antecedents of our own civilization. A superior being was supposed to have taught the Sumerians mathematics, astronomy, agriculture, social and political organization, and written language—all the arts necessary for making the transition from a hunter-gatherer society to the first civilization.

But as provocative as this and similar legends were, I concluded that it was impossible to *demonstrate* extraterrestrial contact from such legends: There are plausible alternative explanations. We can understand why priests might make myths about superior beings who inhabit the skies and give directions to human beings on how to order their affairs. Among other "advantages," such legends permit the priests to control the people.

There is only one category of legend that would be convincing: When information is contained in the legend that could not possibly have been generated by the civilization that created the legend—if, for example, a number transmitted from thousands of years ago as holy turns out to be the nuclear fine structure constant. This would be a case worthy of some considerable attention.

Also convincing would be a certain class of artifact. If an artifact of technology were passed on from an ancient civilization—an artifact that is far beyond the technological capabilities of the originating civilization—we would have an interesting *prima facie* case for extraterrestrial visitation. An example would be an illuminated manuscript, rescued from an Irish monastery, that contains the electronic circuit diagram for a superheterodyne radio receiver. Great care would have to be taken about the provenance of this artifact, just as art collectors are cautious about a newly discovered Raphael. We would make sure that no contemporary Irish prankster was the source of the circuit diagram.

To the best of my knowledge, there are no such legends and no such artifacts. All the ancient artifacts put forward, for example, by Erik von Danniken in his book *Chariots of the Gods* have a variety of plausible, alternative explanations. Representations of beings with large, elongated heads, alleged to resemble space helmets, could equally well be inelegant artistic renditions, depictions of ceremonial head masks or expressions of rampant hydrocephalia. In fact, the expectation that extraterrestrial astronauts would look precisely like American or Soviet astronauts, down to their space suits and eyeballs, is probably less credible than the idea of a visitation itself. Likewise, the idea expressed by von Danniken and others that ancient astronauts erected airfields, employed rockets, and exploded nuclear weapons on Earth is implausible in the extreme, precisely because we ourselves have just developed this technology. A visitor from space will not be so close to us in time. It is as if, framing such an idea in 1870, we concluded that extraterrestrials use hot-air balloons for space exploration. Far from being too daring, such ideas are stodgy in their unimaginativeness. Most popular accounts of alleged contact with extraterrestrials are strikingly chauvinistic.

An American author named Richard Shaver claims that ordinary rocks, sliced fine, contain a set of still photographs left by an ancient civilization, which can be run as a movie film. Just pick up any rock and slice it fine, he says.

In the great high plain of Nasca in Peru, there is a set of enormous geometrical figures. They are quite difficult to discern when standing among them, but quite discernible from the air. It is easy to see how an early human civilization could have made such figures. But why, it is asked, should such constructs be made except for or by an extraterrestrial civilization? If people believe in the existence of gods in the sky, it is not straining credulity to imagine them making messages to communicate with those gods. The markings may be a kind of collective graphical prayer. But they do not necessarily demonstrate the reality of the intended recipient of the prayer.

There are other cases that seem to be quite convincing at first, such as a perfectly machined steel cube, said to reside in the Salzburg Museum and to have been recovered from geological strata millions of years old, or the receipt of the television call signals of a television station off the air for three years. These cases are almost certainly hoaxes.

There are equally provocative archaeological circumstances that the writers of such sensational books have somehow missed. For example, in the frieze of the great Aztec pyramids at San Juan Teotihuacán, outside Mexico City, there is a repeated figure, described as a rain god, but looking for all the world like an amphibious tracked vehicle with four headlights [. . .] . I do not for a moment believe that such amphibious vehicles were indigenous in Aztec times—among other reasons, because they are too close to what we have today rather than too far from it.

These artifacts are, in fact, psychological projective tests. People can see in them what they wish. There is nothing to prevent anyone from seeing signs of past extraterrestrial visitations all about him. But to a person with an even mildly skeptical mind, the evidence is unconvincing. Because the significance of such a discovery would be so enormous, we must employ the most critical reasoning and the most skeptical attitudes in approaching such data. The data do not pass such tests. Pondering wall paintings, for this purpose, like looking for UFOs, remains an unprofitable investment of terrestrial intelligence— if we are truly interested in the quest for extraterrestrial intelligence.

⊱─◆─○─◆─⊰

WALTER SULLIVAN

Looking for the Drift, from
Continents in Motion (1974)

Walter Sullivan (1918–1996) began working as a newspaper journalist for the *New York Times*, and in 1957, science became his specialty. That was the international geophysical year, a period of intense study of the earth. The fall of 1957 also saw the launch of *Sputnik*, the first artificial satellite, and suddenly science was big news. For almost forty years, Sullivan was one of America's leading science writers, covering many changes for the *Times*. As this essay on geology shows, he could make his subjects both vivid and logical. It was easy for readers to follow his trail through the pathways of science.

This selection is drawn from one of Sullivan's many books, in which he alerts the general public to the news that a theory of spreading oceans was confirming Wegener's idea of continental drift. Sullivan does a masterful job of pulling together many observations and threads, and he also shows that despite the widespread scoffing, many scientists were taking Wegener's ideas seriously. They knew just what they needed: an account of a previously unknown geological process that could split apart and push enormous landmasses. Thus, as the evidence of spreading seafloors mounted, a few scientists were like mystery readers who had already peeked at the last chapter. For them, seemingly obscure clues such as changes in the ocean floor's magnetic pattern had great meaning. They were signs of an active seafloor, something powerful enough to accomplish what Wegener had shown must have happened.

One would hardly expect that Etruscan vases have a tale to tell of the past history of the earth's core, but that, in fact, seems to be the case. Late in the nineteenth century, from studies of magnetism it was realized that, when certain substances, including potting clays, cool after high heating, they become imprinted with the magnetic field of the earth. The force lines of this field, as observed at the surface of the earth, resemble those that would be generated by a bar magnet near the center of the earth. The shape of such a magnetic field is often demonstrated in the schoolroom, two-dimensionally, by holding a bar magnet under a sheet of paper sprinkled with iron filings. The filings instantly arrange themselves in a beautifully symmetrical pattern of curved lines that tend to radiate from points nearest the two poles of the magnet.

On the earth's surface the force lines are vertical at the magnetic poles and horizontal at the magnetic equator. Since the compass needle is oriented solely by the horizontal component of the earth's field, at the magnetic poles (where there is no horizontal component) the needle swings aimlessly, whereas at the magnetic equator [. . .] the force controlling a compass needle is strongest (although still several hundred times weaker than that between the poles of a toy horseshoe magnet).

One of the tools of the polar explorer is a dip needle, which is similar to a compass except that it swings in a vertical, instead of horizontal plane. The dip of the needle indicates roughly his distance from the magnetic pole. When it becomes vertical, the explorer knows he is at that pole.

Pots, urns, and vases, old and new, can serve as frozen dip needles—frozen at the time of their firing, because, when they are very hot, the magnetic particles within them become liberated and can orient themselves with the force lines of the earth's magnetic field. When the vase cools, the particles are locked into place and the orientation of the magnetic field, measurable by laboratory methods, is preserved for posterity.

By 1907 magnetic analyses had been done on Etruscan vases of the eighth century B.C., Greek vases of the seventh century B.C. neolithic pots from about 1500 B.C., and various volcanic lavas that also had captured the local magnetic field when they cooled. It was assumed that the pots had all been baked upright and the remanent magnetism within them, therefore, indicated the local dip of the field at the time of their firing. It was thus shown that the local magnetic field, in the area where each of these objects was fired, had changed.

Ever since the seventeenth century navigators have known that there is a slight year-to-year change in the magnetic field, requiring small corrections in compass headings. [. . .] The overall axis of the field drifts slowly westward, taking about 10,000 years to make one complete journey around the earth's spin axis,[37] and the strength of the field, at any one point, may vary by fifty percent within several thousand years. It has long been assumed that these changes manifest some form of activity deep within the planet. As Christopher Hansteen put it, early in the last century, "The earth speaks of its internal movements through the silent voice of the magnetic needle."

Another way in which the earth's magnetic field becomes imprinted for future reference is in the formation of underwater sediments. As the material settles to the bottom, or before loose, wet sediment is compacted and finally turns to rock, the magnetic grains within it are free to align themselves with the earth's field, but from then on they are locked in place.

In the 1940s scientists from the Department of Terrestrial Magnetism of the Carnegie Institution of Washington painstakingly studied a long succession of ice age varved clays from New England. A varved clay is one that has been laid down, layer upon layer, by annual cycles of sedimentation, leaving a record of the past somewhat analogous to that formed by rings in a tree. From the clays it was found that between 15,000 B.C and 9000 B.C magnetic "north," as seen from New England, drifted back and forth to either side of true north.

If clays laid down during the past few millennia carried a record of the earth's magnetic history, what about rocks formed from similar sediments of much greater age? Was it possible that, within the sedimentary rocks that carpet much of the earth, there lurk hidden compasses frozen into place millions of years ago? To test this idea John W. Graham, a young doctoral candidate working with the Carnegie group, collected sedimentary rocks from the Hudson Valley, Colorado, Utah, Wyoming, and Montana. Finally, in sediments that had lain in West Virginia for 200 million years, he found convincing evidence that such a magnetic memory had survived intact. For a distance of fifty miles the formation showed a consistent direction for the magnetic pole—one very different from that of today.

Even more powerful evidence came from lava flows. In some regions of the earth, such as Iceland, California, Oregon, the Aleutians,

---

[37] *Spin axis is the imaginary pole on which the earth is turning.*

Hawaii, and Japan, there have been intermittent eruptions for many millions of years, laying down lava layer upon lava layer, each of which has trapped the magnetic field in existence when it cooled. When the oxides of iron and titanium within such lavas are heated above a certain level [. . .] atoms within these substances align themselves with the local magnetic field. Then, once the rock cools below that level they become locked in their orientation. The resulting "magnetic memory" of the lava is preserved indefinitely. [. . .]

From early studies of remanent magnetism[38] in old lava flows and sedimentary rocks it appeared that locations of the magnetic poles in the distant past were very different from those of today. This was taken by some to indicate slow wandering of the earth's spin axis relative to surface geography. The difficulty in assessing such evidence for polar wandering was the absence of any accepted explanation for the main magnetic field of the earth and its relationship to the earth's spin. After Patrick M. S. Blackett, in 1946, heard a challenging lecture on the subject at the Royal Institution in London, he started to investigate magnetism, including the magnetic properties of rocks on the earth's surface, and the results ultimately led him into the camp of drifters— the first member of that scientific aristocracy, the Nobel Laureates, to become a full-fledged champion of continental drift. [. . .]

Armed with [a] supersensitive magnetic detector, Blackett turned his attention to the magnetic memory within ancient rocks. He saw in the frozen compass needles within rocks of all ages and from all parts of the world a way not only to unravel the history of the earth's magnetic field, but to assess the possibility of continental drift.

In a 1954 lecture in Jerusalem he said: "Major countries will have to study the magnetism of their own rocks just as they do their own geology. I have no doubt at all that the result of this work will, in the next decade, effectively settle the main facts of land movements, and in so doing will have a profound effect on geophysical studies of the earth's crust." The actual time interval was somewhat more than a decade, but this was a prophetic statement.

When Blackett first became interested in continental drift he was professor of physics at the University of Manchester, having taken

---

[38] *Remanent magnetism is the magnetism remaining in an object after the magnetizing force has disappeared.*

over that post from W. Lawrence Bragg, who, with his father, William Henry Bragg, had won a Nobel Prize in 1915 for their discovery that X-rays could be used to determine molecular structure. [. . .] In Manchester the geological community [. . .] was far from hospitable to drift theory, as his predecessor, Bragg, had discovered. One day in 1919 Bragg went for a walk in the nearby Derbyshire hills with Sydney Chapman, an inveterate walker and cyclist. [. . .] Chapman was greatly excited by the idea of moving continents and, on his walk with Bragg, explained the theory at length.

"Chapman's description of it impressed me so much," Bragg wrote later, "that I can remember the exact spot where he began to talk about it and where the idea of the very great movement of the continents involved in the theory dawned on me. I was so thrilled that I wrote to Wegener for an account of his theory, got it translated, and presented it to our Manchester Literary and Philosophical Society." [. . .] But, Bragg said, its geological members were "furious." Until then, he added, he had never known at first hand what it meant to "froth at the mouth." In fact, he said years later, "words cannot describe their utter scorn of anything so ridiculous as this theory, which has now proved so abundantly to be right."

Blackett was not one to be deterred because a concept was unpopular, particularly when he found that his new field of interest, the remanent magnetism in old rocks, might resolve the controversy. His students fanned out over Britain and far more remote parts of the world to collect specimens and determine, from laboratory analysis, the orientation of magnetic field lines that had passed through them thousands or millions of years earlier. If the placement of the sample was carefully noted before it was extracted from its parent formation, the former orientation of the magnetic force lines within it could be determined, not only in terms of their dip, as in the ancient pots, but three-dimensionally (in terms of north-south and east-west vertical planes and the horizontal plane).

One of the first indications of changed geography derived from such analyses came from three members of Blackett's group: J. A. Clegg, Mary Almond and Peter H. S. Stubbs. They found that the "compass needles" in rocks laid down in England during the Triassic Period, 200 million years ago, did not point to the magnetic pole of that period deduced from rocks elsewhere. The apparent explanation, they reported in 1954, was that England, in the interim, had rotated 34 degrees clockwise. Similarly from the continent came evidence that Spain had swung away from France, opening out the Bay of Biscay and crushing the hinge area to form the Pyrenees.

The protégé of Blackett who was to become the chief British ball-carrier for drift theory was S. Keith Runcorn, a stocky, red-haired native of Lancashire who seemed to relish controversy. Runcorn studied lava flows in Oregon and then, in 1954, scrambled down into the Grand Canyon, collecting shale samples "from top to bottom." From these and specimens gathered elsewhere it was possible to reconstruct the migration route of the North Magnetic Pole over the past several hundred million years, from the south east Pacific toward the Philippines (in terms of present geography), then north across China and Siberia into the Arctic Ocean and its present position. Since the work on New England clays indicated that in more recent times the magnetic pole had never wandered very far from the geographic North Pole (which lies on the spin axis of the earth), it appeared to Runcorn that the spin axis, too, had migrated. "We can only suppose from this," he wrote, "that . . . the planet has rolled about, changing the location of its geographic poles. Either mountain building or convection currents in the mantle[39] might account for this rolling." [. . .]

Runcorn remained loyal to the idea of polar wandering (as opposed to continental drift) until he ran into Harry Hess at an Atlantic City meeting of the American Association of Petroleum Geologists. Hess persuaded him that the magnetic findings could better be explained by changes in geography due to sea-floor spreading, and from then on Runcorn was Britain's most vociferous protagonist of the theory.

By 1960 it was evident to Blackett's group that radical changes in the magnetic latitude of various regions had occurred. Europe and North America were near the magnetic equator 300 million years ago, they said. Australia "appears to have moved in latitude in a somewhat complicated way." Some 400 million years ago it was near the Equator, then drifted to within 2300 kilometers (1400 miles) of the South

---

[39] *Convection currents in the mantle are heat flows in the layer of the earth that lies between its outer crust and its inner core.*

Pole 200 million years ago, and finally north to its present latitude. "India has apparently moved farther and faster than any other continent," they wrote, from a latitude farther south than that of Australia 150 million years ago, across the Equator to its present location.

By this time Runcorn had moved to King's College, at Newcastle upon Tyne, and had formed a powerful team of investigators of rock magnetism. Using a map of the globe with today's geography they plotted positions of the North Magnetic Pole as far back as the Precambrian Period, more than 600 million years ago. What struck them was the difference between the path of the migrating pole as indicated by American rocks and that derived from European rocks. The positions based on European rocks always lay to the east of those calculated from American rocks. Did this mean there were two magnetic poles? It struck Runcorn and his colleagues that, if they shifted Europe up against North America, the two indicated paths of polar motion came into coincidence.

While this, to Runcorn and others, was a strong argument for drift, another aspect of the earth's magnetism demonstrated sea-floor spreading in a manner that even the most determined skeptics could not ignore. A long-standing puzzle had been the 1909 discovery by Bernard Brunhes that some ancient lava flows in the Massif Central, the mountains that bisect France, are imprinted with a magnetic polarity exactly opposite that of other rocks of somewhat greater or lesser age. Brunhes assumed that this was a local phenomenon. Then, in 1929, a Japanese scientist, Motonori Matuyama, found that in Japan many rocks laid down some 700,000 years ago also were reversely magnetized. Was it possible that the entire magnetic field of the earth had temporarily flipped over so that a compass needle that normally pointed north would point south? This seemed too preposterous for serious consideration. In the next three decades a number of other formations with reversed polarity were discovered. [. . .]

Furthermore, rocks laid down over the great span of geologic time seemed about equally divided between formations with normal polarity and those with reversed polarity. [. . .]

Such evidence led a trio at the laboratories of the United States Geological Survey in Menlo Park, California, to undertake a major investigation. They were Allan Cox, G. Brent Dalrymple, and Richard R. Doell, all three of whom had begun studying the earth's magnetic history while at the University of California at Berkeley. They realized that the reversals represented a major scientific challenge. "The idea that the earth's magnetic field reverses at first seems so preposterous," they wrote later, "that one immediately suspects a violation of some basic law of physics, and most investigators working on reversals have sometimes wondered if the reversals are really compatible with the physical theory of magnetism." [. . .]

What enabled this trio, with contributions from several other groups, to demonstrate that the earth's field, in fact, does flip over from time to time was the simultaneous occurrence of two processes within rock that is cooling after eruption. One is the freezing of its magnetic "compass needles." The other is the setting in motion of a "stopwatch" that, millions of years later, can be read to determine how much time has elapsed since the cooling occurred and the needles froze into position. [. . .]

By thus correlating magnetic reversals with age it was possible to show that rocks formed at any given time in the past almost all carried the same magnetic polarity, regardless of where in the world they originated. For some periods the polarity resembled that of today. For others it was reversed.

Early in this research Blackett recognized that the timetable of such reversals could be especially useful to geologists. If, he said, "the reversals have been spaced irregularly in time, then it may prove possible to recognize a 'pattern of reversals' at different places on the globe. If reversals are very infrequent, for instance, only once, say, in 100 million years, then the tracing of such a reversal across the globe should be relatively easy, thus enabling the results to be used for geological dating purposes." [. . .]

The present orientation of the earth's magnetism was found to be characteristic of rocks laid down during at least most of the past 700,000 years, and this has been named the Brunhes Epoch in honor of the Frenchman who first detected reversed magnetism. Before that, for some 1.8 million years, was the great Matuyama Epoch of reversed polarity, named for the Japanese pioneer in this research. Earlier than that the field was again "normal," as it is today. [. . .]

The manner in which magnetic reversals helped resolve the question of sea-floor spreading came from a completely unexpected quarter. During World War II the problem of submarine detection from the air had led to the development of an extremely sensitive device called the Magnetic Airborne Detector, or M.A.D. It was housed in a teardrop capsule that could be lowered from a blimp or scouting plane to trail below and behind the craft, well clear of its magnetic disturbance.

The device was effective for submarines at shallow depth, and after the war geologists looking for oil-bearing formations found it useful for aerial magnetic surveys, since it could delineate, in rough terms, regional distributions of igneous and sedimentary rocks. [. . .]

Then, in 1955, the United States Coast and Geodetic Survey undertook an intensive deep-water mapping program off the West Coast. The research ship *Pioneer* was to steam back and forth along a succession of east-west tracks spaced eight kilometers (five miles) apart with orders not to deviate more than 150 meters from this assigned pattern. [. . .] After an initial period of monotonous sailing back and forth, [they] brought their accumulated data back to the laboratory, where a visiting British scientist, Ronald G. Mason, plotted the variations in magnetic intensity as recorded along successive tracks. He then drew the counterpart of a contour map, showing the hills of high magnetic intensity and the valleys of low intensity. When he was through, Raff wrote, "A single glance was enough to show that we had something quite new in geophysics." Over the whole map were parallel, north-south "hills" and "valleys" of magnetic intensity. Although the intensity variations between these "hills" and "valleys" were only a few percent, the features were unmistakable, varying in width from a few kilometers to a few tens of kilometers. "No dry-land surveys," said Raff, "had ever revealed a lineation that approached this one in uniformity and extent." [. . .]

In [. . .] 1962 the British research frigate *H.M.S. Owen*, which was in the Indian Ocean as a participant in the International Indian Ocean Expedition, at Bullard's initiative, did an intensive magnetic survey of an area that measured forty by fifty nautical miles (seventy-four by ninety-three kilometers) spanning the Carlsberg Ridge. The echo sounder showed that the ship was steaming back and forth across a succession of ridges a kilometer or more in height parallel to the main Carlsberg Ridge and its central rift valley. When Drummond H. Matthews of Cambridge University brought home the magnetic measurements made on these crossings of the ridge, he turned them over to a new graduate student named Frederick J. Vine for analysis. "I already believed in continental drift, sea-floor spreading and reversals of the earth's magnetic field," Vine wrote later to this author, "and was particularly looking for some record of drift and spreading within the ocean basins."

What first struck Vine, as he analyzed the data with the aid of a computer, was that the magnetic "signature" of one of the sea-mounts on the Indian Ocean floor indicated it to be reversely polarized. It occurred to him that this could be explained if the sea-mount were a volcanic feature formed during a period when the magnetic field of the earth was reversed.

If that were the case, and if the sea-floor spreading recently proposed by [Harry] Hess were correct, then material erupting along a midocean ridge, as it cooled, would become imprinted with the polarity of the earth's magnetism in effect at that time. After this ribbon of new sea floor had been split and pushed to either side by additional material erupting along the ridge, this newly forming sea floor would also capture the current field; and, if the field by then had reversed itself, the polarity of the imprinted magnetism would be reversed. The effect of this process, after a succession of reversals, would be to produce a series of sea-floor bands parallel to the ridge and alternating in their magnetic polarity.

There seemed a suggestion of such a pattern in the data obtained over the Carlsberg Ridge. [. . .]

In the issue of *Nature* for September 7, 1963, Vine and his thesis supervisor, "Drum" Matthews, published their explanation for the magnetic stripes. Work on the Indian Ocean magnetic survey, they said, had led them to suggest "that some 50 percent of the oceanic crust might be reversely magnetized." In other words, they proposed that oceans throughout the world are paved with such parallel avenues of rock, divided about equally between anomalies indicating normal and reversed magnetism. [. . .]

By 1965 [. . .] the key to confirmation of the Vine-Matthews explanation was in hand. [. . .] Vine and Wilson realized that, if the long magnetic bands, or anomalies, were being manufactured on a ridge, then split and pushed aside as a new band was formed, the resulting magnetic patterns should be symmetrical to either side of the ridge— that is, the sequence of broad and narrow stripes on one side, generated by long and short magnetic epochs, should be a mirror image of that on the other side. It would be as though two conveyor belts, slowly moving away from each side of the ridge, were being painted at the ridge either black or white, depending on the magnetic polarity in force at the time. The two belts would then carry the same sequence of broad or narrow, black and white bands, but in opposite symmetry.

The only detailed magnetic survey then available [. . .] was the original one of the sea floor west of North America. [Wilson] and Vine spread out the magnetic map and studied the vicinity of [a] feature, which they called the Juan de Fuca Ridge. They were immediately struck by the symmetry of the magnetic avenues to either side of it.

By this time, Cox, Doell, and Dalrymple—that team of inveterate charters of the earth's magnetic history—had pieced together a rough timetable of reversals that had occurred about one million, 2.5 million and 3.4 million years ago, with short flip-flops at 1.9 million and possibly at three million years. Vine and Wilson decided to test the hypothesis that this timetable was inscribed in the relative widths of the magnetic avenues on each side of this ridge.

To carry out their test the two men computed the profile that would be recorded if a magnetometer were towed across a ridge that had been manufacturing zones of rock normally and reversely polarized according to the timetable. They realized that the central avenue—the one down the centerline of the ridge—would be double-width relative to the others, not yet having been split in two by the intrusion of material imprinted with a new magnetic reversal. [. . .]

Wilson and Vine found that, if they assumed a flow rate of two centimeters (one inch) per year away from each side of the ridge, the resulting magnetic profile bore a striking resemblance to the one based on actual observations. A similar result was obtained when this test was applied to a magnetic profile across the East Pacific Rise. Thus, said the two scientists, it was possible to account for the striped patterns "without recourse to improbable structures or lateral changes in petrology [rock composition]."

The following November, after Vine had gone to Princeton as a young geology instructor, he attended a meeting of the Geological Society of America in Kansas City, and it was there that the full significance of the magnetic sea-floor patterns became apparent. At the meeting Cox, Doell, and Dalrymple displayed a more extended and detailed timetable of the reversals. [. . .] The timetable now stood out like a great, coded message, inscribed in "dots" and "dashes" (or broad and narrow tree rings) to represent the succession of brief and extended periods of magnetic polarity reaching back at least four million years. This same message, Vine realized, must be written on the floor of every ocean in the world. [. . .]

Particularly clean data had been obtained by the United States research ship *Eltanin* on Leg 19 of its survey of the South Pacific, where the East Pacific Rise curves west. At the memorable 1966 meeting of the American Geophysical Union [. . .] [James] Heirtzler threw on the screen the magnetic profile obtained by the *Eltanin* as it crossed the rise on Leg 19 of its zigzagging journey. To dramatize the symmetry of this profile to either side of the rise, he displayed above it the same profile in reverse sequence—that is, from southeast to northwest instead of the other way around. Then, below it, he showed a computer calculation of what the profile should look like in terms of the known timetable of reversals and an assumed spreading rate of 4.6 centimeters (1.8 inches) a year from each side of the rise (which was at the center of each profile).

All three profiles were strikingly alike. As the session chairman, Allan Cox, commented later, "The profile was beautifully symmetrical on either side of the oceanic rise. I hadn't really believed in seafloor spreading up until then because the magnetic data hadn't been very symmetrical. But suddenly there was the incredible symmetry of the *Eltanin*-19 profile. I remember my reaction: 'Good grief! Vine is right after all.'"

Vine himself was there with copies of a new and comprehensive paper on the implications of the magnetic patterns. The following December 16 this report, by a lowly, newly appointed geology instructor at Princeton, was published, in revised form, as the lead article in *Science*. "Magnetic anomalies," said its subtitle, "may record histories of the ocean basins and Earth's magnetic field for 2×10^8 [200 million] years." [. . .]

Even to the most ardent protagonist of drift, there had been little to indicate just how the continents were once fitted together and by what routes and what timetables they drifted apart. One might reasonably assume that such information was lost forever. But now there was the history of such motions imprinted on the bottom of the sea.

The spreading of the sea floor south of Australia clearly has been pushing that continent and Antarctica apart. The entire Atlantic, north and south, seems to be growing along the Mid-Atlantic Ridge.

But in the North Pacific almost the whole sea floor is moving north-west, sliding along the rim of California and disappearing under Japan and the Aleutians.

"The magnetic lineations in the ocean floor," said Heirtzler, "serve as 'footprints' of the continents, marking their consecutive positions before they reached their present positions." Thus it appeared that, as previously postulated, the present continents originated in two primordial continents: Gondwanaland in the south and Laurasia in the north.

▷─┼─◆▷─○─◁◆─┼─◁

GEORGE SMOOT

Looking for the Big Bang, from
Wrinkles in Time (1994)

The work of George Smoot (1945–) is hidden behind ugly acro-
nyms and technical machines that measure things that people
cannot discern with their senses. Yet this work stems from the same
impulses that moved Herodotus and Marie Curie. Smoot led
NASA's Cosmic Background Explorer (COBE) program. "Cosmic
Background" refers to radiation that comes toward earth from all
directions. According to contemporary theory, this radiation is the
residue of the "Big Bang," the instant—10 billion or 15 billon years
ago—when all matter and energy in the universe appeared and
began to move apart.

The Big Bang therefore contained the seeds of the future universe. Did matter in the universe collect into the clumps that we now
call galaxies? If so, there must have been a predisposition to form
such clumps that was built into the Big Bang itself, and such a predisposition should be evident in the background radiation. But the
background radiation is hard to study, because the earth's atmosphere absorbs and distorts so much of it. Smoot's team sent instruments into space to reach beyond the problems of the atmosphere.
One instrument, for example, was the Far Infrared Absolute Spec-
trophotometer (FIRAS), which measured infrared waves. A second,
the Differential Microwave Radiometer (DMR), compared temperatures between various points in the sky.

Despite the military-sounding jargon, the reason for the COBE
program was exactly the one that sent Herodotus sailing off to
Egypt—the urge to have a look. And like Madame Curie, the researchers had an idea that they hoped to demonstrate. That is, they
wanted to show that even in the background radiation were small
clumps that, over the eons, could grow into galaxies.

We were looking for tiny variations in the smooth background
temperatures, something less than one part in a hundred thousand—
that is something like trying to spot a dust mote lying on a vast,
smooth surface like a skating rink. And, just like a skating rink, there
would be many irregularities on the surface that had nothing to do

with those we sought. These irregularities are like the systematic errors that plagued the differential microwave radiometer data—stray sources of heat, magnetic radiation, artifacts in software analysis, and so on. It is difficult to convey how obsessed we were with trying to eliminate these errors. I had started writing a list of potential things that could fool us back in 1974. Ever since I had continually updated the list, adding new candidates and studying their effects. I drilled everyone to check for errors all the time. And then I would have them double-checked by someone else.

When we'd been through that, I got Al Kogut in to check them through *again*. He would be someone fresh and independent of my years of effort. [. . .] It was grueling work, hardly the stuff of glamorous discovery, but essential. We simply had to do it if we were to have any hope of boiling everything down to a purely cosmological signal, a signal that came from only one place—the edge of space and time. We had seen too many "discoveries" prove to be nothing more than artifacts and did not wish to be caught in the same trap. [. . .]

All of us knew that if we did discover cosmic wrinkles, they wouldn't leap out of the map overnight, fully formed. If present, their image on the map would slowly become more visible, for two reasons: First, the repetition of their exceedingly weak signal would become ever more evident, like the ever-darkening mark left by repeatedly rubbing a pencil lightly over a piece of paper; second, we would steadily clear out the noise in the system that tended to obscure the signal.

There would be four steps in the discovery of wrinkles: First, we would see the cosmic background radiation itself, seemingly uniform from all parts of the universe [. . .]; second, we would observe the dipole, a slight distortion of the background radiation caused by the peculiar motion of our galaxy [. . .]; third, we would detect the quadrupole [. . .], which corresponds to the first cosmic distortion; finally, we would find the wrinkles themselves, the primordial seeds. At the end of 1990, a year after launch, we chalked up the first two steps, but had only set tighter limits on the last two. Now, early in 1991, I thought I could see step three emerging from the data—the picture was getting crisper.

In March 1991, I told the science team what we saw in the data, but cautioned that the noise and our limits on potential errors in the system were still as great as the signal. In other words, it could be wrong. I said the same in June at the next meeting. Ned Wright, who was working on the FIRAS data at the time, looked over the DMR maps and reported that they showed a signal resembling a quadrupole. The rest of the team were much more cautious; they knew how cranky instruments could be and how exciting data could fall flat the next day, so they wanted to continue analyzing more data before making any announcements. I agreed.

We kept absolute secrecy. We couldn't have anything leaking out about what we had found, until we were certain about what we had.

As we worked on, doggedly scrutinizing the data for systematic errors, marking putative anisotropies,[40] tension began to build, both inside and outside the team. Inside, a conflict began to grow over how real our results were, and, more tricky to handle, how soon we should go public with them. Outside, our colleagues were wondering what was going on. Here we were, a year since launch, blessed with the most sensitive instrument for detecting cosmic seeds, and yet we had reported nothing—beyond the fact that, by January 1991, we had seen no evidence of seeds, and that any variations in the cosmic background radiation had to be less than several parts per hundred thousand. Had we failed?

Colleagues often asked me what we had found; I said we were still trying to figure it out. A report by the National Research Council was ominous: "If no variations are found at these increased sensitivities, then theoretical extragalactic astronomy will be thrown into crisis. Something will be seriously wrong—either with our theories of galaxy formation or with our understanding of the cosmic background radiation."

News media made hay out of the "crisis" that allegedly faced big bang theorists. While the articles tended to be fairly balanced, the headlines had a prematurely funereal tone: "The Big Bang: Dead or Alive?" asked a headline in *Sky and Telescope*. A student magazine distributed by *USA Today* ran a story titled "Goodbye to Big Bang Theory?" A *Science News* article was titled "State of the Universe—If

---

[40] Anisotropy *is the technical term for what Smoot calls "wrinkles in time," or fluctuations in the temperature of the background radiation that is everywhere in the universe.*

Not with a Big Bang, Then What?" Said *Astronomy* magazine: "Beyond the Big Bang—New observations cast doubt on conventional theories of how the universe formed." *Science* magazine tried to balance things with the headline "Despite Reports of Its Death, the Big Bang Is Safe," but acknowledged: "Even so, cosmologists are going to have to rethink a lot of what they thought they knew about what happened later." [. . .]

Unbeknownst to the outside world, by the fall of 1991 our celestial map was coalescing ever more clearly, with evidence of wrinkles becoming ever more persuasive, though not conclusively so. We had analyzed our data with new software and gathered a clean map. In September we held a major meeting on systematic errors—again—and resolved yet more potential traps. I could tell that confidence was building, and that worried me. We have to be even more diligent, I insisted, and offered a free airline ticket to anywhere in the world to anyone who could prove the wrinkles were an artifact. I wanted to motivate the team to look for problems rather than assume everything was fine. If you get cocky, you get sloppy. In science, if you want to see an effect, it is all to easy to be seduced into believing you have. To quote Richard Feynman: "The first principle is that you must not fool yourself and you're the easiest person to fool."

Ned Wright, convinced we had a genuine signal, pressed ever harder for us to publish our results and let the cosmological community evaluate them however it wished. On October 9, 1991, he delivered a "preliminary paper" on his views at a Science Working Group meeting. Ned argued that we should offer an article on the data to *Astrophysical Journal*. He thought the team should publish the data so theorists wouldn't waste any more time wandering down blind alleys. David Wilkinson urged caution: He hadn't forgotten the quadrupole he thought he'd detected years earlier, only to see it vanish. [. . .]

In the end, the group turned Ned down; we wouldn't notify the *Astrophysical Journal*, not yet. As a member of the Science Working Group, Ned was bound by its decision to keep things quiet for the time being. We would wait until a slightly revised version of the new software had reprocessed the first year's data, and we could check through a few more potential errors. The new software had more checks and robustness built into it. We felt it was important to assess the results very carefully before taking the fateful step of announcing them. Even a tentative announcement posed all kinds of potential risks. There was a great deal at stake; we saw no virtue in announcing a result that would probably attract massive press attention, then might fall through and humiliate us all. It's difficult to appreciate how instruments can turn around and bite you if you don't watch them every step of the way.

Rumors began to course through the cosmological community, and more and more of our colleagues tried to pry out of us what we had found. At one point Dick Bond came up to me and said, "I hear you have a thirteen-microkelvin quadrupole in your data." I smiled and answered him evasively. Sometimes someone would bait me by asking, "I hear you found anisotropies at such and such a level," and I'd reply, "That's a little big, don't you think?" Likewise, at Goddard, Sasha Kashlinsky worked right down the hall from Chuck Bennett; Sasha frequently walked into Chuck's office and tried to bait him by saying something like, "So, Chuck, I hear you saw a big quadrupole," or, "I hear you see something but *no* quadrupole." Chuck didn't bite the bait. From time to time I sent out electronic-mail messages to the team reminding everyone to keep mum about our results until we were sure of them. [. . .] They say that during World War II, even the great "secret" of the atom bomb project was the subject of rumors throughout U.S. academia. Who knew what people might be saying about the DMR as 1991 drew to a close?

In October 1991 we had the latest version of the map pasted on the wall in the instrument operations room. It looked unprepossessing, with red and blue blotches on a green background. I stood in front of it one evening, no one else around, and thought, Yes, this is it. This is what I've spent eighteen years looking for. [. . .] I felt deep down that we had done everything right with our instrument. I thought the team had worked so well together, the analysis was so careful, and the search for systematic errors so thorough, that we *had* found the wrinkles, the Holy Grail of modern cosmology.

"Those Who Would Judge the Book Must Read It"

Thomas Huxley

JOHANNES KEPLER

I Admit the Moon Has Seas, from
Conversation with the Sidereal Messenger (1610)

Johannes Kepler (1571–1630) is a popular figure among historians of science, and this selection gives a sense of the reason. He wrote with his heart throbbing on his sleeve. The puzzle he most wished to solve is something that people today do not think of as a scientific question at all: why are there seven planets? "Because there are not eight" was Shakespeare's answer, given at that time, and we tend to agree. But Kepler wanted a fuller explanation and spent many years studying the geometry of planetary orbits in the hope of finding a reason. But for all the oddity of his choice of puzzles, he would not accept false logic and made his observations as important as his mathematical reasoning.

Another reason for liking Kepler is his ready willingness to change his mind. This selection shows his quick response to Galileo's *Sidereal Messenger*, also included in this volume. In many ways Kepler's reply reads like modern e-mail. It is an emotional reaction, full of self-citation, but for all his ego, Kepler is saying that Galileo was right and he, Kepler, was wrong. Of course, Kepler's embrace of seas on the moon was wrong, but his admission of error and willingness to accept somebody else's work and build on it is essential to a living, progressing science. We can think again tomorrow only if we are willing to think again today. Popper's essay speaks of the need for boldness in trying to be right. But every bold idea cannot be right, so people who can bring themselves to say, "Oops, that was wrong," make a real contribution.

Three months ago the Most August Emperor[1] raised various questions with me about the spots on the moon. He was convinced that the images of countries and continents are reflected in the moon as though in a mirror. He asserted in particular that Italy with its two adjacent islands seemed to him to be distinctly outlined. He even offered his glass for the examination of these spots on subsequent days, but this was not done. Thus at that very same time, Galileo, when you cherished the abode of Christ our Lord above the mere appellation,[2] you vied in your favorite occupation with the ruler of Christendom (actuated by the same restless spirit of inquiry into nature).

But this story of the spots in the moon is also quite ancient. It is supported by the authority of Pythagoras and Plutarch, the eminent philosopher who also, if this detail helps the cause, governed Epirus with the power of a proconsul under the Caesars. I say nothing about Mästlin and my treatise on "Optics,"[3] published six years ago; these I shall take up later on in their proper place.

Such assertions about the body of the moon are made by others on the basis of mutually self-supporting evidence. Their conclusions agree with the highly illuminating observations which you report on the same subject. Consequently I have no basis for questioning the rest of your book and the four satellites of Jupiter. I should rather wish that I now had a telescope at hand, with which I might anticipate you in discovering two satellites of Mars (as the relationship seems to me to require) and six or eight satellites of Saturn, with one each perhaps for Venus and Mercury.

For this search, so far as Mars is concerned, the most propitious time will be next October,[4] which will show Mars in opposition to the sun and (except for the year 1608) nearest to the earth, with the error in the predicted position exceeding 3°. [. . .]

[It] is a highly agreeable circumstance that I too engaged in observing the spots on the moon, not like you with upturned face,

---

[1] *Rudolf II was the Holy Roman Emperor in 1610. Although considered to be politically incompetent, Rudolf was an eager patron of the sciences, especially astronomy, and he was Kepler's patron. In our day the patronage redounds to Rudolf's honor, but in 1610 it redounded to Kepler's.*

[2] *An indirect pun linking Galileo's name with Jesus of Galilee.*

[3] *In 1604, Kepler published an explanation of how eyeglasses worked.*

[4] *Mars's two satellites were not observed until 1877.*

but with my head down. You will find a sketch of this observation on page 247 of my book.[5] It shows that the moon's limb appeared very bright all around to me also, only the interior of the body being marked by spots.

Hence it occurs to me to compete with you in scrutinizing those small spots first noticed by you in the brighter region. Yet, if this is agreed upon, I hope to get results by my own method of observation, with my back to the moon. I shall admit the light of the moon through an aperture to a sheet of paper mounted on a rod. [. . .]

Let us proceed, Galileo, with the investigation of your phenomena. [. . .]

In your discussion of the spots first noticed by you in the bright region of the moon, you show in a thorough optical analysis of the illumination that they are hollow or sunken cavities in the body of the moon. But you provoke a dispute over the nature of those numerous spots in what has been regarded from ancient times as the bright region of the moon. You compare them with the valleys on our earth. There are some valleys of this kind, I admit, especially in the province of Styria.[6] They are almost round in appearance. Through very narrow passes they admit the Mur River at their upper end, and discharge it at their lower end. Such are the so-called Fields of Graz, Leibniz, and Maribor on the Drava. There are others in other regions. Round about these fields rise the lofty summits of mountains, creating the impression of a bowl, since the height of the surrounding peaks is no small fraction of the width of the fields. For my part I concede the possibility of such lunar valleys, carved by rivers from the curving slopes of the mountains. But these spots are so numerous, you add, that they make the bright part of the body of the moon resemble a peacock's tail, divided into various eye-shaped reflecting surfaces. Hence the question occurs to me whether these spots on the moon indicate something else. For among us on the earth there are some curving valleys, but they are also extensive in length, according to the course of the rivers, and of no mean depth. An example of this

---

[5] *Kepler invented a projection device that let him study an enlarged image of the moon focused onto a piece of paper.*

[6] *Styria is a region in central Austria, known today as Steiermark.*

sort of virtually uninterrupted valley is offered by almost the whole of Austria, which, on account of the Danube, lies deep and, as it were, hidden between the mountains of Moravia and Styria. Why then do you report no such long spots on the moon? Why are they for the most part bounded by a circle? May l indulge in the guess that the moon is like pumice, with very many pores of the largest size opening up on all sides? For you will patiently permit me here to take the opportunity to make some reference to the speculations set forth in Chapter 34, page 175, of my "Commentary on Mars." I pointed out there that the moon is set in motion by the earth with twice the velocity of the outlying areas of the earth at the equator. Hence I inferred that the body of the moon is not very dense and that, possessing the slight resistance of a thin material, it offers no great opposition to the force exerted by the earth.

But these suggestions (about cavities that are below the surface and not cut through the mountains) are not so weighty that I would deem them worthy of a stubborn defense, should they be rendered completely untenable by your subsequent reports. For you have established most firmly by brilliant observations in full accord with the laws of optics that many peaks tower above the body of the moon, throughout the bright region, especially in the lower portion. Like the loftiest mountains on our earth, they are the first to enjoy the light of the sun as it rises for the moon, and are thereby revealed to you when you make use of your telescope.

What shall I say now about your very acute analysis of the ancient spots on the moon? On page 251 of my book I cited the opinion of Plutarch, who regarded those ancient spots on the moon as lakes or seas, and the bright areas as continents. I did not hesitate to oppose him and to reverse his interpretation, by attributing the spots to continents, and the purity of the bright region to the effects of a liquid. Wackher used to give strong approval to my stand on this question. We were deeply engaged in these discussions last summer (I suppose, because nature was seeking the same results through us as it achieved a little later through Galileo). To please Wackher, I even founded a new astronomy for the inhabitants of the moon, as it were; in plain language, a sort of lunar geography. Among its basic propositions was this thesis, that the spots are continents, while the bright areas are seas. My motive in contradicting Plutarch in this regard may be seen on page 251 of my book. I there report an observation which I performed on Mt. Schöckel in Styria. From my vantage point the river below looked bright, and the land darker. But the flaw in my reasoning is indicated in the margin of the very next page. Obviously the river did not, like the land, shine by light received from the sun, but by light reflected from the illuminated air. Hence my analysis of the causes of the phenomenon was also unfortunate. For, in opposition to the doctrine of Aristotle's book "On Colors," I asserted that water partakes of black less than earth does. Yet how could this be true, since earth turns darker when it is soaked with water? But why go on at length? Suppose that the moon, like the island of Crete, is composed of a white soil (as Lucian said that the moon is a cheese-like land). We shall have to admit that the soil shines by sunlight more vividly than the seas, however little they may be tinged with black.

My book, consequently, does not prevent me from agreeing with you, as you adduce mathematical arguments against me in favor of Plutarch with brilliant and irrefutable logic. Certainly the bright areas are broken up by many cavities; the bright areas are bounded by an irregular line; the bright areas contain great peaks, on account of which they light up sooner than the neighboring region. Where they face the sun, they are bright; where they face away from the sun, they are dark. All these characteristics suit a dry, solid, and high material, but not a fluid. On the other hand, the dark spots, known since antiquity, are flat. The dark spots light up later—a fact which proves their low elevation—when the surrounding peaks are already aglow far and wide. When the dark spots are illumined, a certain shadow-like black effect differentiates them from the peaks. The boundary of the illumination in the dark area is a straight line at half-moon. These characteristics, in turn, belong to a liquid, which seeks the lowest levels and on account of its weight settles in a horizontal position.

By these arguments, I say, you have proved your point completely. I admit that the spots are seas, I admit that the bright areas are land.

VOLTAIRE

The Importance of Isaac Newton, from
Letters on the English Nation (1733)

Voltaire (1694–1778) was a propagandist for liberty who understood the connection between science and freedom. In 1726, he was grabbed by henchmen of a leading chevalier, beaten up, dragged to the Bastille, and then shipped forcibly to England. Voltaire's trouble had come because he had no respect for authority. But in England he encountered the most liberal society in the world. Voltaire thus became a strong defender of English ways: He maintained that Shakespeare was as great as the French dramatists and that Locke was the philosopher of freedom. He also hailed British science. He saw at once that by denying kings and clergy the authority to say how nature behaved, modern science was a termite eating away at all the traditional social structures. He liked that subversiveness and declared Newton to be a greater scientist than Descartes.

In 1733 Voltaire published a series of "letters" on the glories of England, devoting four of them to Newton. The point was not to satisfy an inherent curiosity about science but to force his French readers to see that Newton and English liberty had really accomplished something. Voltaire successfully instigated a great European interest in Newton. He followed his letters with a book-length account of Newton's science, and his mistress, Madame de Châtelet, learned English for the specific purpose of translating Newton's works into French.

A Frenchman who arrives in London, will find philosophy, like everything else, very much changed there. He had left the world a plenum, and he now finds it a vacuum. At Paris the universe is seen, composed of vortices of subtle matter, but nothing like it is seen in London. In France, 'tis the pressure of the moon that causes the tides, but in England 'tis the sea that gravitates toward the moon; so that when you think that the moon should make it flood with us, those gentlemen fancy it should be ebb. [. . .]

You'll observe farther that the sun, which in France is said to have nothing to do in the affair, comes in here for very near a quarter of its assistance. According to your Cartesians, everything is performed by an impulsion, of which we have very little notion, and according to Sir Isaac Newton, 'tis by an attraction, the cause of which is as much unknown to us. At Paris you imagine that the earth is shaped like a melon, or of an oblique figure; at London it has an oblate one. A Cartesian declares that light exists in the air, but a Newtonian asserts that it comes from the sun in six minutes and a half. The several operations of your chemistry are performed by acids, alkalis, and subtle matter, but attraction prevails even in chemistry among the English. [. . .]

The discoveries which gained Sir Isaac Newton so universal a reputation, relate to the system of the world, to light, to geometrical infinites, and lastly to chronology, with which he used to amuse himself after the fatigue of his severer studies. [. . .] With regard to the system of our world, disputes were a long time maintained on the cause that turns the planets and keeps them in their orbits, and on those causes which make all bodies here below descend towards the surface of the earth.

The system of Descartes, explained and improved since his time, seemed to give a plausible reason for all those phenomena, and this reason seemed more just, as 'tis simple, and intelligible to all capacities. But in philosophy, a student ought to doubt of the things he fancies he understands too easily, as much as those he does not understand.

Gravity, the falling of accelerated bodies on the earth, the revolution of the planets in their orbits, their rotations round their axis, all this is mere motion. Now motion can't perhaps be conceived any otherwise than by impulsion; therefore all those bodies must be impelled. But by what are they impelled? All space is full, it therefore is filled with a very subtle matter, since this is imperceptible to us; this matter goes from west to east since all the planets are carried from west to east. Thus from hypothesis to hypothesis, from one appearance to another, philosophers have imagined a vast whirlpool of subtle matter, in which the planets are carried round the sun. They also have created another particular vortex which floats in the great one, and which turns daily round the planets. When all this is done, 'tis pretended that gravity depends on this diurnal motion, for, say these, the velocity of the subtle matter that turns round our little vortex must be seventeen times more rapid than that of the earth or, in case its velocity is seventeen times greater than that of the earth, its centrifugal force must be vastly greater, and consequently impel all bodies towards the earth. This is the cause of gravity, according to the

Cartesian system. But the theorist, before he calculated the centrifugal force and velocity of the subtle matter, should first have been certain that it existed.

Sir Isaac Newton seems to have destroyed all these great and little vortices, both that which carries the planets round the sun, as well as the other which supposes every planet to turn on its own axis. [. . .] He proves that there is no such thing as a celestial matter which goes from west to east since the comets transverse those spaces, sometimes from east to west and at other times from north to south.

In fine, the better to resolve, if possible, every difficulty, he proves, and even by experiments, that 'tis impossible there should be a plenum and brings back the vacuum which Aristotle and Descartes had banished from the world.

Having by these and several other arguments destroyed the Cartesian vortices, he despaired of ever being able to discover, whether there is a secret principle in nature which, at the same time, is the cause of the motion of all celestial bodies, and that of gravity on the earth. But being retired in 1666, upon account of the plague, to a solitude near Cambridge, as he was walking one day in his garden, and saw some fruits fall from a tree, he fell into a profound meditation on that gravity, the cause of which had so long been sought, but in vain, by all the philosophers, whilst the vulgar think there is nothing mysterious in it. He said to himself, that from what height soever in our hemisphere, those bodies might descend, their fall would certainly be in the progression discovered by Galileo, and the spaces they run through would be as the square of the times. Why may not this power which causes heavy bodies to descend, and is the same without any sensible diminution at the remotest distance from the center of the earth, or on the summits of the highest mountains, why, said Sir Isaac, may not this power extend as high as the moon? And in case, its influence reaches so far, is it not very probable that this power retains it in its orbit, and determines its motion? But in case the moon obeys this principle (whatever it be) may we not conclude very naturally, that the rest of the planets are equally subject to it? In case this power exists (which besides is proved) it must increase in an inverse ratio of the squares of the distances. All therefore that remains is, to examine how far a heavy body, which should fall upon the earth from a moderate height, would go, and how far in the same time, a body which should fall from the orbit of the moon, would descend. To find this, nothing is wanted but the measure of the earth, and the distance of the moon from it.

Thus Sir Isaac Newton reasoned, but at that time the English had a very imperfect measure of our globe, and depended on the uncertain supposition of mariners, who computed a degree to contain but sixty English miles, whereas it consists in reality of near seventy. As this false computation did not agree with the conclusions which Sir Isaac intended to draw from them, he laid aside this pursuit. A half-learned philosopher, remarkable only for his vanity, would have made the measure of the earth agree, anyhow, with his system. Sir Isaac, however, chose rather to quit the researches he was then engaged in. But after Mr. Picart had measured the earth exactly, by tracing that meridian, which redounds so much to the honor of the French,[7] Sir Isaac Newton resumed his former reflections, and found his account in Mr. Picart's calculation.

A circumstance which has always appeared wonderful to me, is that such sublime discoveries should have been made by the sole assistance of a quadrant and a little arithmetic.

The circumference of the earth is 123,249,600 feet. This, among other things, is necessary to prove the system of attraction.

The instant we know the earth's circumference, and the distance of the moon, we know that of the moon's orbit, and the diameter of this orbit. The moon performs its revolution in that orbit in twenty seven days, seven hours, forty three minutes. 'Tis demonstrated, that the moon in its mean motion makes a 187,960 feet (of Paris) in a minute. 'Tis likewise demonstrated, by a known theorem, that the central force which should make a body fall from the height of the moon, would make its velocity no more than fifteen Paris feet in a minute of time. Now, if the law by which bodies gravitate, and attract one another in an inverse ratio of the squares of the distances be true, if the same power acts, according to that law, throughout all nature, 'tis evident that as the earth is sixty semi-diameters distant from the moon, a heavy body must necessarily fall (on the earth) fifteen feet in the first second, and fifty four thousand feet in the first minute.

Now a heavy body falls, in reality, fifteen feet in the first second, and goes in the first minute fifty four thousand foot, which number is

---

[7] *Jean Picart, an astronomer in Paris, calculated the size of the earth by determining the correct length of 1 degree along a longitudinal meridian.*

the square of sixty multiplied by fifteen. Bodies therefore gravitate in an inverse ratio of the squares of the distances; consequently, what causes gravity on earth, and keeps the moon in its orbit, is one and the same power; it being demonstrated that the moon gravitates on the earth, which is the center of its particular motion, 'tis demonstrated that the earth and the moon gravitate on the sun which is the center of their annual motion.

The rest of the planets must be subject to this general law, and if this law exists, these planets must follow the laws which Kepler discovered. All these laws, all these relations, are indeed observed by the planets with the utmost exactness; therefore, the power of attraction causes all the planets to gravitate towards the sun, in like manner as the moon gravitates towards our globe.

Finally, as in all bodies, reaction is equal to action, 'tis certain that the earth gravitates also towards the moon, and that the sun gravitates towards both, that every one of the satellites of Saturn gravitates towards the other four, and the other four towards it, all five towards Saturn and Saturn towards all. That 'tis the same with regard to Jupiter, and that all these globes are attracted by the sun, which is reciprocally attracted by them.

This power of gravitation acts proportionately to the quantity of matter in bodies, a truth which Sir Isaac has demonstrated by experiments. This new discovery has been of use to show that the sun (the center of the planetary system) attracts them all in a direct ratio of their quantity of matter combined with their nearness. From hence Sir Isaac, rising by degrees to discoveries which seemed not to be formed for the human mind, is bold enough to compute the quantity of matter contained in the sun and in every planet, and in this manner shows, from the simple laws of mechanics, that every celestial globe ought necessarily to be where it is placed.

His bare principle of the laws of gravitation, accounts for all the apparent inequalities in the course of the celestial globes. The variations of the moon are a necessary consequence of those laws. Moreover, the reason is evidently seen why the nodes of the moon[8] perform their revolutions in nineteen years, and those of the earth in about twenty six thousand. The several appearances observed in the

---

[8] Node. *In astronomy, the point where an orbit crosses the plane of the apparent path of the sun through the stars.*

tides are also a very simple effect of this attraction. The proximity of the moon, when at the full, and when it is new, and its distance in the quadratures or quarters combined with the action of the sun, exhibit a sensible reason why the ocean swells and sinks. [. . .]

This is attraction, the great spring by which all nature is moved. Sir Isaac Newton, after having demonstrated the existence of this principle, plainly foresaw that its very name would offend, and therefore this philosopher in more places than one of his books, gives the reader some caution about it. He bids him beware of confounding this name with what the ancients called occult qualities, but to be satisfied with knowing that there is in all bodies a central force which acts to the utmost limits of the universe, according to the invariable laws of mechanics.

'Tis surprising, after the solemn protestations Sir Isaac made, that such eminent men as Mr. Sorin and Mr. de Fontenelle, should have imputed to this great philosopher the verbal and chimerical way of reasoning of the Aristotelians, Mr. Sorin in the Memoirs of the Academy of 1709, and Mr. de Fontenelle in the very elogium of Sir Isaac Newton. Most of the French, the learned and others, have repeated this reproach. These are forever crying out, why did he not employ the word impulsion, which is so well understood, rather than that of attraction, which is unintelligible.

Sir Isaac might have answered these critics thus: First, you have as imperfect an idea of the word impulsion, as that of attraction, and in case you cannot conceive how one body tends towards the center of another body, neither can you conceive by what power one body can impel another.

Secondly, I could not admit of impulsion, for to do this, I must have know that a celestial matter was the agent, but so far from knowing that there is any such matter, I have proved it to be merely imaginary.

Thirdly, I use the word attraction for no other reason, but to express an effect which I discovered in nature, a certain and indisputable effect of an unknown principle, a quality inherent in matter, the cause of which persons of greater abilities than I can pretend to, may, if they can find out.

What have you then taught us? Will these people say further: And to what purpose are so many calculations to tell us what you yourself don't comprehend?

I have taught you, may Sir Isaac rejoin, that all bodies gravitate towards one another in proportion to their quantity of matter, that these central forces alone, keep the planets and comets in their orbits, and cause them to move in the proportion before set down. I demonstrate to you, that 'tis impossible there should be any other cause which keeps the planets in their orbits, than that general phenomenon of gravity. For heavy bodies fall on the earth according to the proportion demonstrated of central forces, and the planets finishing their course according to these same proportions. In case there were another power that acted upon all those bodies, it would either increase their velocity, or change their direction. Now not one of those bodies ever has a single degree of motion or velocity, or has any direction but what is demonstrated to be the effect of the central forces, consequently 'tis impossible there should be any other principle.

Give me leave once more to introduce Sir Isaac speaking: Shall he not be allowed to say, My case and that of the ancients is very different. These saw, for instance, water ascend in pumps, and said, the water rises because it abhors a vacuum. But with regard to myself, I am in the case of a man who should have first observed that water ascends in pumps, but should leave others to explain the cause of this effect. The anatomist who first declared that the motion of the arm is owing to the contraction of the muscles, taught mankind an indisputable truth, but are they less obliged to him because he did not know the reason why the muscles contract? The cause of the elasticity of the air is unknown, but he who first discovered this spring performed a very signal service to natural philosophy. The spring that I discovered was more hidden and more universal, and for that reason mankind ought to thank me the more. I have discovered a new property of matter, one of the secrets of the Creator, and have calculated and discovered the effects of it. After this, shall people quarrel with me about the name I give it?

THOMAS H. HUXLEY

From *The Darwinian Hypothesis* (1859)

T homas Henry Huxley (1825–1895) was a world-renowned advocate of the theory of evolution through natural selection.
Trained as a doctor, Huxley became interested in zoology while serving as ship's surgeon during an expedition to the South Pacific. It
was as a propagandist for Darwinism that Huxley made his biggest
mark. While Darwin stayed at home, "Darwin's bulldog"—as Huxley became known—went out into the world and loudly promoted
Darwin's theory. Before Darwin published *On the Origin of Species*,
he showed a copy to Huxley for comment. So Huxley's published
review of Darwin's book may seem amazingly understated.

This dispassionate tone was all the more striking in Huxley's own
day because Darwin's book immediately sparked a furor. People were
already lining up on both sides, and suddenly Huxley appeared in the
guise of a neutral commentator. He starts quietly, laying out the facts
of biology that seem random and devoid of plan. But his conclusion
seems inevitable and undeniable. The central point is "Those who
would judge the book must read it." Already, Darwin's book was
more widely judged than read.

The hypothesis of which the present work of Mr. Darwin is but
the preliminary outline, may be stated in his own language as follows:—"Species originated by means of natural selection, or through
the preservation of the favoured races in the struggle for life." To render this thesis intelligible, it is necessary to interpret its terms. In the
first place, what is a species? The question is a simple one, but the
right answer to it is hard to find, even if we appeal to those who
should know most about it. It is all those animals or plants which
have descended from a single pair of parents; it is the smallest distinctly definable group of living organisms; it is an eternal and immutable entity; it is a mere abstraction of the human intellect having
no existence in nature. Such are a few of the significations attached to
this simple word which may be culled from authoritative sources; and
if, leaving terms and theoretical subtleties aside, we turn to facts and
endeavour to gather a meaning for ourselves, by studying the things

to which, in practice, the name of species is applied, it profits us little. For practice varies as much as theory. Let two botanists or two zoologists examine and describe the productions of a country, and one will pretty certainly disagree with the other as to the number, limits, and definitions of the species into which he groups the very same things. In these islands, we are in the habit of regarding mankind as of one species, but a fortnight's steam will land us in a country where divines and savants, for once in agreement, vie with one another in loudness of assertion, if not in cogency of proof, that men are of different species; and, more particularly, that the species negro is so distinct from our own that the Ten Commandments have actually no reference to him. [. . .]

The truth is that the number of distinguishable living creatures almost surpasses imagination. At least 100,000 such kinds of insects alone have been described and may be identified in collections, and the number of separable kinds of living things is under-estimated at half a million. Seeing that most of these obvious kinds have their accidental varieties, and that they often shade into others by imperceptible degrees, it may well be imagined that the task of distinguishing between what is permanent and what fleeting, what is a species and what a mere variety, is sufficiently formidable. [. . .]

If, weary of the endless difficulties involved in the determination of species, the investigator, contenting himself with the rough practical distinction of separable kinds, endeavours to study them as they occur in nature [. . .] he finds himself, according to the received notions, in a mighty maze, and with, at most, the dimmest adumbration of a plan. If he starts with any one clear conviction, it is that every part of a living creature is cunningly adapted to some special use in its life. Has not his Paley[9] told him that that seemingly useless organ, the spleen, is beautifully adjusted as so much packing between the other organs? And yet, at the outset of his studies, he finds that no adaptive reason whatsoever can be given for one-half of the peculiarities of vegetable structure. He also discovers rudimentary teeth, which are never used, in the gums of the young calf and in those of

---

[9] *William Paley (1743–1805), in* Natural Theology *(1802), argued that the adaptation of every organ to function was proof of the providence of God.*

the foetal whale; insects which never bite have rudimental jaws, and others which never fly have rudimental wings; naturally blind creatures have rudimental eyes; and the halt have rudimentary limbs. [. . .]

But, if the doctrine of final causes will not help us to comprehend the anomalies of living structure, the principle of adaptation must surely lead us to understand why certain living beings are found in certain regions of the world and not in others. The Palm, as we know, will not grow in our climate, nor the Oak in Greenland. The white bear cannot live where the tiger thrives, nor *vice versâ*, and the more the natural habits of animal and vegetable species are examined, the more do they seem, on the whole, limited to particular provinces. But when we look into the facts established by the study of the geographical distribution of animals and plants it seems utterly hopeless to attempt to understand the strange and apparently capricious relations which they exhibit. One would be inclined to suppose *à priori* that every country must be naturally peopled by those animals that are fittest to live and thrive in it. And yet how, on this hypothesis, are we to account for the absence of cattle in the pampas of South America, when those parts of the New World were discovered? It is not that they were unfit for cattle, for millions of cattle now run wild there; and the like holds good of Australia and New Zealand. It is a curious circumstance, in fact, that the animals and plants of the northern hemisphere are not only as well adapted to live in the southern hemisphere as its own autochthones, but are, in many cases, absolutely better adapted, and so overrun and extirpate the aborigines. Clearly, therefore, the species which naturally inhabit a country are not necessarily the best adapted to its climate and other conditions. [. . .]

But our knowledge of life is not confined to the existing world. Whatever their minor differences, geologists are agreed as to the vast thickness of the accumulated strata which compose the visible part of our earth, and the inconceivable immensity of the time the lapse of which they are the imperfect but the only accessible witnesses. Now, throughout the greater part of this long series of stratified rocks are scattered, sometimes very abundantly, multitudes of organic remains, the fossilized exuviae of animals and plants which lived and died while the mud of which the rocks are formed was yet soft ooze, and could receive and bury them. It would be a great error to suppose that these organic remains were fragmentary relics. Our museums exhibit fossil shells of immeasurable antiquity, as perfect as the day they were formed; whole skeletons without a limb disturbed; nay, the changed flesh, the developing embryos, and even the very footsteps of primaeval organisms. Thus the naturalist finds in the bowels of the earth species as well defined as, and in some groups of animals more numerous than, those which breathe the upper air. But, singularly enough, the majority of these entombed species are wholly distinct from those that now live. Nor is this unlikeness without its rule and order. As a broad fact, the further we go back in time the less the buried species are like existing forms; and, the further apart the sets of extinct creatures are, the less they are like one another. In other words, there has been a regular succession of living beings, each younger set, being in a very broad and general sense, somewhat more like those which now live. [. . .]

Such is a brief summary of the main truths which have been established concerning species. Are these truths ultimate and irresolvable facts, or are their complexities and perplexities the mere expressions of a higher law?

A large number of persons practically assume the former position to be correct. They believe that the writer of the Pentateuch was empowered and commissioned to teach us scientific as well as other truth, that the account we find there of the creation of living things is simply and literally correct, and that anything which seems to contradict it is, by the nature of the case, false. All the phenomena which have been detailed are, on this view, the immediate product of a creative fiat and, consequently, are out of the domain of science altogether.

Whether this view prove ultimately to be true or false, it is, at any rate, not at present supported by what is commonly regarded as logical proof, even if it be capable of discussion by reason; and hence we consider ourselves at liberty to pass it by, and to turn to those views which profess to rest on a scientific basis only, and therefore admit of being argued to their consequences. And we do this with the less hesitation as it so happens that those persons who are practically conversant with the facts of the case (plainly a considerable advantage) have always thought fit to range themselves under the latter category.

The majority of these competent persons have up to the present time maintained two positions—the first, that every species is, within certain defined limits, fixed and incapable of modification; the second, that every species was originally produced by a distinct creative act. The second position is obviously incapable of proof or disproof, the direct operations of the Creator not being subjects of science; and it must therefore be regarded as a corollary from the first, the truth or falsehood of which is a matter of evidence. Most persons imagine that the arguments in favour of it are overwhelming; but to some few minds, and these, it must be confessed, intellects of no small power and grasp of knowledge, they have not brought convictions. Among these minds, that of the famous naturalist Lamarck, who possessed a greater acquaintance with the lower forms of life than any man of his day, Cuvier not excepted, and was a good botanist to boot, occupies a prominent place.

Two facts appear to have strongly affected the course of thought of this remarkable man—the one, that finer or stronger links of affinity connect all living beings with one another, and that thus the highest creature grades by multitudinous steps into the lowest; the other, that an organ may be developed in particular directions by exerting itself in particular ways, and that modifications once induced may be transmitted and become hereditary. Putting these facts together, Lamarck endeavoured to account for the first by the operation of the second. Place an animal in new circumstances, says he, and its needs will be altered; the new needs will create new desires, and the attempt to gratify such desires will result in an appropriate modification of the organs exerted. Make a man a blacksmith, and his brachial muscles will develop in accordance with the demands made upon them, and in like manner, says Lamarck, "the efforts of some short-necked bird to catch fish without wetting himself have, with time and perseverance, given rise to all our herons and long-necked waders."

The Lamarckian hypothesis has long since been justly condemned,[10] and it is the established practice for every tyro to raise his heel against the carcass of the dead lion. But it is rarely either wise or instructive to treat even the errors of a really great man with mere ridicule, and in the present case the logical form of the doctrine stands on a very different footing from its substance. [. . .]

---

[10] *Jean-Baptiste Lamarck (1744–1829) proposed that species evolved as a result of the use or disuse of organs during an animal's lifetime. Giraffes, for example, were said to have developed long necks because they often stretched them while trying to reach high branches.*

Since Lamarck's time, almost all competent naturalists have left speculations on the origin of species to such dreamers as the author of the "Vestiges,"[11] by whose well-intentioned efforts the Lamarckian theory received its final condemnation in the minds of all sound thinkers. Notwithstanding this silence, however, the transmutation theory,[12] as it has been called, has been a "skeleton in the closet" to many an honest zoologist and botanist who had a soul above the mere naming of dried plants and skins. Surely, has such an one thought, nature is a mighty and consistent whole, and the providential order established in the world of life must, if we could only see it rightly, be consistent with that dominant over the multiform shapes of brute matter. But what is the history of astronomy, of all the branches of physics, of chemistry, of medicine, but a narration of the steps by which the human mind has been compelled, often sorely against its will, to recognise the operation of secondary causes in events where ignorance beheld an immediate intervention of a higher power? And when we know that living things are formed of the same elements as the inorganic world, that they act and react upon it, bound by a thousand ties of natural piety, is it probable, nay is it possible, that they, and they alone, should have no order in their seeming disorder, no unity in their seeming multiplicity, should suffer no explanation by the discovery of some central and sublime law of mutual connection?

Questions of this kind have assuredly often arisen but it might have been long before they received such expression as would have commanded the respect and attention of the scientific world, had it not been for the publication of the work which prompted this article. Its author, Mr. Darwin, inheritor of a once celebrated name,[13] won his spurs in science when most of those now distinguished were young men, and has for the last twenty years held a place in the front ranks of British philosophers. After a circumnavigatory voyage, undertaken

---

[11] *Robert Chambers (1802–1871) published anonymously* Vestiges of Natural History *in 1844, in which he presented a picture of evolution. Huxley condemned the book, but in this essay he begins to set forth the many facts Chambers had assembled in his argument.*

[12] *Transmutation theory is the idea that over time, species change into other species.*

[13] *A reference to Erasmus Darwin (1731–1802), the grandfather of Charles and an early proponent of the idea that species change over time.*

solely for the love of his science, Mr. Darwin published a series of researches which at once arrested the attention of naturalists and geologists; his generalisations have since received ample confirmation and
now command universal assent, nor is it questionable that they have
had the most important influence on the progress of science. More recently Mr. Darwin, with a versatility which is among the rarest of
gifts, turned his attention to a most difficult question of zoology and
minute anatomy; and no living naturalist and anatomist has published
a better monograph than that which resulted from his labours. Such a
man, at all events, has not entered the sanctuary with unwashed
hands, and when he lays before us the results of twenty years' investigation and reflection we must listen even though we be disposed to
strike. But, in reading his work, it must be confessed that the attention
which might at first be dutifully, soon becomes willingly, given, so
clear is the author's thought, so outspoken his conviction, so honest
and fair the candid expression of his doubts. Those who would judge
the book must read it: we shall endeavour only to make its line of argument and its philosophical position intelligible to the general reader
in our own way.

The Baker Street Bazaar has just been exhibiting its familiar annual spectacle. Straight-backed, small-healed, big-barrelled oxen, as
dissimilar from any wild species as can well be imagined, contended
for attention and praise with sheep of half-a-dozen different breeds
and styes of bloated preposterous pigs, no more like a wild boar or
sow than a city alderman is like an ourang-outang. The cattle show
has been, and perhaps may again be, succeeded by a poultry show of
whose crowing and clucking prodigies it can only be certainly predi-
cated that they will be very unlike the aboriginal *Phasianus gallus*. If
the seeker after animal anomalies is not satisfied, a turn or two in
Seven Dials will convince him that the breeds of pigeons are quite as
extraordinary and unlike one another and their parent stock, while
the Horticultural Society will provide him with any number of corresponding vegetable aberrations from nature's types. He will learn
with no little surprise, too, in the course of his travels, that the pro-
prietors and producers of these animal and vegetable anomalies regard them as distinct species, with a firm belief, the strength of which
is exactly proportioned to their ignorance of scientific biology, and
which is the more remarkable as they are all proud of their skill in
originating such "species."

On careful inquiry it is found that all these, and the many other artificial breeds or races of animals and plants, have been produced by one method. The breeder—and a skilful one must be a person of much sagacity and natural or acquired perceptive faculty—notes some slight difference, arising he knows not how, in some individuals of his stock. If he wish to perpetuate the difference, to form a breed with the peculiarity in question strongly marked, he selects such male and female individuals as exhibit the desired character, and breeds from them. Their offspring are then carefully examined, and those which exhibit the peculiarity the most distinctly are selected for breeding; and this operation is repeated until the desired amount of divergence from the primitive stock is reached. It is then found that by continuing the process of selection—always breeding, that is, from well-marked forms, and allowing no impure crosses to interfere— a race may be formed, the tendency of which to reproduce itself is exceedingly strong; nor is the limit to the amount of divergence which may be thus produced known; but one thing is certain, that, if certain breeds of dogs, or of pigeons, or of horses, were known only in a fossil state, no naturalist would hesitate in regarding them as distinct species.

But in all these cases we have human interference. Without the breeder there would be no selection, and without the selection no race. Before admitting the possibility of natural species having originated in any similar way, it must be proved that there is in nature some power which takes the place of man, and performs a selection *suâ sponte*. It is the claim of Mr. Darwin that he professes to have discovered the existence and the *modus operandi* of this "natural selection," as he terms it; and, if he be right, the process is perfectly simple and comprehensible, and irresistibly deducible from very familiar but well nigh forgotten facts.

Who, for instance, has duly reflected upon all the consequences of the marvellous struggle for existence which is daily and hourly going on among living beings? Not only does every animal live at the expense of some other animal or plant, but the very plants are at war. The ground is full of seeds that cannot rise into seedlings; the seedlings rob one another of air, light and water, the strongest robber winning the day, and extinguishing his competitors. Year after year, the wild animals with which man never interferes are, on the average, neither more nor less numerous than they were; and yet we know that the annual produce of every pair is from one to perhaps a million young; so that it is mathematically certain that, on the average, as many are killed by natural causes as are born every year, and those only escape which happen to be a little better fitted to resist destruction than those which die.[14] The individuals of a species are like the crew of a foundered ship, and none but good swimmers have a chance of reaching the land.

Such being unquestionably the necessary conditions under which living creatures exist, Mr. Darwin discovers in them the instrument of natural selection. Suppose that in the midst of this incessant competition some individuals of a species (A) present accidental variations which happen to fit them a little better than their fellows for the struggle in which they are engaged, then the chances are in favour, not only of these individuals being better nourished than the others, but of their predominating over their fellows in other ways, and of having a better chance of leaving offspring, which will of course tend to reproduce the peculiarities of their parents. Their offspring will, by a parity of reasoning, tend to predominate over their contemporaries, and there being (suppose) no room for more than one species such as A, the weaker variety will eventually be destroyed by the new destructive influence which is thrown into the scale, and the stronger will take its place. Surrounding conditions remaining unchanged, the new variety (which we may call B)—supposed, for argument's sake, to be the best adapted for these conditions which can be got out of the original stock—will remain unchanged, all accidental deviations from the type becoming at once extinguished, as less fit for their post than B itself. The tendency of B to persist will grow with its persistence through successive generations, and it will acquire all the characters of a new species.

But, on the other hand, if the conditions of life change in any degree, however slight, B may no longer be that form which is best adapted to withstand their destructive, and profit by their sustaining, influence; in which case if it should give rise to a more competent variety (C), this will take its place and become a new species; and

---

[14] *Compare this paragraph with the argument given in Wallace's essay.*

thus, by natural selection, the species B and C will be successively derived from A.

That this most ingenious hypothesis enables us to give a reason for many apparent anomalies in the distribution of living beings in time and space, and that it is not contradicted by the main phenomena of life and organisation appear to us to be unquestionable; and, so far, it must be admitted to have an immense advantage over any of its predecessors. But it is quite another matter to affirm absolutely either the truth or falsehood of Mr. Darwin's views at the present stage of the inquiry. Goethe has an excellent aphorism defining that state of mind which he calls "Thätige Skepsis"—active doubt. It is doubt which so loves truth that it neither dares rest in doubting, nor extinguish itself by unjustified belief; and we commend this state of mind to students of species, with respect to Mr. Darwin's or any other hypothesis, as to their origin. The combined investigations of another twenty years may, perhaps, enable naturalists to say whether the modifying causes and the selective power, which Mr. Darwin has satisfactorily shown to exist in Nature, are competent to produce all the effects he ascribes to them; or whether, on the other hand, he has been led to over-estimate the value of the principle of natural selection, as greatly as Lamarck overestimated his *vera causa* of modification by exercise.

But there is, at all events, one advantage possessed by the more recent writer over his predecessor. Mr. Darwin abhors mere speculation as nature abhors a vacuum. He is as greedy of cases and precedents as any constitutional lawyer, and all the principles he lays down are capable of being brought to the test of observation and experiment. The path he bids us follow professes to be, not a mere airy trek, fabricated of ideal cobwebs, but a solid and broad bridge of facts. If it be so, it will carry us safely over many a chasm in our knowledge, and lead us to a region free from the snares of those fascinating but barren virgins, the Final Causes, against whom a high authority has so justly warned us. "My sons, dig in the vineyard," were the last words of the old man in the fable: and, though the sons found no treasure, they made their fortunes by the grapes.

WILLIAM BATESON

Galton's Genetics, from
Mendel's Principles of Heredity (1909)

William Bateson (1861–1926) is said to be the man who coined the term *genetics*. In 1894 he published a theory that evolution proceeded in spurts rather than in the slow-but-steady way that Darwin asserted. This view is now most closely associated with Stephen Jay Gould and Niles Eldridge. To make his case stronger, Bateson began doing breeding experiments in order to understand how change could be rapid and abrupt. In 1900, while investigating the early literature on the subject, Bateson found Gregor Mendel's paper "Experiment in Plant Hybrids." Mendel by then was dead. He had been an Austrian monk and had published his paper 34 years earlier. Nobody seems to have paid any attention to it. But Bateson immediately accepted Mendel's work, translated his paper into English, and wrote a book entitled *Mendel's Principles of Heredity* (1902).

The selection here is from the second edition of Bateson's book, which appeared after the new heredity had swept most of the field before it. In it, Bateson discusses the most rigorous statement of the traditional view that heredity is a family affair, carrying the traits of all one's ancestors, and thus making rapid evolution much more difficult. Shortly before the discovery of Mendel's work, Francis Galton (1822–1911) published a mathematical description of how offspring inherit the traits of their families. Although Galton's equation sounded learned and rational, it just did not work. Bateson's respectful account of Galton's law is designed to get people to say that they had once agreed with that theory but had changed their minds. I especially like the part where Bateson says that Galton's work "deserves to be long remembered" even as he sweeps it out the door toward the dustbin.

The supreme importance of an exact knowledge of heredity was urged by Galton. [. . .] He pointed out that the phenomena manifested regularity, and he made the first comprehensive attempt to determine the rules they obey. It was through his work and influence that the existence of some order pervading the facts became generally recognized. In 1897 he definitely enunciated his now famous "Law" of heredity, which declared that to the total heritage of the offspring the parents on an average contribute ½, the grandparents ¼, and the great-grandparents ⅛, and so on, the total heritage being taken as unity.[15] To this conclusion he had been led by several series of data, but the evidence upon which he especially relied was that of the pedigrees of Basset Hounds furnished him by the late Sir Everett Millais. In that instance the character considered was the presence or absence of *black* in addition to yellow and white. The colours were spoken of as tri-colour and non-tri-colour, and the truth of the law was tested by the average numbers of the respective colours which resulted from the various matings of dogs of known ancestral composition. These numbers corresponded so well with the expectations given by the law as to leave no reasonable doubt that the results of calculation were in general harmony with natural fact. [. . .]

Though there was admittedly a statistical accord between Galton's theory and some facts of heredity, yet no one familiar with breeding or even with the literature of breeding could possibly accept that theory as a literal or adequate presentation of the facts. Galton himself in promulgating it made some reservations; but in the practice of breeding, so many classes of unconformable phenomena were already known, that while recognizing the value of his achievement, we could not from the first regard it as more than an adumbration of the truth. As we now know, Galton's method failed for want of analysis. His formula should in all probability be looked upon rather as an occasional consequence of the actual laws of heredity than in any proper sense one of those laws. [. . .]

><+·+·O·<+·+·<

The colours of Bassets are two, the first spoken of as *tricolour,* consisting of black and yellow marks on a white ground; the second, *non-tricolour,* which differs from the first in having no black. It is said that dogs which cannot be easily referred to one or other of these two types do not occur, and they must certainly be very rare if they exist at all. [. . .]

---

[15] *Galton's math has subtle differences from the modern equation, but the key difference was in the presence of ancestors. In Mendelian genetics, everything comes from the two parents. In Galton's equation, the grandparents have a direct effect on the grandchildren, quite independent of the parents' contribution.*

This evidence consisted in records giving the number of offspring of each type which had occurred in families of various compositions. It was thus possible to compare the number of tricolour and non-tricolour dogs produced in the families with the number of the respective types distributed among their pedigrees. Galton's figures indicated that there was a close correspondence between these two numbers, so that it was possible, given the ancestral composition of the families, to predict with considerable accuracy the numerical proportions in which the respective types would appear. According to Galton's system the family was regarded as the production of all the ancestors. Each ancestor was supposed to contribute in his or her degree to this total heritage, the more immediate progenitors contributing more, and the remoter progenitors less, according to a definite arithmetical rule. This rule was that the average contribution of each ancestor was to be reckoned

for each parent	$\frac{1}{4}$
for each grandparent	$\frac{1}{16}$
for each great-grandparent	$\frac{1}{64}$

and so on, the total heritage being thus reckoned as unity. It will be observed that this scheme differs entirely from those based on Mendelian principles, inasmuch as every ancestor is, according to the Law of Ancestral Heredity, supposed to have some effect on the composition of each family in its posterity, and each recent progenitor is regarded as having a very sensible influence on these numbers.

Though no one with a knowledge of practical breeding could entertain the supposition that Galton's Law had the universality of application claimed for it, there was on the other hand no doubt that the Law had successfully expressed a variety of facts in which no order at all had been previously detected.

We have now to consider the meaning of this evidence in the light of modern knowledge. At the time that Galton's views were promulgated nothing was known of segregation.[16] The supposition that any individual, whatever its own characters, was capable of carrying on and transmitting to its posterity any of the characters exhibited by its

---

[16] *Mendel's first principle is the law of segregation: each person has two "alleles" per characteristic but contributes only one of those allelles to a sex cell. In this sense, segregation refers simply to separation. Mendel said that during reproduction, the alleles separate.*

immediate progenitors, at all events, was generally received without question by biologists.[17] According to that idea the number of classes of individuals differing in respect of their ancestral composition and transmitting powers is to be regarded as indefinitely large, whereas in all cases of sensible allelomorphism the number of classes of individuals is three only, two being homozygous and one heterozygous.[18] The difference between the two schemes is thus absolute and irreconcilable.

When Mendelian phenomena were first recognized it was naturally supposed that some classes of cases would be found to conform to the Mendelian scheme and others to the Law of Ancestral Heredity. With the progress of research however almost all the cases to which precise analytical methods have been applied have proved to be reducible to terms of Mendelian segregation; and of those which have not already been so elucidated some, we may feel confident, if not all, will be eventually shown to be governed by similar rules. In discussing aberrant phenomena like those alleged in regard to the Bassets the first question to be settled is whether the facts are correctly reported. If either type is recessive[19] we should naturally expect this to be the non-tricolour, which is without black. Unfortunately as the non-tricolours are not fashionable there were comparatively few matings between two parents of that colour. Nevertheless 41 dogs, offspring of such matings, are given, of these 20 being tricolour. Though the records were not made by scientific men or with a scientific purpose directly in view it is almost impossible to imagine that all these cases can depend on mistakes, and pending the production of new and direct evidence we must take the records as correct.

In the *Theory and Practice of Rational Breeding* [. . .] Sir Everett Millais gives one or two more notes bearing on this question. He says

---

[17] *The old theory held that having, say, blue-eyed ancestors meant that you could have blue-eyed descendants. But the Mendelian theory says that you can have blue-eyed descendants only if you have inherited an allele for blue eyes.*

[18] *A homozygous individual has two matching allelles for a characteristic, and a heterozygous person has two different alleles.*

[19] *Recessive qualities are expressed only when a person inherits a recessive allele from each parent. Mendel distinguished between dominant traits, which are expressed when even one allele is present, and recessive traits, which require two alleles before they appear in a person.*

that in England there were then two strains of Bassets, the Couteulx and the Lane. "The Couteulx is as a rule a very perfectly marked tricolour, with the tan and black markings deeply accentuated. The Lane hounds, on the other hand, are very weak in markings if they happen to be tricolour, but as a fact they are far more generally found to be lemon and white." In another place he mentions that "in nearly every litter of pure Couteulx there is generally a lemon and white puppy."

If it were not that the genetic relations of yellow and black pigments are, as we have seen, so complicated and uncertain in other types, we might be inclined to attribute the alleged production of tricolours by non-tricolours to imperfect classification of "weak" tricolours, but in dealing with this group of phenomena no such suggestion can be hazarded with any confidence. [. . .]

At present the Basset phenomena must be regarded as definitely unconformable. Perhaps the most probable view of their nature is that they are an illustration of irregular dominance, but this cannot be asserted with much confidence.

It is curious that the one example to which a partially correct system of analysis was applied before Mendelian methods were rediscovered, should have been of this remarkably exceptional order.

There is little reason to anticipate, as we once did, that a distinct group of cases obeying a Law of Ancestral Heredity will have to be recognized. That principle in certain cases gives an epitome of the consequences of the Mendelian process, and in all likelihood its applicability to any phenomena of natural inheritance is due to this fact. The Law of Ancestral Heredity takes of course no account of dominance, or of segregation with all the consequences it entails; but as describing the results to be witnessed among a population interbreeding at random, its predictions would frequently approximate to the truth. [. . .]

The suggestion that methods based on unanalysed statistics have scientific value in the study of heredity can scarcely mislead those who have examined the facts. Professor Pearson and others committed to these methods have of late defended their position by arguing that there is no fundamental incompatibility between Laws of Ancestral Heredity and the conclusions of Mendelian analysis. The matter would not be worth notice were it not that the same proposition is being freely repeated by several writers seeking some convenient shelter of neutrality. It is to be observed however that the supposition of an underlying harmony between Mendelian and biometrical results was not put forward by the biometricians until every possible means of discrediting the truth of Mendelian facts had been exhausted. Those attacks having failed, we are asked to observe that the Law of Ancestral Heredity was meant as a statement of a statistical consequence, and is not concerned with physiological processes. Mr. Galton's views on this point are well shown in the following passage in which he explicitly appeals to the physiological process of gametogenesis[20] as apparently occurring in the way which his Law requires. For in introducing the Law as applicable to Bassets he wrote:

> It should be noted that nothing in this statistical law contradicts the generally accepted view that the chief, if not the sole, line of descent runs from germ to germ and not from person to person. The person may be accepted on the whole as a fair representative of the germ, and being so, the statistical laws which apply to the persons would apply to the germs also, though with less precision in individual cases. Now this law is strictly consonant with the observed binary subdivisions of the germ cells, and the concomitant extrusion and loss of one-half of the several contributions from each of the two parents to the germ-cell of the offspring. The apparent artificiality of the law ceases on these grounds to afford cause for doubt; its close agreement with physiological phenomena ought to give a prejudice in favour of its truth rather than the contrary.

Had segregation been known to Mr. Galton, the Law of Ancestral Heredity would not have been promulgated. It is obvious that so soon as that phenomenon is recognized and appreciated, all question of useful or direct applicability of the Law of Ancestral Heredity is at an end. That method of representing the phenomena of heredity and all modifications of it are based on the false assumption that any individual can transmit the characteristics of any ancestor, and especially of any recent ancestor. When this conception was shown to be un-

---

[20] *Gametogenesis is the formation of reproductive cells.*

true, the structure which the biometricians have offered to the world as a scientific study of heredity ceased to have meaning or value. Statistical examination of ancestral composition may, as we have seen, occasionally give a prediction in good correspondence with fact, but this is due to coincidence and not to any elements of truth in the ratiocination by which the prediction was reached.

As an attempt to compass the solution of an intricate problem by labour and ingenuity without proper data or equipment Mr. Galton's work deserves long to be remembered. It stands out as a significant and stimulating event in the history of biology.

EDMUND BLAIR BOLLES

Gestalt Psychology, from
A Second Way of Knowing (1991)

More than any other group of psychologists, the "gestalt" school developed techniques for studying the relation between the mind and sensory experiences. How, gestaltists wondered, do the body's individual sensations become unified, meaningful experiences? How, for example, does a mix of red, yellow, and green become the sight of fruits in a bowl? This is an old philosophical question, one asked by Aristotle. More recently, our inability to explain this seemingly simple recognition has proved to be an important barrier to giving computers common sense. Computers may be great at manipulating symbols, but common sense uses perception rather than logic. How does perception work?

The gestalt psychologists approached the question experimentally and, long before the age of the computer, established that perception was no mere metaphysical conundrum. As scientists, the gestalt psychologists recognized the importance of their question to understanding how scientific knowledge itself is possible and how scientific imaginations work. After all, scientists use sensory experiences to move from ignorant curiosity to general understanding. Gestalt psychology understood that Galileo's feat was not simply being the first on his block to point a telescope skyward. Somehow he had recognized the meaning of the data that he had collected. How did such a transformation of data into meaning happen? How did Marie Curie move from noticing an oddity in her measurements to realizing that an unknown element was mixed in with her material? How could Alfred Wegener look at a globe, as many others had done before him, and notice that Africa and South America fit too neatly together?

That question was at the heart of gestalt research, just as it is the organizing theme of this anthology. I have included this selection on gestalt psychology—which I wrote—because it unites the story of a science with the issue of the scientific imagination. Edmund Blair Bolles (1942–)—aka "me"—is not a gestaltist himself/myself, and this piece would not appear in any collection of great science writing that was organized differently. But this anthology traces the trail from sensory wonder to articulate expression, and this essay is about that change.

The nature of sensory input was debated a century ago and was finally resolved in 1910 by a set of experiments that were as tidy as anything Galileo ever did. Max Wertheimer, a psychologist from Prague, proved that perceptions include effects not reducible to the sensory input. Psychology was, in the main, offended by that rejection of common sense and turned its attention elsewhere. It is no accident that the decade following 1910 saw American psychology switch from the study of "mental life" to a behaviorism that ignored the mind altogether. In time, of course, people forgot that a battle over input's role had ever been lost and won.

Central to the old hurly-burly was an "atomic" theory of perception. Atomism traced to the ancient Greeks and had not changed much over the millennia. The tradition's prestige was especially high in the late nineteenth century because of its splendid success at making sense of chemistry. For psychology, atomism held that we see a cow because atoms travel from the cow to stimulate the eye. We perceive *only* the bundles of atoms pounding against our various sense organs. In other words, the argument was nearly identical with the one made today by people who believe that computers, in principle, can perceive as well as humans. Only instead of talking about atoms, modern physicalists say energy carries information to the sense organs. [. . .]

Is Everything Out There?

In 1891 [. . .] an Austrian psychologist/philosopher named Christian von Ehrenfels published a paper titled *On Gestalt Qualities*. "Gestalt" is a German word that does not translate easily into English. Ehrenfels used it to suggest those things in a perception that are not intrinsic to the physical stimulation. Centuries earlier Galileo had said that physics studies a thing's "primary" qualities, its measurable features. From then on physics limited itself to considering objects in terms of their weights, sizes, and times. Gestalt qualities are the "secondary" qualities of an experience that Galileo kicked out of the realm of physics. A melody is an example of a gestalt quality. We hear a melody without thinking about it, but where in the sound's measurable dimensions does the melody lie?

Ehrenfels did some straightforward experiments to show that the melody does not come from the primary qualities of the stimulation.

A melody or tune is hard to define, but people know one when they hear it. The atomist theory held that the separate tones in music produces a melody. Ehrenfels disproved that idea by showing that people could hear a melody in one key and then recognize it when played again in a different key. Although this finding can hardly surprise anyone familiar with music, the atomists could not readily explain it. Ehrenfels said there had to be another quality out there in the atom, but such an explanation points toward Pandora's box. If psychologists start restoring to physics all those qualities that Galileo banished, the age of science will pass into history. We will return to the old days when vague qualitative description was the best people could do.

Some psychologists tried to get around the von Ehrenfels experiment by saying that the melody lay in the relation between the musical sensations, the particular notes did not matter. Further experiments showed that changing the relationships did not always obscure the melody. Even with the input distorted listeners could recognize it.

One possible explanation is that the melody is not objectively out there in the music. Melody might be a subjective invention by the listener, something that reflects subjective facts. Von Ehrenfels did not offer that answer because it breaks with the physicist's model of truth. In physics, no effect occurs without an external cause. The environment shapes everything. If we start saying that minds construct things not found in the physical environment, we break with the physicist's model and say it does not apply to people. Von Ehrenfels had no such radical ambitions, and did not suggest that the melody sings only in the listening human.

Some Things Are Not Out There

It took almost twenty years for a psychologist to suggest that we can perceive things not found in the atoms. Hindsight makes the suggestion seem inevitable. The late nineteenth century saw the beginning of the mass production of illusional entertainment. In particular, cinematic illusion had begun to create a great stir. Even today movies stand as the most familiar example of a perception of something absent from the input. A movie is merely a rapid display of still photographs; the motion we see is not out there on the silver screen. It is a performance in our heads. The routine physicalist "explanation" for such apparent motions relies on a phenomenon known as after-image, an effect in which people continue to see a light after it shuts off. Possibly, afterimages explain why we do not notice the flicker of the projection, but it does nothing to explain how we see still images move. Why doesn't afterimage just turn all movie scenes into a blurry mess?

In the summer of 1910 the psychologist Max Wertheimer boarded a train bound for the Rhine valley and his summer vacation. As he went, he began thinking about the problem of apparent motion. How could you get motion out of those still images? The natural way for any scientist to think about a problem was to break it into parts. As Wertheimer later said, "Isolate the elements, discover their laws, then reassemble them and the problem is solved." As he rode the train, Wertheimer suddenly realized that this analytical approach was hopeless. Apparent motion is not an effect of the parts. It comes from the whole. Wertheimer eventually expressed this insight in the principle that became the keystone of Wertheimer's gestalt theory: The whole is different from the sum of its parts. (Please note the word *different*. Gestalt psychologists are often accused of teaching the whole is *greater* than the sum of its parts. Gestalt psychology did not challenge mathematics. It said that you could not find the nature of the whole merely by adding up the parts. [. . .])

Wertheimer was so taken with his idea that he hopped off his train and checked into a hotel to begin experimenting. He bought a flip book, one of these small tablets with pictures that appear to move when you flip rapidly through its pages, and he began playing with it. Soon he developed a series of experiments to study apparent motion.

The most important of Wertheimer's experiments sat a person in a dark room where he stared into black space. Unbeknownst to the observer, two lights stood in front of him. One was on the left side of the room, the other to the right. Suddenly the left light flashed on and then off again. One-twentieth of a second later the light at the right flashed on and off.

The observers reported that a single light came on at the left and then moved quickly to the right before going off. This perceived motion did not occur out there in the room. There were no atoms that could account for the effect. The motion was a construct of the observer. Wertheimer conducted many other experiments to measure the role of time and space in the effect. He also showed people a light that really did come on, move, and go off. The observers saw no difference between this real motion and the apparent motion.

Anyone who has ever watched a movie knows that these effects are not intellectual deductions, nor are they like some half-remembered incident in which we recall details that "must have been" a particular way. Apparent motion looks and feels just like real motion. People walking across a street in a movie look just like people walking across a street in real life. Yet the motion is not out there in the input; it is a performance going on in the observer's head.

Gestalt Psychology

Wertheimer and two participants in his motion experiments—Wolfgang Kohler and Kurt Koffka—went on to develop the school known as Gestalt psychology. *Gestalt* here is usually translated as a "whole" and refers to the "wholes" that people perceive, think about, and act upon. Koffka used the term "living whole" for groups that people consider meaningful. A living whole is something people perceive as a unit.

In Wertheimer's light experiment, for example, subjects see the individual flashing lights as part of one unified experience. They do not perceive the components of the experience, but give a meaning to the whole input. Much of Gestalt research went into trying to formulate rules that would predict what things people will perceive as wholes. For example, if a person looks at these lines,

Gestalt psychology predicts that you will see narrow pole-like units separated by larger spaces. Some readers may be taken aback by

that "prediction." Surely that is what is really out there! With a little conscious effort you may be able to see the lines as making wide tree-like units separated by narrow spaces. Is that also out there? Are two different images in the same place at the same time? Other units too might be imagined; however, it is usually difficult to see the lines as meaningless individuals, each its own unit, separate and distinct from all other lines. Yet, speaking atomistically, this last description states exactly what is on the page. The only stimulus units are the lines. The larger units are imagined by the observer. We can make the lines appear as individual items if we even out the spaces between them.

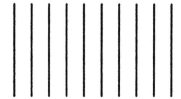

Meaningful organization suddenly seems to disappear.

Readers today have a clearer sense of what is peculiarly alive about seeing these wholes. We have become familiar with the universal product code found on nearly all packaging. These lines look like the sort of things Gestalt psychologists studied,

but meaning never quite seems to jump out at the observer. I often feel that if I just studied these codes a bit more carefully, I could make sense of them. Yet I never do.

In truth, the universal product code does not work the way we perceive things. It has been designed to be read atomistically. Standardized markings signal the beginning and end of input. One at a time, a laser scans the lines and spaces between them. Contrary to our own perceptual instincts, the light spaces carry as much information as the dark lines. Any patterns that we see by looking at the whole code is invisible to the computer as it reads unit by unit.

Wertheimer used a famous metaphor in describing our perception of living wholes. In a 1924 speech before a philosophical society in Berlin he said:

> Suppose the world were a vast plateau upon which were many musicians. I walk about listening and watching the players.

This idea is not as bizarre as it may seem. One of the oldest metaphors in science called nature's ordered beauty the "harmony of the spheres." In his unassuming way, Wertheimer was going for science's jugular by wondering just how we are to know this harmony when we meet it:

> First suppose that the world is a meaningless plurality. Everyone does as he will, each for himself. What happens together when I hear ten players might be the basis for my guessing as to what they are all doing, but this is merely a matter of chance and probability, much as in the kinetics of gas molecules.

[. . .] Of course, this probabilistic approach to experience is not the only kind imaginable. Wertheimer went on:

> A second possibility would be that each time one musician played *c*, another played *f* so and so many seconds later. I work out a theory of blind coupling, but the playing as a whole remains meaningless. This is what many people think physics does, but the real work of physics belies this.

[. . .] Powerful as it is, this analysis still does not find meaning in the pattern. Wertheimer went on:

> The third possibility is [that the musicians play], say, a Beethoven symphony when it would be possible to select one part of the whole and work from that toward an idea of the structural principle motivating and determining the whole. Here the fundamental laws are not those of fortuitous pieces, but concern the very character of the event.

Gestalt psychologists showed in experiment after experiment that people do perceive meanings—motion in lights, order in lines, etc.— that get at the very character of the events. Yet such meaning is not included in the sensory input.

Meaning

Meaning at the perceptual level is such a crude idea it almost sounds like a joke. We are used to looking for profound ideas in meaning. What did Shakespeare mean when he said *blah blah blah*? No. Meaning in Gestalt psychology is more basic. What does it mean when you see a large metallic object with a cannon on the front aimed straight at you and you hear somebody say, "Fire." It means there is a tank in front of you. That is the level of meaning we are talking about here. It is the group of qualities that pulls associations together into a concrete unit. [. . .]

How do we find the meaning? The bulk of Gestalt research probed that question. For example, it asked how we distinguish the meaningful from the rest. We can understand the problem by remembering what it looks like to watch a TV channel without a broadcast. The screen is a series of thousands of dots that flicker randomly from black to white. The sound is a rushing racket. We can see and hear that the dots and sounds keep changing, but their meaninglessness remains constant. The experience of watching such a channel quickly becomes irritating and boring. There is nothing stable to grab our attention. The channel is all noise and no signal, or, to use Gestalt terminology, all ground and no figure.

In Gestalt psychology, *figures* are the things we attend to; *ground* is the array of sensations we ignore. At a cocktail party we can converse with one person. That person's speech is the figure we pay attention to. The roar of the other speakers in the room fades into background and loses its identity. At the same party we look around the room and recognize the face of an old friend. The face suddenly becomes a figure. Perception begins with the sorting of figure from ground. A further step, like recognizing a face at a party, discovers individuality. Gestalt research did little on this issue of the individuality of figures. It focused chiefly on finding identity.

Keep this point clear: Central to discovering an experience's perceptual meaning is the recognition of its *identity* and of its *individuality*.

Perceptual Meaning

An amazing quality of perceptual figures is the way they pop out at us. We first see them whole. We do not need the artificial pauses that computer programmers like to insert between units of input. Publishers do insert spaces between printed words, but speakers seldom bother. An entire book printed without word breaks would be fatiguing, but meaningmakesitsownspaces. Gestalt psychology teaches that perceiving the distinction between figure and ground depends on choice. We select the things that make up a figure and mentally separate it from the ground. Gestalt psychology also teaches that perceiving the distinction between figure and ground depends on automatic choices. We automatically select the things that make up a figure and sort it from the ground.

Gestalt ideas have proven valuable in understanding issues of perception in many real-world contexts. Artists can deliberately use these principles to help viewers see and respond to their images. Critics find that Gestalt principles can help explain effects created by different works of art.

Natural phenomena too have proven understandable in Gestalt terms. My own favorite is the insight they offer into the nature of camouflage. We are all aware of the use of camouflage to fade into the background. Chameleons, lions, and gazelles all use that technique to hide their identity, but Gestalt ideas also explain the camouflage that makes animals easy to see. These work by obscuring individuality.

Zebras are notable examples of animals whose markings keep them from blending into the landscape even at three hundred yards. Indeed, zebra markings provide a near-perfect illustration of the various Gestalt principles that create functional wholes. The *law of similarity* says we will group similar things into wholes. When we perceive black and white stripes we automatically distinguish them from the stripeless background. The *law of good continuation* says we see straight or smoothly curving lines as belonging together. Thus, even if a rock or bush partially obscures the view of a zebra, the grace of their stripes still makes it easy to see the animal. The *law of proximity* says we will group things that are near each other into a figure. The narrowness of the stripes helps us see the zebra as a whole. Lastly, Gestalt psychologists speak of a *law of common fate*. It says we group things that move in the same direction. When a zebra runs, all its stripes move in the same direction and its whole body remains clearly visible. The effect on the observer is very strong. The zebra stands cleanly apart from the plain and sky. In short, evolution has created a heavily-hunted animal whose appearance makes it absurdly easy for predators to see. Everyone wonders why.

A biologist on the Serengeti plains has pointed out a peculiarity of zebras—their calves tag along beside them. Most animal young stay put and use camouflage to blend into the background. From birth zebras trot with the herd and have to blend in with the figures all about them. Although it is easy to distinguish a zebra's upper body—head, rump, back—against the landscape, the lower parts run together in the herd's tangle of black and white contrasts. Young zebras disappear amid the markings of the larger adults. Zebra identity is clear, but zebra individuality is lost. Suddenly all those Gestalt rules for distinguishing figure from background turn into rules for blurring herd members together. Similarity, nearness, good continuation, and common fate each serve to make the herd look like a unit. Calves trot past the watchful hyena without being discovered.

Technicians can use Gestalt principles to create illusory experiences. The Walt Disney people have so developed the craft they have given it a name, imagineering. A typical display at Walt Disney World, Florida, will use a giant screen that shows a movie of some exciting vehicle, like a sled, speeding forward. The audience rides past these screens in little cars that slowly move sideways. The true motion of the observer, therefore, is from left to right, yet the perceived motion goes forward. The images on the movie screen give people the feeling that they are swooping down a mountain in a sled. Intellectually, of course, riders know they are not sliding down the Matterhorn, but even intellectually it is difficult to accept the truth that everyone is really moving sideways. Atomism is lost in the face of such contradictory perceptions, but Gestalt theory takes them in stride. It says we perceive whole meanings; we do not add up meaningless bits.

BERTRAND RUSSELL

What Einstein Did, from
The ABC of Relativity (1925)

Bertrand Russell (1872–1970) wrote the best popular book on the theory of relativity. He succeeded because he brought all of his strengths to his task. First, as a philosopher, he put relativity into context, showing what Einstein had solved and what he had changed. Russell gave his readers a sense of why Einstein was important. Second, Russell understood the mathematics supporting Einstein's theory. Russell had become an expert on non-Euclidean geometry even before the twentieth century was born and won a fellowship at Cambridge University on the basis of his analysis of geometry's foundations. Third, he could set vanity aside. Many people would find it impossible to resist showing off their mathematical knowledge, even at the price of confusing their readers, but Russell was careful to tell his readers only as much as they needed to know to understand why Einstein mattered. Finally, Russell appreciated his readers' needs. He knew that the greatest difficulty in understanding Einstein was not the intensity of his logic but the freedom of his imagination. Russell worked to loosen the imagination of his readers. The result was a classic account of Einstein's accomplishment and why it mattered.

Everybody knows that Einstein has done something astonishing, but very few people know exactly what it is that he has done. It is generally recognized that he has revolutionized our conception of the physical world, but his new conceptions are wrapped up in mathematical technicalities. It's true that there are innumerable popular accounts of the theory of relativity, but they generally cease to be intelligible just at the point where they begin to say something important. [. . .] What is demanded [to understand Einstein] is a change in our imaginative picture of the world—a picture which has been handed down from remote, perhaps pre-human, ancestors, and has been learned by each one of us in early childhood. A change in our imagination is always difficult, especially when we are no longer young. The same sort of change was demanded by Copernicus, when he taught that the earth is not stationary and the heavens do not revolve about it once a day. To us now there is no difficulty in this idea, because we learned it before our mental habits had become fixed. Einstein's ideas, similarly, will seem easy to a generation which has grown up with them; but for our generation a certain effort of imaginative reconstruction is unavoidable. [. . .]

In studying the heavens, we are debarred from all senses except sight. We cannot touch the sun, or travel to it; we cannot walk round the moon, or apply a foot rule to the Pleiades. Nevertheless, astronomers have unhesitatingly applied the geometry and physics which they found serviceable on the surface of the earth, and which they had based upon touch and travel. In doing so, they brought down trouble on their heads, which it has been left for Einstein to clear up. [. . .]

An illustration may help us to understand how much is impossible to the astronomer as compared to the man who is interested in things on the surface of the earth. Let us suppose that a drug is administered to you which makes you temporarily unconscious, and that when you wake you have lost your memory but not your reasoning powers. Let us suppose further that while you were unconscious you were carried into a balloon, which, when you come to, is sailing with the wind in a dark night—the night of the fifth of November if you are in England, or of the fourth of July if you are in America. You can see fireworks which are being sent off from the ground, from trains, and from aeroplanes traveling in all directions, but you cannot see the ground or the trains or the aeroplanes because of the darkness. What sort of picture of the world will you form? You will think that nothing is permanent: there are only brief flashes of light, which, during their short existence, travel through the void in the most various and bizarre curves. You cannot touch these flashes of light, you can only see them. Obviously your geometry and your physics and your metaphysics will be quite different from those of ordinary mortals. If an ordinary mortal is with you in the balloon, you will find his speech unintelligible. But if Einstein is with you, you will understand him more easily than the ordinary mortal would, because you will be free from a host of preconceptions which prevent most people from understanding him.

The theory of relativity depends, to a considerable extent, upon getting rid of notions which are useful in ordinary life but not to our drugged balloonist. Circumstances on the surface of the earth, for various more or less accidental reasons, suggest conceptions which turn out to be inaccurate, although they have come to seem like necessities of thought. The most important of these circumstances is that most objects on the earth's surface are fairly persistent and nearly stationary from a terrestrial point of view. If this were not the case, the idea of going on a journey would not seem so definite as it does. If you want to travel from King's Cross to Edinburgh, you know that you will find King's Cross where it always has been, that the railway line will take the course that it did when you last made the journey, and that Waverley Station in Edinburgh will not have walked up to the Castle. You therefore say and think that you have traveled to Edinburgh, not that Edinburgh has traveled to you, though the latter statement would be just as accurate. The success of this common-sense point of view depends on a number of things which are really of the nature of luck. Suppose all the houses in London were perpetually moving about, like a swarm of bees; suppose railways moved and changed their shapes like avalanches; and finally suppose that material objects were perpetually being formed and dissolved like clouds. There is nothing impossible in these suppositions: something like them must have been verified when the earth was hotter than it is now. But obviously what we call a journey to Edinburgh would have no meaning in such a world. You would begin, no doubt, by asking the taxi-driver: "Where is King's Cross this morning?" [. . .]

The effects in regard to time are even more strange. This matter has been explained with almost ideal lucidity by Eddington in *Space, Time and Gravitation*. He supposes an aviator traveling, relatively to the earth, at a speed of 161,000 miles a second, and he says:

"If we observed the aviator carefully we should infer that he was unusually slow in his movements; and events in the conveyance moving with him would be similarly retarded—as though time had forgotten to go on. His cigar lasts twice as long as ours. I said 'infer' deliberately; we should *see* a still more extravagant slowing down of time; but that is easily explained, because the aviator is rapidly increasing his distance from us and the light impressions take longer and longer to reach us. The more moderate retardation referred to remains after we have allowed for the time of transmission of light.

But here again reciprocity comes in, because in the aviator's opinion it is we who are traveling at 161,000 miles a second past him; and when he has made all allowances, he finds that it is we who are sluggish. Our cigar lasts twice as long as his."

What a situation for envy! Each man thinks that the other's cigar lasts twice as long as his own. It may, however, be some consolation to reflect that the other man's visits to the dentist also last twice as long. [. . .]

[Thus,] on logical grounds, [. . .] Newton's law of gravitation cannot be quite right.

Newton said that between any two particles of matter there is a force which is proportional to the product of their masses and inversely proportional to the square of their distance. That is to say, [. . .] if there is a certain attraction when the particles are a mile apart, there will be a quarter as much attraction when they are two miles apart, a ninth as much when they are three miles apart, and so on: the attraction diminishes much faster than the distance increases. Now, of course, Newton, when he spoke of the distance, meant the distance at a given time: He thought there could be no ambiguity about time. But [. . .] this was a mistake. What one observer judges to be the same moment on the earth and the sun, another will judge to be two different moments. "Distance at a given moment" is therefore a subjective conception, which can hardly enter into a cosmic law. Of course, we could make our law unambiguous by saying that we are going to estimate times as they are estimated by Greenwich Observatory. But we can hardly believe that the accidental circumstances of the earth deserve to be taken so seriously. And the estimate of distance, also, will vary for different observers. We cannot, therefore, allow that Newton's form of the law of gravitation can be quite correct, since it will give different results according to which of many equally legitimate conceptions we adopt. This is as absurd as it would be if the question whether one man had murdered another were to depend upon whether they were described by their Christian names or their surnames. It is obvious that physical laws must be the same whether distances are measured in miles or in kilometers, and we are concerned with what is essentially only an extension of the same principle. [. . .] Newton *must* be wrong and that something [. . .] *must* be substituted.

The arguments in favor of the relativity of motion are [. . .] quite conclusive. In daily life, when we say that something moves, we mean that it moves relatively to the earth. In dealing with the motions of the planets, we consider them as moving relatively to the sun, or to the center of mass of the solar system. When we say that the solar system itself is moving, we mean that it is moving relatively to the stars. There is no physical occurrence which can be called "absolute motion." Consequently the laws of physics must be concerned with relative motions, since these are the only kind that occur. [. . .]

Newton says that the force of gravitation between two bodies is proportional to the product of their masses [. . .] [and] that in a given gravitational situation, all bodies behave exactly alike. As regards the surface of the earth, this was one of the first discoveries of Galileo. Aristotle thought that heavy bodies fall faster than light ones; Galileo showed that this is not the case, when the resistance of the air is eliminated. In a vacuum, a feather falls as fast as a lump of lead. As regards the planets, it was Newton who established the corresponding facts. At a given distance from the sun, a comet, which has a very small mass, experiences exactly the same acceleration towards the sun as a planet experiences at the same distance. Thus the way in which gravitation affects a body depends only upon where the body is, and in no degree upon the nature of the body. This suggests that the gravitational effect is a characteristic of the locality, which is what Einstein makes it. [. . .]

We have another indication as to what sort of thing the law of gravitation *must* be, if it is to be a characteristic of a neighborhood, as we have seen reason to suppose that it is. It must be expressed in some law which is unchanged when we adopt a different kind of co-ordinates. [. . .] We must not, to begin with, regard our co-ordinates as having any physical significance: they are merely systematic ways of naming different parts of space-time. Being conventional, they cannot enter into physical laws. That means to say that, if we have expressed a law correctly in terms of one set of co-ordinates, it must be expressed by the same formula in terms of another set of co-ordinates. Or, more exactly, it must be possible to find a formula which expresses the law, and which is unchanged however we change the co-ordinates. It is the business of the theory of tensors to deal with such formulae. And the theory of tensors shows that there is one formula which obviously suggests itself as being possibly the law of gravitation. When this possibility is examined, it is found to give the right results; it is here that the empirical confirmation comes in. But if Einstein's law had not been found to agree with experience, we could not have gone back to Newton's law. [. . .]

A poet might say that water runs down hill because it is attracted to the sea, but a physicist or an ordinary mortal would say that it moves as it does, at each point, because of the nature of the ground at that point, without regard to what lies ahead of it. Just as the sea does not cause the water to run toward it, so the sun does not cause the planets to move round it. The planets move round the sun because that is the easiest thing to do—in the technical sense of "least action." It is the easiest thing to do because of the nature of the region in which they are, not because of an influence emanating from the sun.

⊳─┼─◆─○─◆─┼─◁

J. ROBERT OPPENHEIMER

A Science in Change, from
Science and the Common Understanding (1953)

J. Robert Oppenheimer (1904–1967) will be forever remembered as the science administrator of the Manhattan Project, the construction of the first atomic bomb. Although his success made him an instant American hero, less than 10 years later, Oppenheimer lost his security clearance. The immediate cause was his opposition to developing a hydrogen bomb, but the larger reason was the atmosphere of fear and vague accusations of communist "sympathies" that characterized the late 1940s and early 1950s. During the cold war, for a nuclear physicist to lose his security clearance was tantamount to losing his license to do science.

In late 1953, just as his security standing was reaching its crisis, Oppenheimer gave a series of lectures on the BBC about twentieth-century physics. The program was greeted as a model of clarity, and the lectures were reprinted in the following year in book form. By then, however, Oppenheimer had been pronounced a security risk, and his reputation was too bound up in the political passions of the day for people to accept the book at face value. That was unfortunate, as this selection proves that Oppenheimer had an unusual ability to write clearly, even about something as puzzling as quantum physics. He does an excellent job of showing that the famous quantum paradoxes come out of observation and that the old-fashioned Newtonian view of matter will just not work at this level.

Our understanding of atomic physics, of what we call the quantum theory of atomic systems, had its origins at the turn of the century and its great synthesis and resolutions in the nineteen-twenties. It was a heroic time. It was not the doing of any one man; it involved the collaboration of scores of scientists from many different lands, though from first to last the deeply creative and subtle and critical spirit of Niels Bohr guided, restrained, deepened, and finally transmuted the enterprise. It was a period of patient work in the laboratory, of crucial experiments and daring action, of many false starts and many untenable conjectures. It was a time of earnest correspondence and hurried conferences, of debate, criticism, and brilliant mathematical improvisation. [. . .]

Most of us are convinced that today, in our present probings in the sub-atomic and sub-nuclear world, we are laying the groundwork for another such time for us and for our sons. The great growth of physics, the vast and increasingly complicated laboratories of the mid-twentieth century, the increasing sophistication of mathematical analysis, have altered many of the conditions of this new period of crisis. We do not think that they will have altered its heroic and creative character.

When quantum theory was first taught in the universities and institutes, it was taught by those who had participated, or had been engaged spectators, in its discovery. Some of the excitement and wonder of the discoverer was in their teaching; now, after two or three decades, it is taught not by the creators but by those who have learned from others who have learned from those creators. It is taught not as history, not as a great adventure in human understanding, but as a piece of knowledge, as a set of techniques, as a scientific discipline to be used by the student in understanding and exploring new phenomena in the vast work of the advance of science, or its application to invention and to practical ends. It has become not a subject of curiosity and an object of study but an instrument of the scientist to be taken for granted by him, to be used by him, to be taught to him as a mode of action, as we teach our children to spell and to add.

What we must attempt to do in these talks is wholly different. This is no school to learn the arts of atomic physics. Even those prior arts—the experimental tools, the mathematical powers, the theories, inventions, instruments, and techniques which defined the problems of atomic physics, which established the paradoxes, described the phenomena, and underlay the need for synthesis—are not known to us of our own experience. We must talk of our subject not as a community of specialized scientists but as men concerned with understanding, through analogy, description, and an act of confidence and trust, what other people have done and thought and found. So men listen to accounts of soldiers returning from a campaign of unparalleled hardship and heroism, or of explorers from the high Himalayas, or of tales of deep illness, or of a mystic's communion with his God. Such stories tell little of what the teller has to tell. They are the threads which bind us in community and make us more than separate men.

Here, then, we have our atoms. Their ingredients have been made manifest by Rutherford and his α-particles, as have the forces that act between the ingredients, and by probing with electrons and with light as well as with α-particles. There is the nucleus, with almost the whole of the atom's mass and almost none of its size, and with a charge which is measured by the atomic number, equal to the number of electrons that surround the nucleus in the normal atom. We have the simple laws of attraction and repulsion, familiar from the large-scale, everyday experience with electricity. Unlike charges attract and like repel; and the forces, like Newton's, decrease inversely with the square of their separation.

In Rutherford's day it seemed reasonable, as it no longer entirely does today in facing our modern physics, to subdivide the problem of atomic structure into three questions: what are the ingredients of the atom; what are the forces, and the laws of force, acting between these ingredients; how in response to those forces do the ingredients move? [. . .]

The atom, then, has a massive charged nucleus; the atom as a whole is neutral and 10,000 to 100,000 times as far across as its tiny nuclear core. The rest of the atom is composed of electrons and electric fields—electrons that are the universal ingredients of matter, the determinants of almost all its chemical properties and of most of its familiar physical properties as well. There will be as many electrons in the atom as the atomic number, the nuclear charge; this makes the atom as a whole neutral. There will be one electron in hydrogen and thirteen in aluminum and ninety-two in uranium. These are the ingredients; and the laws of force, complex only in the last refinements, are basically simple. The electron feels an attractive Coulomb force exerted by the nucleus, attractive since the electron and nucleus are oppositely charged, and once again falling off with distance in the same way as gravitational forces according to Newton's law. For hydrogen, this means a simple situation: two bodies with a force between them identical in structure with that which the sun exerts on the planets; two bodies small enough compared to the atom's size so that they almost never touch, and the properties of their contact can have little influence. The law of forces has been verified not only by probing with particles, by which it was originally discovered, but by probing with electrons themselves, in the first instances by the beta rays of naturally radioactive substances. For other atoms there is in addition the electrical repulsion between the several electrons, balancing to some extent the nuclear attraction. And there is, further, the well-known mathematical complication of describing quantitatively the behavior of a system with many particles.

But with hydrogen this should not be so. Here we have essentially a single light body moving in a simple and well-known force. The description of this system should be a perfect example of Newtonian dynamics, and should, in its refinements, be intelligible in terms of all that the nineteenth century had discovered about the behavior of charged particles in motion and the electromagnetic radiation produced when they are accelerated.

But it did not turn out that way. To what appeared to be the simplest questions, we will tend to give either no answer or an answer which will at first sight be reminiscent more of a strange catechism than of the straightforward affirmatives of physical science. If we ask, for instance, whether the position of the electron remains the same, we must say "no"; if we ask whether the electron's position changes with time, we must say "no"; if we ask whether the electron is at rest, we must say "no"; if we ask whether it is in motion, we must say "no." The Buddha has given such answers when interrogated as to the conditions of a man's self after his death; but they are not familiar answers for the tradition of seventeenth- and eighteenth-century science.

Let us review, then, what a hydrogen atom should be like if we could apply Newton's laws and the whole classical picture of matter in motion to the simple model. The electron is held to its nucleus as the earth is to the sun, or as is Venus. It should revolve in an ellipse, as Kepler found and Newton explained. The size of the ellipse could be varied from atom to atom as the orbits of the planets are different, depending on how it was formed and what its history, and so should the shape of the orbits, whether they are narrow or round. There should be no fixed size for a hydrogen atom and no fixed properties; and when we disturb one by one of our probings, or when it is disturbed in nature, we would not expect it to return to a size and shape at all similar to that from which it started. This is not all—there are more recondite points. When a charge moves in anything but a straight line, it should send out electromagnetic radiation. This is what we see in every radio antenna. As far as our model goes, this radiation should in time sap the energy of the electron to make up for the energy that has been sent out in the form of light waves; and the ellipses on which the electron moves should get smaller and smaller as it gets nearer to its attractive sun and loses its energy. For a system about the size of the hydrogen atom as we know it in nature, a few hundredths of a millionth of an inch across, this process should go very rapidly; and the atom should become far, far smaller than atomic dimensions in very much less than a millionth of a second. The color of the light that the electron radiates should be determined by the period of its revolution; it too should be random, differing from orbit to orbit, differing from time to time as the orbits shrink and alter. This is the picture which classical physics—Newtonian physics—predicts for the hydrogen atom, if Rutherford's model is right.

It could hardly be further from the truth. By all we know, hydrogen atoms if undisturbed are all identical. They are the same size and each has the same properties as any other, whatever its history, provided only that it has had a chance to recover from any disturbance. They last indefinitely. We think of them, rightly, as completely stable and unchanging. When they are undisturbed, they do not radiate light or any other electromagnetic radiation, as indeed they could not if they are to remain unaltered. When they are disturbed, they sometimes do radiate, but the color of the light that they emit is not random and continuous but falls in the sharp lines of the hydrogen spectrum. The very stability, extent, and definiteness is not at all understandable on the basis of classical physics; and indeed on the basis of classical physics there is no length that we can define in terms of the masses and charges of the ingredients of the atom, and that is even roughly of the actual dimensions of the atom.

In other respects, too, the atomic system shows a peculiar lack of continuity wholly at variance with the properties of Newtonian dynamics. If we probe atoms with a stream of electrons, for instance, the electrons will typically lose some of their initial energy, but these losses are not random in amount. They correspond to definite, well-defined energy gaps, characteristic for the atom in question, reproducible and not too hard to measure. When an atom is irradiated by light, an electron will be ejected, if and only if the energy of that light exceeds a certain minimum known as the photo-electric threshold. Indeed, it was this discovery which led Einstein in the early years of the century to a finding about light almost equally revolutionary for our understanding of light and for our understanding of atomic systems. This finding, to be more precise, is that as one alters the frequency of the light that shines on a body, the energy of the electrons ejected increases linearly with the frequency; linearly—that means proportionally. The constant of proportionality, which connects energy with frequency, is the new symbol of the atomic domain. It is called Planck's constant, or the quantum of action, and it gives a measure of energy in terms of frequency. It is the heraldic symbol over the gateway to the new world; and it led Einstein to the bold, though at the time hardly comprehensible, conclusion that light, which we know as an electromagnetic disturbance of rapidly changing electrical fields, which we know as a continuous phenomenon propagating from point to point and from time to time like a wave, is also and is nevertheless corpuscular, consisting of packets of energy determined by the frequency of the light and by Planck's constant. When a material system absorbs light, it absorbs such a packet, or quantum, of energy, neither less nor more; and the discontinuous nature of the energy exchanges between an atom and an electron is paralleled by the discontinuous nature of the energy exchanged when radiation is absorbed or emitted.

We shall have to come back more than once to light as waves, and light as quanta; but how radical a problem of understanding this presents can be seen at once from all of classical optics, from the work of Huygens and its mathematical elaboration by Fresnel, and even more completely from its electromagnetic interpretation by Maxwell. We know that light waves interfere. We know, that is, that if there are two sources of light, the intensity of the light to be found at some other place will not necessarily be just composed of the sum of that which comes from the two sources; it may be more and it may be less. We know from unnumbered attempts how to calculate, and how to calculate correctly, what the interference of the sources will turn out to be. If we have light impinging on a screen which is opaque and there are two holes in the screen, not too large and not too far apart in the terms of the wave length, the wavelets that come from one of the holes will be added to those that come from the other. Where two crests of these wavelets coincide, we shall have more light than the sum of the two. Where a crest and a trough coincide, we shall have less; and so we observe and understand and predict and are quite confident of these phenomena of interference.

Try for a moment to describe this in the terms of the passage of particles of quanta. If one of those quanta which characterize both the emission of light at the source and its detection—let us say, by the eye or by the photographic plate or photo-cell on the far side of the screen—if a quantum passes through one of the holes, how can the presence of the other hole through which it did not pass affect its destiny? How can there be any science or any prediction if the state of affairs remote from the trajectory of the quantum can determine its behavior? Just this question and our slow answer to it will start us on the unravelling of the physics of the atomic world.

The first great step, taken long before the crisis of quantum theory, was [. . .] Bohr's first theory. It has given us the symbol of the atomic world: the nucleus and a series of circles and ellipses represent in a pictorial way the states of the atom. We use it today, though we know in far more detail and far more completely what Bohr knew when he proposed it, that it could be at best a temporary and partial analogy. This was Bohr's first postulate: that in every atom there were stationary states whose stability and uniqueness could not be understood in terms of classical dynamics. The lowest one, the one with the least energy, the ground state, is truly stable. Unless we disturb it, it will last unaltered. The others are called excited states, and they may be excited by collision or radiation or other disturbance. They, too, are stable in a sense incomprehensible in terms of Newton's theory. Their stability is not absolute though. Just as these states could be reached by transition induced by collision or disturbance, so an atom may return to states of lower energy, whether by further collision or spontaneously. In these spontaneous changes it gives out that radiation which is the analog of the radiation which in classical theory would make all motion unstable. [. . .] The rules which Bohr laid down for determining the character of the orbits that would correspond to stationary states, the so-called quantum conditions, were from the first recognized by him as incomplete and provisional. We know now that the states are in fact nothing like orbits at all; that the element of change with time, which is inherent in an orbit, is missing from these states; and that in fact the very notion of an orbit can be applied to the motion of matter only when the stationary state is not defined, and that a stationary state can exist only when there is no possibility of describing an orbit at all.

That was the first rule. And what is the second? The second rule is that an atom can change only by passing from state to state. [. . .]

But what are we to think of the transitions themselves? Do they take place suddenly? Are they very quick motions, executed in going from one orbit to another? Are they causally determined? Can we say, that is, when an atom will pass from one of its states to another as we disturb it; and can we find what it is that determines that time? To all these questions, the answer would turn out to be "no." What we learned to ask was what determined not the moment of the transition but the probability of the transition. What we needed to understand was not the state of affairs during the transition but the impossibility of visualizing the transition—an even more radical impossibility than with the states themselves—in terms of the notion of matter. We learned to accept, as we later learned to understand, that the behavior of an atomic system is not predictable in detail; that of a large number of atomic systems with the same history, in, let us say, the same state, statistical prediction was possible as to how they would act if they were let alone and how they would respond to intervention; but that nowhere in our battery of experimental probings would we find one to say what one individual atom would in fact do. We saw in the very heart of the physical world an end of that complete causality which had seemed so inherent a feature of Newtonian physics.

How could all this be and yet leave the largely familiar world intact as we knew it? Large bodies are, of course, made up of atoms. How could causality for bullets and machines and planets come out of acausal atomic behavior? How could trajectories, orbits, velocities, accelerations, and positions re-emerge from this strange talk of states, transitions, and probabilities? For what was true yesterday would be true still, and new knowledge could not make old knowledge false. Is there a possible unity between the two worlds and what is its nature?

This is the problem of correspondence. Whatever the laws which determine the behavior of light or of electrons in atoms or other parts of the atomic world, as we come closer and closer to the familiar ground of large-scale experience, these laws must conform more and more closely to those we know to be true. This is what we call the principle of correspondence. In its formulation the key is the quantum of action, whose finiteness characterizes the new features of atomic physics. And so the physicist says that, where actions are large compared to the quantum of action, the classical laws of Newton and Maxwell will hold. What this tends to mean in practice is that when mass and distances are big compared to those of the electron and the atom's size, classical theory will be right. Where energies are large and times long compared to atomic energies and times, we shall not need to correct Newton. Where this is so, the statistical laws of atomic physics will lead to probabilities more and more like certitudes, and the acausal features of atomic theory will be of no moment, and in fact lost in the lack of precision with which questions about large events will naturally be put.

In Bohr's hands and those of the members of his school, this correspondence principle was to prove a powerful tool. It did not say what the laws of atomic physics were, but it said something about them. They must in this sense be harmonious with, and ultimately reducible to, those of large-scale physics. And when to this principle was added the growing conviction that the laws of atomic physics must deal not with the Newtonian position, velocity, and acceleration that characterized a particle but with the observable features of atoms—the energies and properties of stationary states, the probabilities of transitions between these states—the groundwork was laid for the discovery of quantum mechanics.

The principle of correspondence—this requirement that the new laws of atomic mechanics should merge with those of Newtonian mechanics for large bodies and events—thus had great value as an instrument of discovery. Beyond that, it illustrates the essential elements of the relation of new discovery and old knowledge in science; the old knowledge, as the very means for coming upon the new, must in its old realm be left intact; only when we have left that realm can it be transcended.

A discovery in science, or a new theory, even when it appears most unitary and most all-embracing, deals with some immediate element of novelty or paradox within the framework of far vaster, unanalyzed, unarticulated reserves of knowledge, experience, faith, and presupposition. Our progress is narrow; it takes a vast world unchallenged and for granted.

This is one reason why, however great the novelty or scope of new discovery, we neither can, nor need, rebuild the house of the mind very rapidly. This is one reason why science, for all its revolutions, is conservative. This is why we will have to accept the fact that no one of us really will ever know very much. This is why we shall have to find comfort in the fact that, taken together, we know more and more.

"Somehow the Wave Had to Exist"

Fred Alan Wolf

>─┤─◆─○─◆─┤─<

CHAPTER EIGHT

SIR JAMES JEANS

The End of the Universe, from
The Universe Around Us (1929)

Sir James Hopwood Jeans (1877–1946) was a mathematician, astronomer, physicist, and the author of many popular works of science. Although he pursued science his entire life, he retired from university life in 1912 after marrying a wealthy wife. In the late 1920s and early 1930s he produced a series of astronomy books that built on established scientific grounds and asked what the implications of these ideas might be. The selection here, from his hugely successful book *The Universe Around Us* (Rachel Carson's book *The Sea Around Us* played off its title), describes the winding down of the universe like an old clock. The image became famous, although it was widely greeted with dismay. The selection by Fred Hoyle refers to this passage, in a tone that assumes the reader is familiar with it. When reading it now, we can see why. Jeans was a stylist who could make deep space as vivid as a country landscape.

If we ask what is the underlying cause of all the varied animation we see around us in the world, the answer is in every case, energy—the chemical energy of the fuel which drives our ships, trains and cars, or of the food which keeps our bodies alive and is used in muscular effort, the mechanical energy of the earth's motion which is responsible for the alternations of day and night, of summer and winter, of high

tide and low tide, the heat energy of the sun which makes our crops grow and provides us with wind and rain.

The first law of thermodynamics, which embodies the principle of "conservation of energy," teaches that energy is indestructible; it may change about from one form to another, but its total amount remains unaltered through all these changes, so that the total energy of the universe remains always the same. As the energy which is the cause of all the life of the universe is indestructible, it might be thought that this life could go on for ever undiminished in amount.

The second law of thermodynamics rules out any such possibility. Energy is indestructible as regards its amount, but it continually changes in form, and generally speaking there are upward and downward directions of change. It is the usual story—the downward journey is easy, while the upward is either hard or impossible. As a consequence, more energy passes in one direction than in the other. For instance, both light and heat are forms of energy, and a million ergs of light-energy can be transformed into a million ergs of heat with the utmost ease; let the light fall on any cool, black surface, and the thing is done. But the reverse transformation is impossible; a million ergs which have once assumed the form of heat, can never again assume the form of a million ergs of light. [. . .]

It may be objected that the everyday act of lighting a fire disproves all this. Has not the sun's heat been stored up in the coal we burn, and cannot we produce light by burning coal? The answer is that the sun's radiation is a mixture of both light and heat, and indeed of radiation of all wavelengths. What is stored up in the coal is primarily the sun's light and other radiation of still shorter wavelength. When we burn coal we get some light, but not as much as the sun originally put into the coal; we also get some heat, and this is more than the amount of heat which was originally put in. On balance, the net result of the whole transaction is that a certain amount of light has been transformed into a certain amount of heat.

All this shews that we must learn to think of energy, not only in terms of quantity, but also in terms of quality. Its total quantity remains always the same; this is the first law of thermodynamics. But its quality varies, and tends to vary always in the same direction. Turnstiles are set up between the different qualities of energy; the passage is easy in one direction, impossible in the other. A human crowd may contrive to find a way round without jumping over turnstiles, but in nature there is no way round; this is the second law of thermodynamics. Energy flows always in the same direction, as surely as water flows downhill. [. . .]

Although this is the main part of the downward path, it is not the whole of it. Thermodynamics teaches that all the different forms of energy have different degrees of "availability" and that the downward path is always from higher to lower availability.

And now we may return to the question with which we started [. . .]: "what is it that keeps the varied life of the universe going?" Our original answer "energy" is seen to be incomplete. Energy is no doubt essential, but the really complete answer is that it is the transformation of energy from a more available to a less available form; it is the running downhill of energy. To argue that the total energy of the universe cannot diminish, and therefore the universe must go on for ever, is like arguing that as a clockweight cannot diminish, the clock-hand must go round and round for ever.

Energy cannot run downhill for ever, and, like the clockweight, it must touch bottom at last. And so the universe cannot go on for ever; sooner or later the time must come when its last erg of energy has reached the lowest rung on the ladder of descending availability, and at this moment the active life of the universe must cease. The energy is still there, but it has lost all capacity for change; it is as little able to work the universe as the water in a flat pond is able to turn a waterwheel. We are left with a dead, although possibly a warm, universe—a "heat-death."

Such is the teaching of modern thermodynamics. There is no reason for doubting or challenging it, and indeed it is so fully confirmed by the whole of our terrestrial experience that it is difficult to see at what point it could be open to attack. It disposes at once of any possibility of a cyclic universe in which the events we see are as the pouring of river water into the sea, while events we do not see restore this water back to the river. The water of the river can go round and round in this way, just because it is not the whole of the universe; something extraneous to the river-cycle keeps it continually in motion—namely, the heat of the sun. But the universe as a whole cannot so go round and round. Short of postulating continuous action from outside the universe, whatever this may mean, the energy of the universe must continually lose availability; a universe in which the energy had no further availability to lose would be dead already. Change can occur only in the one direction, which leads to the heat-death. With universes as with mortals, the only possible life is progress to the grave.

Even the flow of the river to the sea, which we selected as an obvious instance of true cyclic motion, is seen to illustrate this, as soon as all the relevant factors are taken into account. As the river pours seaward over its falls and cascades, the tumbling of its waters generates heat, which ultimately passes off into space in the form of heat radiation. But the energy which keeps the river pouring along comes ultimately from the sun in the form mainly of light; shut off the sun's radiation and the river will soon stop flowing. The river flows only by continually transforming light-energy into heat-energy, and as soon as the cooling sun ceases to supply energy of sufficiently high availability the flow must cease.

The same general principles may be applied to the astronomical universe. There is no question as to the way in which energy runs down here. It is first liberated in the hot interior of a star in the form of quanta of extremely short wave-length and excessively high energy. As this radiant energy struggles out to the star's surface, it continually adjusts itself, through repeated absorption and re-emission, to the temperature of that part of the star through which it is passing. As longer wave-lengths are associated with lower temperatures [. . .], the wave length of the radiation is continually lengthened; a few energetic quanta are being transformed into numerous feeble quanta. Once these are free in space, they travel onward unchanged until they meet dust particles, stray atoms, free electrons, or some other form of interstellar matter. Except in the highly improbable event of this matter being at higher temperature than the surfaces of the stars, these encounters still further increase the wave-length of the radiation, and the final result of innumerable encounters is radiation of very great wave-length. The quanta have increased enormously in numbers, but have paid for their increase by a corresponding decrease in individual strength. In all probability, the original very energetic quanta had their source in the annihilation of protons and electrons, so that the main process of the universe consists in the energy of exceedingly high availability which is bottled up in electrons and protons being transformed into heat energy at the lowest level of availability.

Many, giving rein to their fancy, have speculated that this low-level heat energy may in due course reform itself into new electrons and protons. As the existing universe dissolves away into radiation, their imagination sees new heavens and a new earth coming into being out of the ashes of the old. But science can give no support to such fancies. Perhaps it is as well; it is hard to see what advantage could accrue from an eternal reiteration of the same theme, or even from endless variations of it.

The final state of the universe will, then, be attained when every atom which is capable of annihilation has been annihilated, and its energy transformed into heat-energy wandering for ever round space, and when all the weight of any kind whatever which is capable of being transformed into radiation has been so transformed.

We have mentioned Hubble's estimate that matter is distributed in space at an average rate of 1.5×10^{-31} grammes per cubic centimetre. The annihilation of a gramme of matter liberates 9×10^{20} ergs of energy, so that the annihilation of 1.5×10^{-31} grammes of matter liberates 1.35×10^{-10} ergs of energy. It follows that the total annihilation of all the substance of the existing universe would only fill space with energy at the rate of 1.35×10^{-10} ergs per cubic centimetre. This amount of energy is only enough to raise the temperature of space from absolute zero to a temperature far below that of liquid air; it would only raise the temperature of the earth's surface by a 6000th part of a degree Centigrade. The reason why the effect of annihilating a whole universe is so extraordinarily slight is of course that space is so extraordinarily empty of matter; trying to warm space by annihilating all the matter in it is like trying to warm a room by burning a speck of dust here and a speck of dust there. As compared with any amount of radiation that is ever likely to be poured into it, the capacity of space is that of a bottomless pit. Indeed, so far as scientific observation goes, it is entirely possible that the radiation of thousands of dead universes may even now be wandering round space without our suspecting it.

FREDERIC C. BARTLETT
Imagery in Thought, from
Remembering (1932)

T he British psychologist Frederic C. Bartlett (1887–1969) did
pioneering work on the study of memory. Whereas classic memory research concerns the way that people learn meaningless lists,
Bartlett developed a series of techniques for studying the memory of
meaningful things. His ideas have been especially influential in artificial intelligence projects that ask computers to turn information
into something meaningful. Bartlett's interests carried him close to
nature's border. He was interested in subjective experiences, and
although there is nothing unnatural about it, the subjective is difficult to study at first hand. He also wanted to understand the biological function, or role, of what he studied. Again, there is nothing unnatural about function, but it is difficult to state objectively and
unambiguously. Even something like a bird's wing has multiple functions, such as flying, signaling, keeping eggs warm, and probably
20 other things known only to the birds.

Unless a person can put an observation into the context of some
independent understanding of nature—for example, the shape of a
wing and the laws of aerodynamics or the markings of a wing and
natural selection—there is always something a bit speculative about
an explanation of function. Speculation, however, is often all we
have and all we can use to try a bit of reverse engineering. The laws
of aerodynamics, for example, were worked out by studying birds'
wings and speculating about what in the wing might support flight.
Bartlett was trying something similar for the mind. By observing the
effects of subjectivity and what it seems to accomplish, he is laying
the groundwork for a fuller understanding of it.

Images have fundamentally important parts to play in mental life.
This hardly appears to be a popular view in modern psychology. It is
not only the extreme behaviourist who is tempted to think that psychology can gain little advantage from a close study of images. In less
radical circles also it is often held that, since images are generally
vague, fleeting, and variable from person to person and from time to

time, hardly any statement can be made about them which will not be immediately contradicted on very good authority. This view has arisen because most statements that have been made about images in traditional psychology concern their nature rather than their functions, what they are rather than what they make it possible to do; and many of the controversies that have raged about them have been concerned primarily with their epistemological status. Undoubtedly considerations about the nature of imagery, and its relation, on the one hand to a world of external objects, and on the other to immediate sensory perception, raise important, interesting and singularly intractable problems. It is not these with which we are now concerned, but simply with the question of what, exactly, images have to do in the general growth of mental life. [. . .]

At one time or another practically every one of the large number of observers who have helped me in my experiments have confidently affirmed the presence of images. These have been described in very various terms, and have been given very various properties, but their actual occurrence has been reported as confidently as anything ever has been in the whole series of experiments, and it seems to be very arbitrary to reject this particular verbal report and to accept any number of others having the same source and precisely the same justification. Moreover, although the images are described with great diversity, they possess one or two marks which are invariably present. Every observer means, when he reports the appearance of an image, that there is something affecting his response which is not referable to an immediate external sensory source, and further that this is certainly not identical with the description that he can give of it. He means, that is to say, that what he is calling an image is not merely reducible to a word formula. In its attempt to discover the exact nature of images, assuming them to be different in some way from the words which are used to describe them, traditional psychology has been mainly preoccupied with apparent divergences between perceiving and imaging, or between percepts and images. It is said that the image, as compared with the percept, is relatively fleeting, fragmentary, flickering and of a low order of intensity; and that it 'strikes upon the mind' in a different manner. The differences between imaging and verbalising have been much less carefully studied, but it is these that become prominent the moment a functional point of view is adopted, and it is somewhere in the realm between direct sensorial response and the bare use of words, that images have their most important functions.

Images and Words

Words very often seem to be the direct expression of meanings: images, however, are somehow there, and it is only when specific meanings are developed out of them that words can flow. This statement may appear to be very indefinite, but it is in fact exactly what many observers say, [. . .] and it seems to me to contain much psychological significance. Consider the course of an experiment. [. . .] A subject comes to his test, and immediately some interest is aroused, some more or less definite orientation set up. The general experimental conditions and the social relation of experimenter and subject will always produce these. From whatever is his starting-point the subject goes on describing in words whatever is saliently related to [. . .] his predominant interest or attitude. Another time, or with another subject, the process begins in precisely the same way, but there is a check; perhaps a clash of interests occurs, or something disturbs the smooth development of the subject's attitude. Suddenly he seems to diverge slightly—sometimes markedly—from the straight path of description. There is an image, and meaning has to be tacked on to that, or, perhaps more accurately, has to flow out of it, or emerge from it, before words can carry the process further. So the course of description, when images abound, is apt to be more exciting, more varied, more rich, more jerky, and, from a merely logical point of view, a little more difficult to follow than when meaning flows directly into words. Also, and this too is a matter of the greatest psychological interest, it is less subject to convention, and is apt to appear more definitely original. Images are, in fact, so much a concern of the individual that, as everybody knows, whenever, in psychological circles, a discussion about images begins, it very soon tends to become a series of autobiographical confessions. [. . .]

Distance Reactions

All important biological advances are marked by two outstanding characteristics: increase in the diversity of reactions to match more nearly the variety of external exciting conditions, and a growth in the capacity to deal with situations at a distance. Sometimes one of these characteristics is particularly prominent and sometimes another, but together they mark the superiority of the higher forms of organic life. The less developed organisms have but a few ways of coming into touch with the external world, and these all result in a few stereotyped responses. Slowly there grows up a multiplicity of sensory reactions, each more or less specialised to its appropriate stimuli. Among these, those which have to do with distance responses take the lead. [. . .]

The capacity to be influenced by past reactions, often—but very likely somewhat inaccurately—called 'modification by experience,' on the whole conflicts with the demand, issued by a diverse and constantly changing environment, for adaptability, fluidity and variety of response. Its general effect is twofold: to lead to stereotyped behaviour and to produce relatively fixed serial reactions. Even on a high level of behaviour the unwinding of responses in a fixed chronological order is very common. We all tend to drop into serial reactions when we are tired, delirious, slightly intoxicated, or when, for any reason, critical keenness is relaxed. It seems as if the processes of organic adjustment are all the time striving to set up serial reactions, and as if, consciousness apart, the series is of greater weight than the elements making it up; so that again and again, if the series fails at any point, there is a tendency for complete collapse to set in, or else for the whole series to begin over again.

The operation of past situations 'in the mass' and the overdetermination by fixed chronological series are both biologically uneconomical and to some extent unsound; uneconomical, because to resume a whole series is often a shocking waste of time, and unsound, because they blur the diversity of actual environmental facts. To surmount these difficulties, the method of images is evolved. I do not think that anything very definite can be said as to the exact mechanism of the process. But in this respect the emergence, or the discovery, of images is in no unique position. Again and again it appears that the integration of functions gives rise to new reactions, although the exact mechanism by which the new reactions are produced remains inexplicable. [. . .] In general, images are a device for picking bits out of schemes, for increasing the chance of variability in the reconstruction of past stimuli and situations, for surmounting the chronology of presentations. By the aid of the image, and particularly of the visual image—for this, like the visual sense, is the best of all our distance mechanisms of its own type—a man can take out of its setting something that happened a year ago, reinstate it with much if not all of its individuality unimpaired, combine it with something that happened yesterday, and use them both to help him to solve a problem with which he is confronted to-day. [. . .]

It has often been pointed out that, in any process of recall, images are particularly liable to arise when any slight check occurs, and that in the absence of this the whole process is likely to run on to completion simply in terms of language or of manipulative activity of some suitable kind. This is just what would be expected. A difference in present circumstances, as compared with those suitable to an automatic response, sets up in the agent a conflict of attitude or of interest. The appropriate response is then temporarily held up, and the image, an item from some scheme, comes in to help to solve the difficulty. The solution, however, very often seems to be an odd one to the onlooker, and may appear unsuitable to the agent. [. . .]

How Images May Fail

The device of images has several defects that are the price of its peculiar excellences. Two of these are perhaps the most important: the image, and in particular the visual image, is apt to go farther in the direction of the individualisation of situations than is biologically useful; and the principles of the combination of images have their own peculiarities and result in constructions which are relatively wild, jerky and irregular, compared with the straightforward unwinding of a habit, or with the somewhat orderly march of thought.

The justification of Napoleon's statement—if, indeed, he ever made it—that those who form a picture of everything are unfit to command, is to be found in the first of these defects. A commander who approaches a battle with a picture before him of how such and such a fight went on such and such an occasion, will find, two minutes after the forces have joined, that something has gone awry. Then his picture is destroyed. He has nothing in reserve except another individual picture, and this also will not serve him for long. Or it may be that, when his first pictured forecast is found to be inapplicable, he has so multifarious and pressing a collection of pictures that equally he is at a loss what practical adjustment to make. Too great individuality of past reference may be very nearly as embarrassing as no individuality of past reference at all. To serve adequately the demands of a constantly varying environment, we have not only to pick items out of their general setting, but we must know what parts of them may flow and alter without disturbing their general significance and functions. This seems to demand just that kind of analysis which words are peculiarly adapted to subserve, but to which the image method is by itself inadequate.

More important still is the fact that images have their own peculiar modes of combination, and that these are less well adapted to the needs of social adjustment than are those of words. The normal psychologist has given much less consideration than is desirable to the ways in which images are associated. Of the commonly recognised forms of association those that go into any considerable detail are nearly always derived from a classification of word-associations and are, as would be expected, distinctly logical in their trend; while those which deal with general categories, such as the traditional association by similarity, contiguity and succession, remain too wide to throw much light on any particular problem. [. . .] Turning once more to the experimental evidence, we find it many times demonstrated that this 'weighting' of detail, this preponderance of certain elements, is primarily a matter of personal factors of the nature of bias or interest. Consequently, images readily flow into one another, condense and combine in one mental life which in another may appear remarkably ill-assorted and incoherent. No reference to principles of similarity, contiguity, or succession will adequately describe such forms of individual association. Still less can these forms be satisfactorily placed into logical categories of relations of subordination, superordination, co-ordination and the like.

The most typical case for the emergence of images is where personal interests or attitudes cross and combine. [. . .] The train of recall, or of construction, which consists largely of a passage from image to image, may be seen to be coherent enough after all. Indeed, just because of the variety and unexpectedness of its concrete imaginal constituents it may be precisely what is called 'brilliantly coherent.' But if the significance is not worked out, and the successive images are merely described, or named, the train will always tend to appear to a second person fantastic, jerky, illogical, insubordinate to the common principles of association. These characteristics are seen at their extreme in the welter of images of the schizophrenic patient, but they cannot be wholly removed from any process of image-laden thought. [. . .]

No doubt many contributory conditions may be at work, but the experiments here and there offer confirmation of what has often been suggested: that prominent among these conditions is the affective colouring of the interests which function simultaneously. There is a good general example in Thomas Hardy's novel, *Desperate Remedies*. Cytheraea Grange has just witnessed her father's fall from a high scaffolding. The shock is tremendous, for she knows that he is probably seriously injured at the very least, even if he is not killed. "The next impression of which Cytheraea had any consciousness was of being carried from a strange vehicle across the pavement to the steps of her own house by a brother and an older man. Recollection of what had passed evolved itself an instant later, and just as they entered the door . . . her eyes caught sight of the south-western sky, and, without heeding, saw white sunlight slanting in shaft-like lines from a rift in a slaty cloud. Emotions will attach themselves to scenes that are simultaneous—however foreign in essence the scenes may be—as chemical waters will crystallise in twigs and trees. Ever after that time any mental agony brought less vividly to Cytheraea's mind the scene from the Town Hall windows, than sunlight slanting in shaftlike lines." Everybody will at once recognise that this expresses accurately a common psychological event. It is not merely the juxtaposition of events in space or time that brings them together in vivid imaginal form, but the fact that, however near or remote, however dissimilar or alike they may be, they are overspread by common emotions or common interests. This is, in fact, that *mémoire affective* concerning which many French psychologists in particular have written with great insight.

From this consideration two or three important conclusions may be drawn. First, we now have before us one of the main reasons why attempts to formulate the significance of images which occur in the course of the working-out of some special interest often seem to their authors to fail to say just what demands to be said. That affective character which gives the image its position and its main justification defies adequate expression. For there is no working vocabulary which comes anywhere near matching the delicate distinctions of affective response of which men are capable. Consequently, in many instances, the images are simply described, or named; their significance is left to tacit understanding, and it almost seems as if they merely provide a kind of aesthetic luxury.

Secondly, this affective character, combining interests of very different nature and origin, may help to explain the great wealth and variety of images which may often be observed in the typical visualiser. The differences of affective response at a human level undoubtedly far outrun in number their available expression in descriptive language, but at the same time they must fall enormously short of the number of discriminable features of external environment. Thus, in every life, exceedingly varied cognitive materials must be capable of being combined, because they have all served the development of interests whose expression was coloured by a fairly constant affective accompaniment.

Finally, it seems not fantastic to suppose that there are some people who are by temperament peculiarly responsive to the subtle affective background of their interests. These are the people whose interests most readily flow and coalesce, or, again, under other conditions, are most severely kept apart. Others, however, are unable to combine realms of interest not conventionally put together, until some reason that can be formulated has been found by themselves or by some different person. The former fore-run reasons, feeling a connexion for which there is no good vocabulary. The latter must await analysis and logic. Since always the image tends to retain its individualising functions, those people who can combine vivid images without waiting for formulated reasons will be persons who deal primarily with concrete representations, metaphor, practical problems. Among them may be found the poet, the inventor, the discoverer. Among them also, possessing the sensitivity which responds to hidden connexions, but lacking orderliness and persistence, are to be found those people who wander at haphazard from one topic to another in a manner that appears wholly inconsequential both to others and to themselves.

Imaging, like all other organic or psychological reactions, is always tending to drop to the level of habit. In general, there seems to me no doubt that the image is one of the answers of a conscious organism to the challenge of an external environment which partially changes and in part persists, so that it demands a variable adjustment, yet never permits an entirely new start. When once the image method has been adopted and practiced, however, it tends itself to become a habit. Whether because of affective conditions, or for any other reason, a typical visualiser, for example, often seems to have a great wealth of images, and these have a great variety of characters, so that he is tempted to stop and describe them—often to his own and others' aesthetic enjoyment—instead of concentrating upon the problem that they are there to help him to solve. Before long his images themselves get into ruts, repeat well-known characters, become conventionalised in a peculiar personal sense, and lose their touch with just those characteristically novel features of a given environment without which the method of images would originally never have developed at all.

Imaging and Thinking

[. . .] If we keep within the indications of the experiments, it would seem as if there must be some kind of supplementary relation between word and image formation, and it is worth while to try to see how this operates. There is a very general agreement that thinking and the use of language are closely connected, though nobody is justified in reducing the one to the other. In one important respect, words and censorial images are alike: both act as signs indicating something else which need not be perceptually present at the moment. Thus they are both instruments of the general function of dealing with situations or objects at a distance. Words have the obvious additional advantage of being social, and they constitute the most direct manner of communicating meaning. The image, to be communicated, has itself to be expressed in words, and we have seen that this can often be done only in a most halting and inadequate manner. Words differ from images in another even more important respect: they can indicate the qualitative and relational features of a situation in their *general* aspect just as directly as, and perhaps even more satisfactorily than, they can describe its peculiar individuality. This is, in fact, what gives to language its intimate relation to thought processes. For thinking, in the proper psychological sense, is never the mere reinstatement of some suitable past situation produced by a crossing of interests, but is the utilisation of the past in the solution of difficulties set by the present. Consequently it involves that amount of formulation which shows, at least in some degree, what is the nature of the relation between the instances used in the solution and the circumstances that set the problem. [. . .] For carrying out this formulation, for utilising the general qualitative and relational features of the situation to which reference is more or less openly made, words appear to be the only adequate instruments so far discovered or invented by man. Used in this way, they succeed just where we have seen that images tend most conspicuously to break down: they can name the general as well as describe the particular, and since they deal in formulated connexions they more openly bear their logic with them.

Thinking, if I am right, is biologically subsequent to the image-forming process. It is possible only when a way has been found of breaking up the 'massed' influence of past stimuli and situations, only when a device has already been discovered for conquering the sequential tyranny of past reactions. But though it is a later and a higher development, it does not supersede the method of images. It has its own drawbacks. Contrasted with imaging it loses something of vivacity, of vividness, of variety. Its prevailing instruments are words, and, not only because these are social, but also because in use they are necessarily strung out in sequence, they drop into habit reactions even more readily than images do. Their conventions are social, the same for all, and far less a matter of idiosyncrasy. In proportion as we lose touch with the image method we run greater and greater risk of being caught up in generalities that may have little to do with actual concrete experience. [. . .] Only in abnormal cases are such risks pushed to their extreme, for, just because images and the language processes of thinking are commonly combined, each method has taken over some of the peculiarities of the other, and images, as in what are often called the 'generic' kind, seem to be striving after some general significance and framework, while language often builds its links from case to case upon elaborate and detailed individual description. Broadly, each method, however closely the two are related, retains its own outstanding character. The image method remains the method of brilliant discovery, whereby realms organised by interests usually kept apart are brought together; the thought-word method remains the way of rationalisation and inference, whereby this connecting of the hitherto unconnected is made clear and possible for all, and the results which follow are not merely exhibited, but demonstrated.

Lying in wait for both these processes is the common fate that may overtake all human effort: they may become mere habit. That very sequence and mass determination which they were developed to surmount may overwhelm them in the end. Then a man will take facility of images and variety of words to be satisfying things in themselves, and it may appear as if images and words are merely luxuries, to be enjoyed.

⊳─◂▸─○─◃▸─◅

GORDON W. ALLPORT

The Mature Personality, from
Personality: A Psychological Interpretation (1937)

G ordon Willard Allport (1897–1967) spent his career at Harvard University studying personality, meaning, roughly, the way a person's individual traits are organized into a whole. He wanted to understand psychology's biggest issue: how people become who they are. He agreed with Freud that personality begins with infantile traits and needs, but Allport believed that some people can get beyond those childish traits, and he wondered how they did it. In this selection he describes mature adults and lists the inadequacies of more than a dozen schools of psychology for explaining the fact that there are at least a few mature personalities. In an essay like this we see the scientific imagination approaching the end of the available road. It recognizes the facts to be explained but finds no guidance in discovering the causes of those facts. At this point, a dishonest investigator would begin to deny the facts, but Allport could not do that. The neurotic investigator would espouse one of the inadequate schools' theories and hope for a miracle, but Allport could not do that either. The pragmatic researcher would switch subjects, and the genius would imagine a new way to cast light on the facts and extend the research. But Allport was neither of those, it seems. After the selection given here, Allport wrote an admirable humanist appreciation of the mature personality.

Nothing requires a rarer intellectual heroism than willingness to see one's equation written out.

—SANTAYANA

The distinctive richness and congruence of a fully mature personality are not easy to describe. There are as many ways of growing up as there are individuals who grow, and in each case the end-product is unique. But if general criteria are sought whereby to distinguish a fully developed personality from one that is still unripe, there are three differentiating characteristics that seem both universal and indispensable.

In the first place, the developed person is one who has a variety of autonomous interests: that is, he can lose himself in work, in contemplation, in recreation, and in loyalty to others. He participates with warmth and vigor in whatever pursuits have for him acquired value. Egocentricity is not the mark of a mature personality. Contrast the garrulous Bohemian, egotistical, self-pitying, and prating of self-expression, with the man of confident dignity who has identified himself with a cause that has won his devotion. Paradoxically, "self-expression" requires the capacity to lose oneself in the pursuit of objectives, *not* primarily referred to the self. Unless directed outward toward socialized and culturally compatible ends, unless absorbed in causes and goals that outshine self-seeking and vanity, any life seems dwarfed and immature.

Whenever a definite objective orientation has been attained, pleasures and pains of the moment, setbacks and defeats, and the impulse for self-justification fade into the background, so that they do not obscure the chosen goals. These goals represent an *extension of the self* which may be said to be the first requirement for maturity in personality.

The second requirement is a curiously subtle factor complementing the first. We may call it *self-objectification*, that peculiar detachment of the mature person when he surveys his own pretensions in relation to his abilities, his present objectives in relation to possible objectives for himself, his own equipment in comparison with the equipment of others, and his opinion of himself in relation to the opinion others hold of him. This capacity for self-objectification is *insight*, and it is bound in subtle ways with the *sense of humor*, which as no one will deny, is, in one form or another, an almost invariable possession of a cultivated and mature personality.

Since there is an obvious antithesis between the capacity for losing oneself in vigorous participation and the capacity for standing off, contemplating oneself, perhaps with amusement, a third, integrative, factor is required in the mature personality, namely, a *unifying philosophy of life*. Such a philosophy is not necessarily articulate, at least not always articulate in words. The preacher, by virtue of his training, is usually more articulate than the busy country doctor, the poet more so than the engineer, but any of these personalities, if actually mature, participates and reflects, lives and laughs, according to some embracing philosophy of life developed to his own satisfaction and representing to himself his place in the scheme of things.

These are the three conditions for the optimum development of personality; each needs more detailed consideration. But before settling to this discussion it is necessary to examine the competence of psychology to deal with such complex mental conditions as are here involved.

When the academic psychologist attempts to account for such intricate formations as these he faces a dilemma. He himself lives and works primarily in professional circles, among persons trained to use their minds and abilities. It is too easy to assume that such men are representative of the majority of personalities, and to overlook those with gross limitations, occurring more frequently in an unselected population. He might easily romanticize the situation by compounding a representative personality on too high and subtle an emotional and mental level. He must not, in his interest in mature personalities, be unrealistic and forget the restrictions on the development of personality resulting from low intelligence, uncontrolled emotion, infantilism, regression, dissociation, stereotypes, autism, suggestibility, and many other entirely human, but none the less abortifacient conditions.

On the other hand, even though the number of completely mature personalities tested by these three criteria may be few, it is still necessary to account *adequately* for such as do exist, and for the multitude of others well along on the road to maturity. This necessity leads us to the second horn of the dilemma, the one on which most psychologists, without realizing it, are already impaled.

They are impaled because they apply the (as yet) crude tools of psychology to material too delicate to be cut and shaped with their aid. For instance, methods and concepts well enough designed to explain the automatic reactions of decorticated cats or simple skin reflexes, are often superimposed upon the vast pattern of mature personality, and declared to overspread it exactly. The results are ridiculous, and are responsible for the view of so many educated people that psychology is a sappy science.

Even the psychologist who honestly desires not to underestimate the complexities of personality finds himself limited by the crudity of the tools within his professional store. As a result he puts the entire strain of investigation upon a few inadequate implements where it would be better to forge new ones.

That the available store of concepts and methods is actually insufficient can be demonstrated by a cursory review of the major limitations of the various branches of psychology concerned with personality. Here it is not a question of the validity of the concepts nor of their suitability for *special* problems, but rather of their *adequacy* in dealing with the subtle characteristics of genuinely mature personalities.

Physiological psychology manifestly fails to specify neural equivalents for complex personal functions, for such subtle processes, for example, as ambition, loyalty, self-criticism, or humor.

Behaviorism of the classical order provides at best a blue-print of non-related excursions and meaningless movements of a mindless organism. Its concepts are better adapted to segmental responses than to fully integrated patterns. Whether the newer "operational behaviorism" can improve the situation remains for it to demonstrate.

Structural psychology of the older introspective order, since it is not interested in conation, can account for no single *dynamic* event in the entire sphere of personality.

Functional psychology, like structural, is preoccupied with mind-in-general, and though it treats instincts, adaptation, the stream of thought, and other vital functions, it does not do so in a personalized, concrete way.

Gestalt psychology, for all its advantageous emphasis on wholeness, has as yet dealt chiefly with *momentary* patterns of conduct, and tends to neglect the problems of lasting structure.

Mathematical psychology, with the aid of highly trained and subtle minds, produces only a caricature of such minds by holding that they can all be reduced to a few basic and *common* factors, or can be regarded as measurable deviations from one standard pattern.

Differential psychology, likewise occupied with the distribution of single qualities in a population, cannot treat the patterning of these differences into individual dynamic formations.

Freudian psychology never regards an adult as truly adult.

Dynamic psychology often commits this error of Freudianism, and usually regards adult motives as variations upon one monotonous standard pattern.

Hormic psychology, promising though it is in its recognition of sentiments, is cramped by preoccupation with the uniform instincts presumed to underlie these sentiments.

Psychiatry and other practical arts of dealing with *whole* men come closer to adequacy, but as yet they are deficient in conceptual formulation, and do not advance the theoretical psychology of personality. Such formulations as they have are derived principally from the study of disease rather than health.

Personalistic psychology also puts its emphasis on totality, but as yet its theories are broad and philosophical, failing to provide specific implements for bringing the concrete complexities of single personalities under inspection. [. . .]

Verstehende psychology is the only school of psychological thought that *flatters* human personality, finding it just as sublime as the ideal types produced in the minds of pre-Hitler German professors. It thereby weakens its own effectiveness; and commits in addition the fault of bifurcating psychology, refusing to utilize the results of any investigations other than its own. [. . .]

This uncomplimentary survey does not deny that each of these branches of psychology has its own distinctive merit in the study of personality, but it does deny that any one of them is at present fully equipped to handle the problems of maturity in personality as defined [here]. The psychology of personality cannot at present come to roost under any one of them. In time some may expand and show greater adequacy (promising signs exist especially in the last five and in *Gestalt* theory). But the required progress will come only if criticisms concerning *inadequacy* are taken to heart.

FRED HOYLE

The Expanding Universe, from
The Nature of the Universe (1950)

In the 1950s and early 1960s, Fred Hoyle (1915–), an astronomer, was extremely popular. Any student in those days who was even slightly interested in science knew that Hoyle was the most articulate proponent of the theory that matter was continuously appearing out of nothing. Although the idea seems unlikely, he insisted that it was no more absurd than the "Big Bang" notion that everything had appeared in an instant. In the mid-1960s, however, Hoyle announced that he had changed his mind. Evidence for the rival theory was mounting fast. In particular, the background radiation from the Big Bang (discussed in the selection by George Smoot) was right where the theory said it should be.

This selection shows a scientist looking for a way to move on. Astronomers needed to explain the discovery that the universe was expanding. One explanation was that the universe was created in an instant and has been expanding ever since. That idea seemed incredible and carried the disconcerting implication of an exhausted universe, as described in the selection by Jeans. Hoyle sought another explanation, but he looked in the wrong place. Both the Big Bang and the continuous creation theories depend on an as yet unexplained and unseen process, the creation of matter. Who could say that instantaneous or continuous creation was the more likely? Hoyle's image of the universe expanding forever is a beautiful work of imagination, and it is to his credit that when the evidence pointed in the other direction, he, like Kepler, put aside his old ideas and moved along the opening road.

By looking at any part of the sky that is distant from the Milky Way you can see right out of the disk that forms our Galaxy. What lies out there? Not just scattered stars by themselves, but in every direction space is strewn with whole galaxies, each one like our own. Most of these other galaxies—or extragalactic nebulae as astronomers often call them—are too faint to be seen with the naked eye, but vast numbers of them can be observed with a powerful telescope. When I say that these other galaxies are similar to our Galaxy, I do not mean that they are exactly alike. Some are much smaller than ours, others are not disk-shaped but nearly spherical in form. The basic similarity is that they are all enormous clouds of gas and stars, each one with anything from 100,000,000 to 10,000,000,000 or so members. [. . .]

Some of them are indeed very like our Galaxy even so far as details are concerned. By good fortune one of the nearest of them, only about 700,000 light years away, seems to be practically a twin of our Galaxy. You can see it for yourself by looking in the constellation of Andromeda. With the naked eye it appears as a vague blur, but with a powerful telescope it shows up as one of the most impressive of all astronomical objects. On a good photograph of it you can easily pick out places where there are great clouds of dust. These clouds are just the sort of thing that [. . .] stops us seeing more than a small bit of our own Galaxy. If you want to get an idea of what our Galaxy would look like if it were seen from outside, the best way is to study this other one in Andromeda. If the truth be known I expect that in many places there living creatures are looking out across space at our Galaxy. They must be seeing much the same spectacle as we see when we look at their galaxy. [. . .]

How many of these gigantic galaxies are there? Well, they are strewn through space as far as we can see with the most powerful telescopes. Spaced apart at an average distance of rather more than 1,000,000 light years, they certainly continue out to the fantastic distance of 1,000,000,000 light years. Our telescopes fail to penetrate further than that, so we cannot be certain that the galaxies extend still deeper into space, but we feel pretty sure that they do. [. . .] I find myself wondering whether somewhere among them there is a cricket team that could beat the Australians. [. . .]

At one time our Galaxy was a whirling disk of gas with no stars in it. Out of the gas, clouds condensed, and then in each cloud further condensations were formed. This went on until finally stars were born. Stars were formed in the other galaxies in exactly the same way. But we can go further than this and extend the condensation idea to include the origin of the galaxies themselves. Just as the basic step in explaining the origin of the stars is the recognition that a tenuous gas pervades the space within a galaxy, so the basic step in explaining the origin of the galaxies is the recognition that a still more tenuous gas fills the whole of space. It is out of this general background material, as I shall call it, that the galaxies have condensed. [. . .]

What is the present density of the background material? The average density is so low that a pint measure would contain only about one atom. But small as this is, the total amount of the background material exceeds about a thousandfold the combined quantity of material in all the galaxies put together. This may seem surprising but it is a consequence of the fact that the galaxies occupy only a very small fraction of the whole of space. [. . .]

If in your mind's eye you take the average galaxy to be about the size of a bee—a small bee, a honeybee, not a bumblebee—our Galaxy, which is a good deal larger than the average, would be roughly represented in shape and size by the fifty-cent piece, and the average spacing of the galaxies would be about three yards, and the range of telescopic vision about a mile. So sit back and imagine a swarm of bees spaced about three yards apart and stretching away from you in all directions for a distance of about a mile. Now for each honeybee substitute the vast bulk of a galaxy and you have an idea of the Universe that has been revealed by the large American telescopes. [. . .]

This colossal swarm is not static: it is expanding. [. . .] If the Universe were static on a large scale [. . .] the background material would [still] condense into galaxies, and after a few thousand million years this process would be completed—no background would be left. Furthermore, the gas out of which the galaxies were initially composed would condense into stars. When this stage was reached hydrogen would be steadily converted into helium. After several hundreds of thousands of millions of years this process would be everywhere completed and all the stars would evolve toward the black dwarfs. [. . .] So finally the whole Universe would become entirely dead. This would be the running down of the Universe that was described so graphically by Jeans.

One of my main aims will be to explain why we get a different answer to this when we take account of the dynamic nature of the Universe. [. . .]

My nonmathematical friends often tell me that they find it difficult to picture this expansion. Short of using a lot of mathematics I cannot do better than use the analogy of a balloon with a large number of dots marked on its surface. If the balloon is blown up the distances between the dots increase in the same way as the distances between the galaxies. Here I should give a warning that this analogy must not be taken too strictly. There are several important respects in which it is definitely misleading. For example, the dots on the surface of a balloon would themselves increase in size as the balloon was being blown up. This is not the case for the galaxies, for their internal gravitational fields are sufficiently strong to prevent any such expansion. A further weakness of our analogy is that the surface of an ordinary balloon is two dimensional—that is to say, the points of its surface can be described by two co-ordinates; for example, by latitude and longitude. In the case of the Universe we must think of the surface as possessing a third dimension. [. . .]

The balloon analogy brings out a very important point. It shows we must not imagine that we are situated at the center of the Universe, just because we see all the galaxies to be moving away from us. For, whichever dot you care to choose on the surface of the balloon, you will find that the other dots all move away from it. In other words, whichever galaxy you happen to be in, the other galaxies will appear to be receding from you. [. . .]

Galaxies lying at only about twice the distance of the furthest ones that actually can be observed with the new telescope at Mount Palomar would be moving away from us at a speed that equalled light itself. Those at still greater distances would have speeds of recession exceeding that of light. Many people find this extremely puzzling because they have learned from Einstein's special theory of relativity that no material body can have a speed greater than light. This is true enough in the special theory of relativity which refers to a particularly simple system of space and time. But it is not true in Einstein's general theory of relativity, and it is in terms of the general theory that the Universe has to be discussed. The point is rather difficult, but I can do something toward making it a little clearer. The further a galaxy is away from us the more its distance will increase during the time required by its light to reach us. Indeed, if it is far enough away the light never reaches us at all because its path stretches faster than the light can make progress. This is what is meant by saying that the speed of recession exceeds the velocity of light. Events occurring in a galaxy at such a distance can never be observed at all by anyone inside our Galaxy, no matter how patient the observer and no matter how powerful his telescope. All the galaxies that we actually see are ones that lie close enough for their light to reach us in spite of the expansion of space that's going on. [. . .]

As you will easily guess, there must be intermediate cases where a galaxy is at such a distance that, so to speak, the light it emits neither gains ground nor loses it. In this case the path between us and the galaxy stretches at just such a rate as exactly compensates for the velocity of the light. The light gets lost on the way. It is a case, as the Red Queen remarked to Alice, of "taking all the running you can do to keep in the same place." We know fairly accurately how far away a galaxy has to be for this special case to occur. The answer is about 2,000,000,000 light years, which is only about twice as far as the distances that we expect the giant telescope at Mount Palomar to penetrate. This means that we are already observing about half as far into space as we can ever hope to do. If we built a telescope a million times as big as the one at Mount Palomar we could scarcely double our present range of vision. So what it amounts to is that owing to the expansion of the Universe we can never observe events that happen outside a certain quite definite finite region of space. We refer to this finite region as the observable Universe. The word "observable" here does not mean that we actually observe, but what we could observe if we were equipped with perfect telescopes. [. . .]

What causes the expansion? Does the expansion mean that as time goes on the observable Universe is becoming less and less occupied by matter? Is space finite or infinite? How old is the Universe? To settle these questions we shall now have to consider new trains of thought. These will lead us to strange conclusions.

First I will consider the older ideas—that is to say, the ideas of the nineteen-twenties and the nineteen-thirties—and then I will go on to offer my own opinion. Broadly speaking, the older ideas fall into two groups. One of them is distinguished by the assumption that the Universe started its life a finite time ago in a single huge explosion. On this supposition the present expansion is a relic of the violence of this explosion. This big bang idea seemed to me to be unsatisfactory even before detailed examination showed that it leads to serious difficulties [. . .] when we try to reconcile the idea of an explosion with the requirement that the galaxies have condensed out of diffuse background material. The two concepts of explosion and condensation are obviously contradictory, and it is easy to show, if you postulate an explosion of sufficient violence to explain the expansion of the Universe, that condensations looking at all like the galaxies could never have been formed. [. . .]

The second group of theories [. . .] work by monkeying with the law of gravitation. The conventional idea that two particles attract each other is only accepted if their distance apart is not too great. At really large distances, so the argument goes, the two particles repel each other instead. On this basis it can be shown that if the density of the background material is sufficiently small, expansion must occur. But once again there is a difficulty in reconciling all this with the requirement that the background material must condense to form the galaxies. For once the law of gravitation has been modified in this way the tendency is for the background material to be torn apart rather than for it to condense into galaxies. [. . .]

What would be the fate of our observable universe if any of these older theories had turned out to be correct? According to them every receding galaxy will eventually increase its distance from us until it passes beyond the limit of the observable universe—that is to say, they will move to a distance beyond the critical limit of about 2,000,000,000 light years that I have already mentioned. When this happens they will disappear—nothing that then occurs within them can ever be observed from our Galaxy. So if any of the older theories were right we should end in a seemingly empty universe, or at any rate in a universe that was empty apart perhaps from one or two very close galaxies that became attached to our Galaxy as satellites. Nor would this situation take very long to develop. Only about 10,000,000,000 years—that is to say, about a fifth of the lifetime of the Sun—would be needed to empty the sky of the 100,000,000 or so galaxies that we can now observe there.

My own view is very different. Although I think there is no doubt that every galaxy we observe to be receding from us will in about 10,000,000,000 years have passed entirely beyond the limit of vision of an observer in our Galaxy, yet I think that such an observer would still be able to see about the same number of galaxies as we do now. By this I mean that new galaxies will have condensed out of the background material at just about the rate necessary to compensate for those that are being lost as a consequence of their passing beyond our observable universe. At first sight it might be thought that this could not go on indefinitely because the material forming the background would ultimately become exhausted. The reason why this is not so, is that new material appears to compensate for the background material that is constantly being condensed into galaxies. This is perhaps the most surprising of all the conceptions of the New Cosmology. For I find myself forced to assume that the nature of the Universe requires continuous creation—the perpetual bringing into being of new background material.

The idea that matter is created continuously represents our ultimate goal in this book. It would be wrong to suppose that the idea itself is a new one. I know of references to the continuous creation of matter that go back more than twenty years, and I have no doubt that a close inquiry would show that the idea, in its vaguest form, goes back very much further than that. What is new about it is this: it has now been found possible to put a hitherto vague idea in a precise mathematical form. It is only when this has been done that the consequences of any physical idea can be worked out and its scientific value assessed. [. . .]

The most obvious question to ask about continuous creation is this: Where does the created material come from? It does not come from anywhere. Material simply appears—it is created. At one time the various atoms composing the material do not exist, and at a later time they do. This may seem a very strange idea and I agree that it is, but in science it does not matter how strange an idea may seem so long as it works—that is to say, so long as the idea can be expressed in a precise form and so long as its consequences are found to be in agreement with observation. Some people have argued that continuous creation introduces a new assumption into science—and a very startling assumption at that. Now I do not agree that continuous creation is an additional assumption. It is certainly a new hypothesis, but it only replaces a hypothesis that lies concealed in the older theories, which assume, as I have said before, that the whole of the matter in the Universe was created in one big bang at a particular time in the remote past. On scientific grounds this big bang assumption is much the less palatable of the two. For it is an irrational process that cannot be described in scientific terms. Continuous creation, on the other hand, can be represented by precise mathematical equations whose consequences can be worked out and compared with observation. On philosophical grounds too I cannot see any good reason for preferring the big bang idea. Indeed it seems to me in the philosophical sense to be a distinctly unsatisfactory notion, since it puts the basic assumption out of sight where it can never be challenged by a direct appeal to observation.

Perhaps you may think that the whole question of the creation of the Universe could be avoided in some way. But this is not so. To avoid the issue of creation it would be necessary for all the material of the Universe to be infinitely old, and this it cannot be for a very practical reason. For if this were so, there could be no hydrogen left in the Universe. [. . .] Hydrogen is being steadily converted into helium throughout the Universe and this conversion is a one-way process—that is to say, hydrogen cannot be produced in any appreciable quantity through the breakdown of the other elements. How comes it then that the Universe consists almost entirely of hydrogen? If matter were infinitely old this would be quite impossible. So we see that the Universe being what it is, the creation issue simply cannot be dodged. And I think that of all the various possibilities that have been suggested, continuous creation is easily the most satisfactory.

Now what are the consequences of continuous creation? Perhaps the most surprising result of the mathematical theory is that the average density of the background material must stay constant. The new material does not appear in a concentrated form in small localized regions but is spread throughout the whole of space. The average rate of appearance of matter amounts to no more than the creation of one atom in the course of about a year in a volume equal to that of a moderate-sized skyscraper. As you will realize, it would be quite impossible to detect such a rate of creation by direct experiment. But although this seems such a slow rate when judged by ordinary ideas, it is not small when you consider that it is happening everywhere in space. The total rate for the observable universe alone is about a hundred million, million, million, million, million tons per second. Do not let this surprise you because, as I have said, the volume of the observable universe is very large. Indeed I must now make it quite clear that here we have the answer to our question, Why does the Universe expand? For it is this creation that drives the Universe. The new material produces an outward pressure that leads to the steady expansion. But it does much more than that. With continuous creation the apparent contradiction between the expansion of the Universe and the requirement that the background material shall be able to condense into galaxies is completely overcome. For it can be shown that once an irregularity occurs in the background material a galaxy must eventually be formed. Such irregularities are constantly being produced by the gravitational effect of the galaxies themselves. For the gravitational field of the galaxies disturbs the background material and causes irregularities to form within it. So the background material must give a steady supply of new galaxies. Moreover, the created material also supplies unending quantities of atomic energy, since by arranging that newly created material should be composed of hydrogen we explain why in spite of the fact that hydrogen is being consumed in huge quantities in the stars, the Universe is nevertheless observed to be overwhelmingly composed of it.

We must now leave this extraordinary business of continuous creation for a moment to consider the question of what lies beyond the observable part of the Universe. In the first place you must let me ask, Does this question have any meaning? According to the theory it does. Theory requires the galaxies to go on forever, even though we cannot see them. That is to say, the galaxies are expanding out into an infinite space. There is no end to it all. And what is more, apart from the possibility of there being a few freak galaxies, one bit of this infinite space will behave in the same way as any other bit. [. . .]

Perhaps you will allow me a short diversion here to answer the question: How does the idea of infinite space fit in with the balloon analogy that I mentioned earlier? Suppose you were blowing up a balloon that could never burst. Then it is clear that if you went on blowing long enough you could make its size greater than anything I cared to specify, greater for instance than a billion billion miles or a billion billion billion miles and so on. This is what is meant by saying that the radius of the balloon tends to infinity. If you are used to thinking in terms of the balloon analogy, this is the case that gives you what we call an infinite space.

Now let us suppose that a film is made from any space position in the Universe. To make the film, let a still picture be taken at each instant of time. This, by the way, is what we are doing in our astronomical observations. We are actually taking the picture of the Universe at one instant of time—the present. Next, let all the stills be run together so as to form a continuous film. What would the film look like? Galaxies would be observed to be continually condensing out of the background material. The general expansion of the whole system would be clear, but though the galaxies seemed to be moving away from us there would be a curious sameness about the film. It would be only in the details of each galaxy that changes would be seen. The overall picture would stay the same because of the compensation whereby the galaxies that were constantly disappearing through the expansion of the Universe were replaced by newly forming galaxies. A casual observer who went to sleep during the showing of the film would find it difficult to see much change when he awoke. How long would our film show go on? It would go on forever.

There is a complement to this result that we can see by running our film backward. Then new galaxies would appear at the outer fringes of our picture as faint objects that come gradually closer to us. For if the film were run backward the Universe would appear to contract. The galaxies would come closer and closer to us until they evaporated before our eyes. First the stars of a galaxy would evaporate back into the gas from which they were formed. Then the gas in the galaxy would evaporate back into the general background from which it had condensed. The background material itself would stay of constant density, not through matter being created, but through matter disappearing. How far could we run our hypothetical film back into the past? Again according to the theory, forever. After we had run backward for about 5,000,000,000 years our own Galaxy itself would disappear before our eyes. But although important details like this would no doubt be of great interest to us there would again be a general sameness about the whole proceeding. Whether we run the film backward or forward the large-scale features of the Universe remain unchanged.

It is a simple consequence of all this that the total amount of energy that can be observed at any one time must be equal to the amount observed at any other time. This means that energy is conserved. So continuous creation does not lead to nonconservation of energy as one or two critics have suggested. The reverse is the case for without continuous creation the total energy observed must decrease with time.

We see, therefore, that no large-scale changes in the Universe can be expected to take place in the future. But individual galaxies will change and you may well want to know what is likely to happen to our Galaxy. This issue cannot be decided by observation because none of the galaxies that we observe can be much more than 10,000,000,000 years old as yet, and we need to observe much older ones to find out anything about the ultimate fate of a galaxy. The reason why no observable galaxy is appreciably older than this is that a new galaxy condensing close by our own would move away from us and pass out of the observable region of space in only about 10,000,000,000 years. So we have to decide the ultimate fate of our Galaxy again from theory, and this is what theory predicts. It will become steadily more massive as more and more background material gets pulled into it. After about 10,000,000,000 years it is likely that our Galaxy will have succeeded in gathering quite a cloud of gas and satellite bodies. Where this will ultimately lead is difficult to say with any precision. [. . .]

Without continuous creation the Universe must evolve toward a dead state in which all the matter is condensed into a vast number of dead stars. The details of the way this happens are different in the different theories that have been put forward, but the outcome is always the same. With continuous creation, on the other hand, the Universe has an infinite future in which all its present very large-scale features will be preserved.

LOREN C. EISELEY
Little Men and Flying Saucers, from
Harper's (1953)

L oren C. Eiseley (1907–1977) had an unusual gift in the history
of science prose. He was both creative and literary, not just in his
language, but also in his ability to invent literary forms. The selection here is a good example, and because in this case, form is critical
to following Eiseley's thought, I have cut nothing from it. Readers
today will be less surprised than midcentury readers were by Eiseley's "collage" style of argumentation, in which he follows one line
of observation and then turns to a different, seemingly unrelated
idea. Finally, he joins the parts together. Eiseley needed this complex
structure because he had something complicated to say. On one level
he is talking about science from the Jeans point of view, working out
the implications of established theory. On another level he is talking
about why people are fascinated by the notion of little men from
space. (In the early 1950s there had been a rash of "flying saucer"
incidents in which unknown objects were said to have been seen
soaring through the night sky.) Eiseley was also putting this into the
context of the early cold war when, with images of World War II's
bombings fresh in people's minds, they were anxious about what the
sky might bring. He weaves all these themes into this piece.

Eiseley himself was an anthropologist who often wrote about
biological and evolutionary subjects. He was a regular contributor to
Harper's magazine and wrote excellent books about evolution and
nature. He also was an environmentalist before there was an environmental movement.

Today as never before, the sky is menacing. Things seen indifferently last century by the wandering lamplighter now trouble a generation that has grown up to the wail of air-raid sirens and the ominous
expectation that the roof may fall at any moment. Even in daytime,
reflected light on a floating dandelion seed, or a spider riding a wisp
of gossamer in the sun's eye can bring excited questions from the
novice unused to estimating the distance or nature of aerial objects.

Since we now talk, write and dream endlessly of space rockets, it is no surprise that this thinking yields the obverse of the coin: that the rocket or its equivalent may have come first to us from somewhere "outside." As a youth, I may as well confess, I waited expectantly for it to happen. So deep is the conviction that there must be life out there beyond the dark, one thinks that if they are more advanced than ourselves they may come across space at any moment, perhaps in our generation. Later, contemplating the infinity of time, one wonders if perchance their messages came long ago, hurtling into the swamp muck of the steaming coal forests, the bright projectile clambered over by hissing reptiles, and the delicate instruments running mindlessly down with no report.

Sometimes when young, and fossil hunting in the western badlands, I had thought it might yet be found, corroding and long dead, in the Tertiary sod that was once green under the rumbling feet of titanotheres. Surely, in the infinite wastes of time, in the lapse of suns and wane of systems, the passage if it were possible, would have been achieved. But the bright projectile has not been found and now, in sobering middle age, I have long since ceased to look. Moreover, the present theory of the expanding universe has made time, as we know it, no longer infinite. If the entire universe was created in a single explosive instant a few billion years ago, there has not been a sufficient period for all things to occur even behind the star shoals of the outer galaxies. In the light of this fact it is now just conceivable that there may be nowhere in space a mind superior to our own.

If such a mind could exist, there are many reasons why it could not reside in the person of a little man. There is, however, a terrible human fascination about the miniature, and one little man in the hands of the spinner of folk tales can multiply with incredible rapidity. Our unexplainable passion for the small is not quenched at the borders of space, nor, as we shall see, in the spinning rings of the atom. The flying saucer and the much publicized little men from space equate neatly with our own projected dreams.

II. When I first heard of the little man there was no talk of flying saucers, nor did his owner ascribe to him anything more than an earthly origin. It has been almost a quarter of a century since I encountered him in a bone hunter's camp in the West. A rancher had brought him to us in a box. "I figured you'd maybe know about him," he said. "He'll cost you money, though. There's money in that little man."

"Man?" we said.

"Man," he countered. "What you'd call a pygmy or a dwarf, but smaller than any show dwarf I ever did see. A mummy, too, a little dead mummy. I figure it was some kind of bein' like us, but little. They put him in the place I found him. Maybe it was a thousand years ago. You'll likely know."

Our heads met over the box. The last paper was withdrawn. The creature emerged on the man's palm. I've seen a lot of odd things in the years since, and fakes by the score, but that little fellow gave me the creeps. He might have been two feet high in a standing posture— not more. He was mummified in a crouching position, arms folded. The face with closed eyes seemed vaguely evil. I could have sworn I was dreaming.

I touched it. There was a peculiar, fleshy consistency about it, still. It was not a dry mummy. It was more like what you would expect a natural cave mummy to be like. It had no tail. I know because I looked. And to this day the little man sits on there, in my brain, and as plain as yesterday I can see the faint, half-smirk of his mouth and the tiny black hands at his sides.

"You can have it for two hundred bucks," said the man. We glanced at each other, sighed, and shook our heads. "We aren't in the market," he said. "We're collecting, not buying, and we're staying with our bones."

"Okay," said the man, and gave us a straight look, closing his box. "I'm going to the carnival down below tonight. There's money in him. There's money in that little man."

I think it may have been just as well for us that we made no purchase. I have never liked the little man, nor the description of the carnival to which he and his owner were going. It may be, I used to think, that I will yet encounter him before I die, in some little colored tent on a country midway. Once, in the years since, I have heard a description that sounded like him in another guise. It involved a fantastic tale of some Paleozoic beings who hunted among the tree ferns when the world was ruled by croaking amphibians. The story did not impress me; I knew him by then for what he was: an anomalous mummified still-birth with an undeveloped brain.

I never expected to see him emerge again in books upon flying saucers, or to see the "little men" multiply and become so common that columnists would take note of them. Nor, though I should have known better, did I expect to live to hear my little man ascribed to an extra-planetary origin. There is a story back of him, it is true, but it is a history of this earth, and of all unlikely things it involves that great man of science, Charles Darwin, though by a curious, lengthy, and involved route.

III. Men have been men for so long that they tend to take the fact for granted. All their experience tells them that their children will precisely resemble themselves; that kittens will become cats and cats will have kittens, and that even caterpillars, though the pattern seems a little odd, will become butterflies, and butterflies will produce caterpillars. It is so habitual an event that we do not stop to ask why this happens, nor to consider that this amazing precision in results implies a strange ordering of life in a world we often think is chanceful and meaningless.

A few wise men since the time of the Greeks have found it a source of wonder, but they have been a minority. Most people have shrugged and spoken indifferently of the gods, or contented themselves, as the Christian world for so long did, with the idea of special creation of each species. Nevertheless the wise ones kept on wondering.

They found, as they began their first groping attempts to classify and arrange the living world, that in spite of the assumed individual creation of every living species by the supernatural intervention of divine power, a basic similarity of structure existed among many forms of life. This was a remarkable thing to find among supposedly individual creations. Offhand one would say that a much greater degree of spontaneous novelty would have been possible. In fact, man once innocently believed himself part of such a creation. The fabulous animals of the ancient bestiaries, the mermaids, griffins, and centaurs, not to mention the men whose ears were so large that their owners slept in them, would have been the natural, spontaneous products of such uncontrolled, creative, whimsy.

But there was the pattern: the ape and the man with their bone-by-bone correspondence. The very fact that one can add a plural to the word *reptile* and so suggest anything from a Brontosaurus to a garter snake shows that a pattern exists. Birds all have feathers, wings, and claws; they are a common class in spite-of their diversities. They have been pulled into many shapes, but there is still an eternal "bird-liness" about them. They are built on a common plan just as I share mammalian characters with a small mouse who inhabits my desk drawer. This is hard to account for in a disordered world, so that recently, when I came upon this mouse, trapped and terrified in the wastebasket, his similarity to myself rendered me helpless, and out of sheer embarrassment I connived in his escape.

Now so long as these remarkable patters could be observed only in the living world around us they occasioned no great alarm. Even after Cuvier, in 1812, made a magnificent attempt to reduce the forms of animal life to four basic blueprints or "archetypes" of divergent character, no one was particularly disturbed—least of all from the religious point of view. In the words of one great naturalist, Louis Agassiz, "This plan of creation . . . has not grown out of the necessary action of physical laws, but was the free conception of the Almighty Intellect, matured in his thought before it was manifested in tangible external forms."

It was not long, however, before *pattern*, the divine blueprint, first recognized in the existing world, was extended by the geologist across the deeps of time. The animal world of the past was in the process of discovery. It proved to be a world of strange and unknown beasts, and it was a world without man. Curiously enough, it was soon learned that extinct animals could be fitted into the broad classifications of the existing world. They were mammals or amphibia or reptiles, as the case might be. Though no living eye had beheld them, they seemed to mark the continuation of the divine abstraction, the eternal patterns, across the enormous time gulf of the past.

The second fact, that man had not been discovered, was a cause for dismay. In the man-centered universe of the time, one can appreciate the anguish of the Reverend Mr. Kirby discovering the Age of Reptiles: "Who can think that a being of unbounded power, wisdom, and goodness, should create a world merely for the habitation of a race of monsters, without a single, rational being in it to serve and glorify him?" This is the wounded outcry of the human ego as it fails to discover its dominance among the beasts of the past. Even more tragically, it learns that the world supposedly made for its enjoyment has existed for untold eons entirely indifferent to its coming. The chill vapors of time and space are beginning to filter under the closed door of the human intellect.

It was in these difficult straits, in the black night of his direst foreboding, that the doctrine of geologic prophecy was evolved by man. For fifty years it would hold time at bay and in one last great effort, its proponents, by clever analogies, would attempt to extend the human drama across the infinite worlds of space; it echoes among us still in the shape of the little men of the flying saucers. No braver mythos was ever devised under the cold eye of science.

In an old book from my shelves, Hugh Miller's *The Testimony of the Rocks*, I find this passage: *"Higher still in one of the deposits of the Trias we are startled by what seems to be the impression of a human hand of an uncouth massive shape, but with the thumb apparently set in opposition, as in man, to the other fingers."*

There is only one way to understand this literature. The biologists of the first half of the nineteenth century had recognized that the unity of animal organization descends into past ages and is observable in forms no living eye has beheld. It was, they believed, an immaterial, a supernatural line of connection. They refused to see in this unity of plan an actual physical relationship. Instead they read the past as a successive series of creations and extinctions upon a divinely modifiable but consistent plan. "Geology," said one writer, "unrolls a prophetic scroll, in which the earlier animated creation points on to the later."

It is in the light of this philosophy that the hand, "massive" and of "uncouth shape," must be interpreted. It foreshadows, out of that slimy concourse of sprawling amphibians and gaping lizards, the eventual emergence of man. Splayed, monstrous, and mud-smeared, it haunts the future. That it is the footprint of some wandering salamandrine beast of the coal swamps may be granted, but it is also a vertebrate. Its very body forecasts the times to come.

It would be erroneous, however, to conceive of salamanders as being the major preoccupation of our geological prophets. They scanned the anatomy of fishes, birds, and reptiles, seeking in their skeletons anticipations of the more perfect structure of man. If they found footprints of fossil bipeds it was a "sign" foretelling man. All things led in his direction. Prior to his entrance the stage was merely under preparation. In this way the blow to the human ego had been softened. The past was only the prologue to the Great Play. Man was at the heart of things after all.

It was a strange half century, as one looks back upon it—that fifty years before the publication of Darwin's *Origin of Species*. It was dominated by a generation that saw the world as a complex symbolic system pointing in the direction of man, who was foreknown and prefigured from the beginning. Man, who comes last, is the end of this strange cycle. With him, in the eyes of many of these thinkers, the process ceases and no further changes in the world of life are to be expected. Since the transcendental "evolutionists" were man-centered, questions involving divergent evolution and adaptation did not come easily to their minds. Working with an immaterial and abstract Platonic concept, it was inevitable that they should seek to extend their doctrine across the wilds of space. Because the pattern was capable of modification, the possibility of the existence of small men, large men, or men of different colors upon other planets did not trouble them, but men they ought to be. There was little comprehension of the fact that man had acquired his particular bodily structure and upright posture through a peculiar set of evolutionary circumstances, not easily to be duplicated.

IV. The theory of the plurality of worlds is a very ancient one; that is, the notion that the lights seen elsewhere in space may be bodies like that which we inhabit. After the rise of the Copernican astronomy and the growing realization that our earth is part of a planetary system revolving around a central sun, it was often contended by philosophers that the other stars seen in space must be similar suns with similar planetary satellites.

Quarrels arose between those who believed God's power infinitely and creatively extended among the stars, and those who regarded it as heresy and dangerous to Christian belief to imply that the infinite mind might be concerned with more than the beings of this planet. It was a struggle heightened by an enormous extension of man's vision into the worlds of the infinitely far and the infinitely small, the telescope and the microscope having momentarily stunned the human imagination. Some clung frantically to the little tight-fenced world of the Middle Ages, refusing to acknowledge what these instruments revealed. Others, with greater willingness to accept the new, tried, nevertheless to equate what they saw with old beliefs and to elaborate an "astrotheology."

In the fifties of the last century there was a great outburst of interest in the possibility of life on other worlds. The recently discovered life history of our own planet and improvements in astronomical apparatus had all excited great interest on the part of a public wavering in its loyalty between old religious dogmas and the new revelations of science. Speculation, in many instances, was roaming far in advance of actual observation.

"The inhabitants of Jupiter," wrote William Whewell in 1854, "must . . . it would be seem, be cartilaginous and glutinous masses. If life be there it does not seem in any way likely, that the living things can be anything higher in the scale of being, than such boneless watery pulpy creatures. . . ."

This remark is not intended as merely innocent theorizing. In this work, the *Plurality of Worlds*, Whewell indicates his definite opposition to the idea that the other planets, or the more remote worlds in other galaxies, are inhabited. At best he is willing to grant the existence of a few gelatinous creatures such as he mentions in the above passage, but that man is to be found elsewhere, he denies. He argues that there are superior and inferior regions of space. Man, preceded by endless eons of lower creatures in time, is yet a superior being. He calls attention to the fact that "the intelligent part of creation is thrust into the compass of a few years, in the course of myriads of ages; why not then into the compass of a few miles, in the expanse of systems?" On this earth a "supernatural interposition" has introduced man; the planet is unique.

Whewell's essay generated a storm of discussion. His was not the popular side of the controversy. Sir David Brewster countered with a volume significantly titled *More Worlds Than One*, in which he bluntly asserts: "The function of one satellite must be the function of all the rest. The function of our Moon, to give light to the Earth, must be the function of the other twenty-two moons of the system; and the function of the Earth, *to support inhabitants*, must be the function of all other planets." He dwells on the "grand combination" of *"infinity of life, with infinity of matter."*

Brewster, moreover, calls attention to the invisible domain revealed by the microscope and argues from this that God has all along been attentive to forms of life of which we had no knowledge. So intriguing became the relativity of size that one author even produced a work whose subtitle bore the query *Are Ultimate Atoms Inhabited Worlds?* Stories like Fitz-James O'Brian's "The Diamond Lens," or Ray Cummings' "The Girl in the Golden Atom," stem from such thought.

Another writer, William Williams, in *The Universe No Desert, the Earth No Monopoly*, strikes more directly at the heart of the argument. He invokes geological prophecy and extends it directly across space: "The archetypal idea of man, revealed in the lower vertebrated animals, proves God's foreknowledge of man's existence; and it equally applies to vertebrates on Jupiter or Neptune as to those on the Earth; and still farther, to the Universe, as these animals were within its precincts."

Williams was not the first nor the last man to utter these sentiments but he did so with a fierce singleness of purpose. The life plans were immanent, prophetic, and immaterial. They could thus be projected across space. Why, he argues with the same horror that the Reverend Mr. Kirby had exhibited toward the Age of Reptiles, should God "banish his own image to one diminutive enclosure and surround . . . the residue of His immense Person with unintelligent, half-formed, crude monsters?" If man is regarded as a good production here he must be found in endless duplication throughout the worlds. The pattern in the rocks of this earth is the pattern of the whole.

The shattering of this scheme of geological prophesy was the work of many men, but it was Charles Darwin who brought the event to pass, and who engineered what was to be one of the most dreadful blows that the human ego has ever sustained: the demonstration of man's physical relationship to the world of the lower animals. It is quite apparent, however, that there is an aspect of Darwin's discoveries which has never penetrated to the mind of the general public. It is the fact that once undirected variation and natural selection are introduced as the mechanisms controlling the development of plants and animals, the evolution of every world in space becomes a series of unique, historical events. The precise accidental duplication of a complex form of life is extremely unlikely to occur in even the same environment, let alone in the different background and atmosphere of a far-off world.

In the modern literature on space travel I have read about cabbage men and bird men; I have investigated the loves of the lizard men and the tree men, but in each case I have labored under no illusion. I have been reading about a man, *Homo sapiens*, that common earthling, clapped into an ill-fitting coat of feathers and retaining all his basic human attributes including an eye for the pretty girl who has just emerged from the space ship. His lechery and miscegenating proclivities have an oddly human ring, and if this is all we are going to find on other planets, I, for one, am going to be content to stay at home. There is quite enough of that sort of thing down here, without encouraging it throughout the starry systems.

V. The truth is that man is a solitary and peculiar development. I do not mean this in any irreverent or contemptuous sense. I want merely to point out that when Charles Darwin and his colleagues established the community of descent of the living world, and observed the fact of divergent evolutionary adaptation, they destroyed forever the concept of geological prophecy. They did not eliminate the possibility of life on other worlds, but the biological principles which they established have totally removed the likelihood that our descendants, in the next few decades, will be entertaining little men from Mars. I would be much more willing to consider the possibility of sitting down to lunch with a purple polyp, but even that has unfortunate connotations with the life of this planet.

Geologic prophecy was based on two things: first a belief, as we have seen, in the man-centered nature of the universe, and second the assumption that since the animals of the past had no physical connection with those of the present, some kind of abstract, immaterial plan in the mind of the Creator linked the forms of the past with those of the present day. The early nineteenth-century thinkers perceived a genuine relationship, but their attachment to the idea of special creation prevented them from recognizing that the relationship arose out of simple biological "descent with modification."

Man could not be proved preordained or predestined from the beginning simply because he showed certain affinities to Paleozoic vertebrates. Instead, he was merely one of many descendants of the early vertebrate line. A moose or a mongoose would have had equally good reason to contend that as a modern vertebrate he had been "prefigured from the beginning," and that the universe had been organized with him in mind.

The situation is something like that of walking through a hall of trick mirrors and being pulled out of shape. The mirror of time does that to all things living, and the distortions stay. Nevertheless, there is a pattern of sorts so that if you have come by the mirror that makes men, and somewhere behind you there is a mirror that makes black cats, you can still see the pattern. You and the cat are related; the shreds of the original shape are in your bones and the shreds of primordial thought patterns move in the eyes of both of you and are understood by both. But somewhere there must be an original pattern; somewhere cat and man and weasel must leap into a single shape. That shape lies inconceivably remote from us now, far back along the time stream. It is historical. In that sense, and in that sense only, the archetype did indeed exist.

Darwin saw clearly that the succession of life on this planet was not a formal pattern imposed from without, or moving exclusively in one direction. Whatever else life might be it was adjustable and not fixed. It worked its way through difficult environments. It modified and then, if necessary, it modified again, along roads which would never be retraced. Every creature alive is the product of a unique history. The statistical probability of its precise reduplication on another planet is so small as to be meaningless. Life, even cellular life, may exist out yonder in the dark. But high or low in nature, it will not wear the shape of man. That shape is the evolutionary product of a strange, long wandering through the attics of the forest roof, and so great are the chances of failure, that no pseudo-human thing is likely ever to come that way again.

The picture of the little man of long ago rises before me as I write. As I have hinted, he was simply a foetal monster, long since scientifically diagnosed and dismissed. The small skull that lent the illusion of maturity to the mummified infant contained a brain which had failed to develop. The describers of two-foot men forget that a normal human brain cannot function with a capacity, at the very minimum, of less than about 900 cubic centimeters of capacity. A man with a hundred-cubic-centimeter brain will not be a builder of flying saucers; he will be less intelligent than an ape. In any case he does not exist.

In a universe whose size is beyond human imagining, where our world floats like a dust mote in the void of night, men have grown inconceivably lonely. We scan the time scale and the mechanisms of life itself for portents and signs of the invisible. As the only thinking mammals on the planet—perhaps the only thinking animals in the entire sidereal universe—the burden of consciousness has grown heavy upon us. We watch the stars but the signs are uncertain. We uncover the bones of the past and seek for our origins. There is a path there, but it appears to wander. The vagaries of the road may have a meaning however; it is thus we torture ourselves.

Lights come and go in the night sky. Men, troubled at last by the things they build, may toss in their sleep and dream bad dreams, or lie awake while the meteors whisper greenly overhead. But nowhere in all space or on a thousand worlds will there be men to share our loneliness. There may be wisdom; there may be power; somewhere across space great instruments, handled by strange, manipulative organs, may stare vainly at our floating cloud wrack, their owners yearning as we yearn. Nevertheless, in the nature of life and in the principles of evolution we have had our answer. Of men elsewhere, and beyond, there will be none forever.

>-+-◆→-O-◆+-◄

WERNER HEISENBERG

Atomic Physics and Causal Law, from
The Physicist's Conception of Nature (1958)

Werner Heisenberg (1901–1976) may have been the most alarming scientist since Galileo. His ideas scared Einstein, who understood them, and have terrified many more people who do not understand them. In 1927 Heisenberg produced the famous "uncertainty principle" that says that observers can know some things about a subatomic particle, but not everything. That is, learning one thing makes it impossible to learn others. Instead of a classically mechanical universe shaped by cause and effect, Heisenberg's universe is shaped by probabilities. At the level of the individual electron, one cannot predict exactly what will happen. Physics seemed to be doing away with the explanatory powers that had made it so powerful. People feared that scientists had become so advanced that they were becoming irrational. Looking at the reaction to Heisenberg, it is easier to understand why Galileo so terrified the people of his time. He, too, seemed to be turning all knowledge into putty, by means of an incomprehensible set of equations. Now when we read Galileo, we see him as a prophet of understanding, and the optimists among us can hope the same holds true for Heisenberg.

This selection, from a series of nonmathematical talks by Heisenberg, makes that optimism seem more reasonable. He is discussing the question of cause and effect. Like the essays by Jeans and Eiseley, Heisenberg is considering what the latest accepted theories of nature tell us about nature itself. Although his writing is notoriously difficult, this essay is astonishingly clear. His ideas were so clear to him that he knew exactly how to make them clear to others. Instead of labeling twentieth-century reasoning as irrational, he shows that the overconfidence of the nineteenth century was where the unreason lay. It is, by the way, interesting to contrast his tone with that of Oppenheimer's essay. Whereas Oppenheimer suggests that nothing all that radical has happened, Heisenberg thinks otherwise.

Some of the most interesting general effects of modern atomic physics are the changes which it has brought about in the concept of natural laws. In the last few years many people have stated that

modern atomic physics has abolished the law of cause and effect, or
at least that it has shown the latter to be partially inoperative, and
that we can no longer properly speak of processes determined by natural laws. Occasionally these statements simply assert that the principle of causality is no longer compatible with modern atomic theory.
Such assertions are always vague if the concept of causality or natural
law is not adequately defined. I should therefore like to [. . .] discuss
the consequences of quantum theory and the development of atomic
physics in the last few years. Very little of this development has
reached the public so far, but it looks very much as if this development also will have repercussions in the sphere of philosophy.

The Concept of Causality

The use of the concept of causality for describing the law of cause
and effect is of relatively recent origin. In previous philosophies the
word *causa* had a very much more general significance than it has
today. [. . .]

The transformations of *causa* into the modern concept of cause
have taken place in the course of centuries, in close connection with
the changes in man's conception of reality and with the creation of science at the beginning of the modern age. As material processes became
more prominent in man's conception of reality, the word *causa* was
used increasingly to refer to the particular material event which preceded, and had in some way caused, the event to be explained. Thus
even Kant, who frequently did what at root amounted to drawing
philosophic consequences from the developments in science since
Newton's time, already used the word 'causality' in a nineteenth-
century sense. When we experience an event we always assume that
there was another event preceding it from which the second has followed according to some law. Thus the concept of causality became
narrowed down, finally, to refer to our belief that events in nature are
uniquely determined, or, in other words, that an exact knowledge of
nature or some part of it would suffice, at least in principle, to determine the future. Newton's physics was so constructed that the future
motion of a system could be calculated from its particular state at a
given time. The idea that nature really was like this was perhaps enunciated most generally and most lucidly by Laplace when he spoke of a

demon, who at a given time, by knowing the position and motion of every atom, would be capable of predicting the entire future of the world. When the word 'causality' is interpreted in this very narrow sense, we speak of 'determinism', by which we mean that there are immutable natural laws that uniquely determine the future state of any system from its present state.

Statistical Laws

From its very beginnings atomic physics evolved concepts which do not really fit this picture. True, they do not contradict it basically, but the approach of atomic physics was by its very character different from that of determinism. Even in the ancient atomic theory of Democritus and Leucippus it was assumed that large-scale processes were the results of many irregular processes on a small scale. That this is basically the case is illustrated by innumerable examples in everyday life. Thus, a farmer need only know that a cloud has condensed and watered his fields. He does not bother about the path of each individual drop of rain. To give another example, we know precisely what is meant by the word 'granite,' even when we are ignorant of the form, colour, and chemical composition of each small constituent crystal. Thus we always use concepts which describe behaviour on the large scale without in the least bothering about the individual processes that take place on the small scale.

This notion of the statistical combination of many small individual events was already used in ancient atomic theory as the basis for an explanation of the world, and was generalized in the concept that all the sensory qualities of matter were indirectly caused by the position and movements of the atoms. Thus Democritus wrote that things only *appeared* to be sweet or bitter, and only appeared to have colour, for in reality there existed only atoms and empty space. Now, if the processes which we can observe with our senses are thought to arise out of the interactions of many small individual processes, we must needs conclude that all natural laws may be considered to be only statistical laws. [. . .]

Yet we must remind the reader that in everyday life all of us encounter statistical laws with every step we take, and make these laws the basis of our practical actions. Thus, when an engineer is constructing a dam he always bases his calculations on the average yearly rainfall, although he cannot have the faintest idea when it will rain and how much of it at a time. [. . .]

From the very beginning of modern times attempts have been made to explain both qualitatively and also quantitatively the behaviour of matter through the statistical behaviour of atoms. Robert Boyle demonstrated that we could understand the relations between the pressure and the volume of a gas if we looked upon pressure as the many thrusts of the individual atoms on the walls of the vessel. Similarly thermodynamical phenomena have been explained by the assumption that atoms move more violently in a hot body than in a cold one. This statement could be given a quantitative mathematical formulation and in this way the laws of heat could be understood.

This application of the concept of statistical laws was finally formulated in the second half of the last century as the so-called *statistical mechanics*. In this theory, which is based on Newton's mechanics, the consequences that spring from an incomplete knowledge of a complicated mechanical system are investigated. Thus in principle it is not a renunciation of determinism. While it is held that the details of events are fully determined according to the laws of Newton's mechanics, the condition is added that the mechanical properties of the *system* are not fully known.

Gibbs and Boltzmann[1] managed to formulate this kind of incomplete knowledge mathematically, and Gibbs was able to demonstrate that, in particular, our conception of temperature is closely related to the incompleteness of our knowledge. [. . .] Gibbs was the first to introduce a physical concept which can only be applied to an object when our knowledge of the object is incomplete. If for instance the motion and position of each molecule in a gas were known, then it would be pointless to continue speaking of the temperature of the gas. The concept of temperature can only be used meaningfully when the system is not fully known and we wish to derive statistical conclusions from our incomplete knowledge.

---

[1] *Willard Gibbs (1839–1903) and Ludwig Boltzmann (1844–1906) developed statistical mechanics.*

The Statistical Character of Quantum Theory

Although the discoveries of Gibbs and Boltzmann made an incomplete knowledge of a system part of the formulation of physical laws, nevertheless determinism was still present in principle until Max Planck's famous discovery ushered in quantum theory. Planck, in his work on the theory of radiation, had originally encountered an element of uncertainty in radiation phenomena. He had shown that a radiating atom does not deliver up its energy continuously, but discreetly in bundles. This assumption of a discontinuous and pulse-like transfer of energy, like every other notion of atomic theory, leads us once more to the idea that the emission of radiation is a statistical phenomenon. However, it took two and a half decades before it became clear *that quantum theory actually forces us to formulate these laws precisely as statistical laws* and to depart radically from determinism. Since the work of Einstein, [Niels] Bohr and [Arnold] Sommerfeld, Planck's theory has proved to be the key with which the door to the entire sphere of atomic physics could be opened. Chemical processes could be explained by means of the Rutherford-Bohr atomic model, and since then, chemistry, physics and astrophysics have been fused into unity. With the mathematical formulation of quantum-theoretical laws pure determinism had to be abandoned.

Since I cannot speak of the mathematical methods here, I should merely like to mention some aspects of the strange situation confronting the physicist in atomic physics.

We can express the departure from previous forms of physics by means of the so-called uncertainty relations. It was discovered that it was impossible to describe simultaneously both the position and the velocity of an atomic particle with any prescribed degree of accuracy. We can either measure the position very accurately—when the action of the instrument used for the observation obscures our knowledge of the velocity, or we can make accurate measurements of the velocity and forego knowledge of the position. The product of the two uncertainties can never be less than Planck's constant. This formulation makes it quite clear that we cannot make much headway with the concepts of Newtonian mechanics, since in the calculation of a mechanical process it is essential to know simultaneously the position and velocity at a particular moment, and this is precisely what quantum theory considers to be impossible.

Another formulation is that of Niels Bohr, who introduced the *concept of complementarity*. By this he means that the different intuitive pictures which we use to describe atomic systems, although fully adequate for given experiments, are nevertheless mutually exclusive. Thus, for instance, the Bohr atom can be described as a small-scale planetary system, having a central atomic nucleus about which the external electrons revolve. For other experiments, however, it might be more convenient to imagine that the atomic nucleus is surrounded by a system of stationary waves whose frequency is characteristic of the radiation emanating from the atom. Finally, we can consider the atom chemically. We can calculate its heat of reaction when it becomes fused with other atoms, but in that case we cannot simultaneously describe the motion of the electrons. Each picture is legitimate when used in the right place, but the different pictures are contradictory and therefore we call them mutually complementary. The uncertainty that is attached to each of them is expressed by the uncertainty relation, which is sufficient for avoiding logical contradiction between the different pictures.

Even without entering into the mathematics of quantum theory these brief comments might have helped us to realize *that the incomplete knowledge of a system must be an essential part of every formulation in quantum theory.* Quantum theoretical laws must be of a statistical kind. To give an example: we know that the radium atom emits alpha-radiation. Quantum theory can give us an indication of the probability that the alpha-particle will leave the nucleus in unit time, but it cannot predict at what precise point in time the emission will occur, for this is uncertain in principle. [. . .]

In large-scale processes this statistical aspect of atomic physics does not arise, generally because statistical laws for large-scale processes lead to such high probabilities that to all intents and purposes we can speak of the processes as determined. Frequently, however, there arise cases in which a large-scale process depends on the behaviour of one or of a few atoms alone. In that case, the large-scale process also can only be predicted statistically. I should like to illustrate this by means of a well-known but unhappy example, that of the atom bomb. In an ordinary bomb the strength of the explosion can be predicted from the mass of the explosive material and its chemical composition. In the atom bomb we can still indicate an upper and a lower limit of the strength of the explosion but we cannot make exact calculations of this strength in advance. This is impossible in principle since it depends on the behaviour of only a few atoms at the moment of firing.

Similarly, there may be biological processes—and Jordan, especially, has drawn our attention to this—in which large-scale events are set off by processes in individual atoms; this would appear to be the case particularly in the mutation of genes during hereditary processes. These two examples were meant to illustrate the practical consequences of the statistical character of quantum theory. This development, too, was concluded over two decades ago, and we cannot possibly assume that the future will see any basic changes in this field.

The History of More Recent Atomic Physics

However, very recently, a new point of view has been added to the problem of causality, which, as I have said at the beginning, stems from the latest developments in atomic physics. [. . .] At the beginning of the modern age the atomic concept was closely linked with that of chemical elements. An element was characterized by the fact that it could not be decomposed further by chemical means, and by the fact that a particular kind of atom belonged to a particular element. Thus, pure carbon consists of carbon atoms only, and pure iron of iron atoms only, and one was forced to assume that there were just as many sorts of atoms as there were chemical elements. Since ninety-two different kinds of chemical elements were eventually known, one assumed the existence of ninety-two kinds of atoms.

However, this conception is not satisfying if we approach the problem from the basic premises of atomic theory. Originally atoms were introduced to explain matter qualitatively through their movements and structure. This conception can have a true explanatory value only if all the atoms are equal, or if there are but a few kinds of atoms—in other words, if the atoms themselves have no qualities at all. If, now, one is forced to assume the existence of ninety-two qualitatively different atoms, this is no great gain over the simple assertion that there are things which differ qualitatively. Thus the assumption of ninety-two different basic particles has long been felt to be unsatisfactory and attempts were made to reduce these ninety-two kinds of atoms to a smaller number of elementary particles. Quite early on it

was thought that chemical atoms themselves might well be composed
of a very small number of basic building-stones. After all, even the
oldest attempts to change one chemical substance into another must
have been based on the assumption that in the final analysis all matter was one.

The last fifty years have shown that all chemical atoms are composed of only three basic building-stones, which we call protons, neutrons and electrons. The atomic nucleus consists of protons and neutrons and is surrounded by a number of electrons. Thus, for instance,
the nucleus of the carbon atom consists of six protons and six neutrons and is surrounded by six electrons at a relatively great distance.
Thanks to the development of nuclear physics in the 1930's we now
have these three different kinds of particles, instead of the ninety-two
different kinds of atoms, and in this respect atomic theory has followed the very path which its basic assumptions had suggested. When
it became clear that all the atoms were composed of only three kinds
of basic building-stones, there arose the practical possibility of changing chemical elements into one another. We know that this physical
possibility was soon to be followed by its technical realization. [. . .]

However, during the last two decades, the picture has once more
become a little confused. In addition to the three elementary particles
already mentioned, the proton, neutron and electron, new elementary
particles were discovered in the thirties and during the last few years
their number has increased most disturbingly. In contrast to the three
basic building-stones, these new particles are always unstable and
have very short lifetimes. Of these so-called *mesons*, one type has a
lifetime of about a millionth of a second, another lives for only one
hundredth part of that time, and a third, which has no electrical
charge, for only a hundred billionth of a second. However, apart from
their instability, these three new elementary particles behave very similarly to the three stable building-stones of matter. At first glance it
looks as if, once more, we were forced to assume the existence of a
great number of qualitatively different particles, which would be most
unsatisfactory in view of the basic assumptions of atomic physics.
However, during experiments in the last few years, it has become clear
that these elementary particles can change into one another during
their collisions, with great changes of energy. When two elementary
particles collide with a great energy of motion, new elementary parti-

cles are created and the original particles, together with their energy, are changed into new matter.

This state of affairs is best described by saying that all particles are basically nothing but different stationary states of one and the same stuff. Thus even the three basic building-stones have become reduced to a single one. *There is only one kind of matter but it can exist in different discrete stationary conditions.* Some of these conditions, *i.e.*, protons, neutrons and electrons, are stable while many others are unstable.

Relativity Theory and the Dissolution of Determinism

[. . .] When dealing with collisions of high-energy elementary particles we must consider the space-time structure of special relativity theory. This space-time structure was not very important in the quantum theory of the atomic shell, since in it the electrons move relatively slowly. Now, however, we are dealing with elementary particles which move almost with the velocity of light, and whose behaviour can therefore only be described with the help of relativity theory. Fifty years ago Einstein discovered that the structure of space and time was not quite as simple as we imagine it to be in everyday life. If we describe all those events as past of which, at least in principle, we can obtain some knowledge, and as future all those events on which, at least in principle, we can still have some influence, then according to our naïve conception we believe that between these two types of events there is but one infinitely short moment which we call the present. This was just the conception on which Newton had based his mechanics. Since Einstein's discovery in 1905, we know that between what I have just called 'future' and 'past' there exists an interval whose extension in time depends on the distance in space between an event and its observer. Thus, the present is not limited to an infinitely short moment in time.

Relativity theory assumes that in principle no effect can be propagated faster than the velocity of light. Now this trend in relativity theory leads to difficulties in connection with the uncertainty relations of quantum theory. According to relativity theory the only effects possible are in that part of space-time limited by the so-called light-cone, *i.e.*, those points in space-time which can be reached by a lightwave emanating from the effective point. This region in space-time is thus—

and this must be stressed—very strictly limited. On the other hand, we have found that in quantum theory a clear determination of position—in other words, a sharp delimitation of space—presupposes an infinite uncertainty of velocity and thus also of momentum and energy. This state of affairs has as its practical consequence the fact that in attempting to arrive at a mathematical formulation of the interactions of the elementary particles, we shall always encounter infinite values for energy and momentum, preventing a satisfactory mathematical statement.

➤⊷❍⊶◄

RICHARD FEYNMAN

The Distinction of Past and Future, from
The Character of Physical Law (1965)

Richard Feynman (1918–1988) became nationally famous when he dropped part of a rocket, called an O-ring, into a glass of ice water and showed that it became hard almost immediately. The implication was that on the cold morning of the space shuttle *Challenger*'s launch, the ring had grown stiff and thus failed. This experiment became the symbol of what went wrong when the *Challenger* blew up shortly after launch and also was a classic Feynman ploy. His whole scientific life was devoted to clarifying obscure technical issues, or at least making them clear enough so that other technically trained people could understand them. He also wrote several books for laypeople. His book *Surely You're Joking, Mr. Feynman!* is reputed to be the only humorous work on the subject of modern physics.

In this selection Feynman tackles a problem that sounds hopelessly metaphysical: why is the past past? He suggests that it is a perfectly understandable phenomenon, one that flows inevitably from the facts of physics. It shows that science, even on a theoretical level, can still be fun.

It is obvious to everybody that the phenomena of the world are evidently irreversible. I mean things happen that do not happen the other way. You drop a cup and it breaks, and you can sit there a long time waiting for the pieces to come together and jump back into your hand. If you watch the waves of the sea breaking, you can stand there and wait for the great moment when the foam collects together, rises up out of the sea, and falls back farther out from the shore—it would be very pretty!

The demonstration of this in lectures is usually made by having a section of moving picture in which you take a number of phenomena, and run the film backwards, and then wait for all the laughter. The laughter just means this would not happen in the real world. But actually that is a rather weak way to put something which is as obvious and deep as the difference between the past and the future; because even without an experiment our very experiences inside are completely different for past and future. We remember the past, we do not remember the future. We have a different kind of awareness about what might happen than we have of what probably has happened. The past and the future look completely different psychologically, with concepts like memory and apparent freedom of will, in the sense that we feel that we can do something to affect the future, but none of us, or very few of us, believe that there is anything we can do to affect the past. Remorse and regret and hope and so forth are all words which distinguish perfectly obviously the past and the future.

Now if the world of nature is made of atoms, and we too are made of atoms and obey physical laws, the most obvious interpretation of this evident distinction between past and future, and this irreversibility of all phenomena, would be that some laws, some of the motion laws of the atoms, are going one way—that the atom laws are not such that they can go either way. There should be somewhere in the works some kind of a principle that uxles only make wuxles, and never vice versa, and so the world is turning from uxley character to wuxley character all the time—and this one-way business of the interactions of things should be the thing that makes the whole phenomena of the world seem to go one way.

But we have not found this yet. That is, in all the laws of physics that we have found so far there does not seem to be any distinction between the past and the future. The moving picture should work the same going both ways, and the physicist who looks at it should not laugh.

Let us take the law of gravitation as our standard example. If I have a sun and a planet, and I start the planet off in some direction, going around the sun, and then I take a moving picture, and run the moving picture backwards and look at it, what happens? The planet goes around the sun, the opposite way of course, keeps on going around in an ellipse. The speed of the planet is such that the area swept out by the radius is always the same in equal times. In fact it just goes exactly the way it ought to go. It cannot be distinguished from going the other way. So the law of gravitation is of such a kind that the direction does not make any difference; if you show any phenomenon involving only gravitation running backwards on a film it will look perfectly satisfactory. You can put it more precisely this way.

If all the particles in a more complicated system were to have every one of their speeds reversed suddenly, then the thing would just unwind through all the things that it wound up into. If you have a lot of particles doing something, and then you suddenly reverse the speed, they will completely undo what they did before.

This is in the law of gravitation, which says that the velocity changes as a result of the forces. If I reverse the time, the forces are not changed, and so the changes in velocity are not altered at corresponding distances. So each velocity then has a succession of alterations made in exactly the reverse of the way that they were made before, and it is easy to prove that the law of gravitation is time-reversible.

The law of electricity and magnetism? Time reversible. The laws of nuclear interaction? Time reversible as far as we can tell. [. . .] In other words, we believe that most of the ordinary phenomena in the world, which are produced by atomic motions, are according to laws which can be completely reversed. So we will have to look some more to find the explanation of the irreversibility.

If we look at our planets moving around the sun more carefully, we soon find that all is not quite right. For example, the Earth's rotation on its axis is slightly slowing down. It is due to tidal friction, and you can see that friction is something which is obviously irreversible. If I take a heavy weight on the floor, and push it, it will slide and stop. If I stand and wait, it does not suddenly start up and speed up and come into my hand. So the frictional effect seems to be irreversible. But a frictional effect, as we discussed at another time, is the result of the enormous complexity of the interactions of the weight of the wood, the jiggling of the atoms inside. The organized motion of the weight is changed into disorganized, irregular wiggle-waggles of the atoms in the wood. So therefore we should look at the thing more closely.

As a matter of fact, we have here the clue to the apparent irreversibility. I will take a simple example. Suppose we have blue water with ink, and white water, that is without ink, in a tank, with a little separation, and then we pull out the separation very delicately The water starts separate, blue on one side and white on the other side. Wait a while. Gradually the blue mixes up with the white, and after a while the water is "luke blue," I mean it is sort of fifty-fifty, the colour uniformly distributed throughout. Now if we wait and watch this for a long time, it does not by itself separate. (You could do something to

get the blue separated again. You could evaporate the water and condense it somewhere else, and collect the blue dye and dissolve it in half
the water, and put the thing back. But while you were doing all that
you yourself would be causing irreversible phenomena somewhere
else.) By itself it does not go the other way.

That gives us some clue. Let us look at the molecules. Suppose
that we take a moving picture of the blue and white water mixing. It
will look funny if we run it backwards, because we shall start with
uniform water and gradually the thing will separate—it will be obviously nutty. Now we magnify the picture, so that every physicist can
watch, atom by atom, to find out what happens irreversibly—where
the laws of balance of forward and backward break down. So you
start, and you look at the picture. You have atoms of two different
kinds (it's ridiculous, but let's call them blue and white) jiggling all the
time in thermal motion. If we were to start at the beginning we should
have mostly atoms of one kind on one side, and atoms of the other
kind on the other side. Now these atoms are jiggling around, billions
and billions of them, and if we start them with one kind all on one
side, and the other kind on the other side, we see that in their perpetual irregular motions they will get mixed up, and that is why the water
becomes more or less uniformly blue.

Let us watch any one collision selected from that picture, and in
the moving picture the atoms come together this way and bounce off
that way. Now run that section of the film backwards, and you find
the pair of molecules moving together the other way and bouncing off
this way. And the physicist looks with his keen eye, and measures
everything, and says, "That's all right, that's according to the laws of
physics. If two molecules came this way they would bounce this way."
It is reversible. The laws of molecular collision are reversible.

So if you watch too carefully you cannot understand it at all,
because every one of the collisions is absolutely reversible, and yet the
whole moving picture shows something absurd, which is that in the
reversed picture the molecules start in the mixed condition—blue,
white, blue, white, blue, white—and as time goes on, through all the
collisions, the blue separates from the white. But they cannot do
that—it is not natural that the accidents of life should be such that the
blues will separate themselves from the whites. And yet if you watch
this reversed movie very carefully every collision is O.K.

Well you see that all there is to it is that the irreversibility is caused by the general accidents of life. If you start with a thing that is separated and make irregular changes, it does get more uniform. But if it starts uniform and you make irregular changes, it does not get separated. It could get separated. It is not against the laws of physics that the molecules bounce around so that they separate. It is just unlikely. It would never happen in a million years. And that is the answer. Things are irreversible only in a sense that going one way is likely, but going the other way, although it is possible and is according to the laws of physics, would not happen in a million years. It is just ridiculous to expect that if you sit there long enough the jiggling of the atoms will separate a uniform mixture of ink and water into ink on one side and water on the other.

Now if I had put a box around my experiment, so that there were only four or five molecules of each kind in the box, as time went on they would get mixed up. But I think you could believe that, if you kept watching, in the perpetual irregular collisions of these molecules, after some time—not necessarily a million years, maybe only a year—you would see that accidentally they would get back more or less to their original state, at least in the sense that if I put a barrier through the middle, all the whites would be on one side and all the blues on the other. It is not impossible. However, the actual objects with which we work have not only four or five blues and whites. They have four or five million, million, million, million, which are all going to get separated like this. And so the apparent irreversibility of nature does not come from the irreversibility of the fundamental physical laws; it comes from the characteristic that if you start with an ordered system, and have the irregularities of nature, the bouncing of molecules, then the thing goes one way. [. . .]

One of the rules of the world is that the thing goes from an ordered condition to a disordered. Incidentally, this word order, like the word disorder, is another of these terms of physics which are not exactly the same as in ordinary life. The order need not be interesting to you as human beings, it is just that there is a definite situation, all on one side and all on the other, or they are mixed up—and that is ordered and disordered.

The question, then, is how the thing gets ordered in the first place, and why, when we look at any ordinary situation, which is only partly ordered, we can conclude that it probably came from one which was more ordered. If I look at a tank of water, in which the water is very dark blue on one side and very clear white on the other, and a sort of bluish colour in between, and I know that the thing has been left alone for twenty or thirty minutes, then I will guess that it got this way because the separation was more complete in the past. If I wait longer, then the blue and white will get more intermixed, and if I know that this thing has been left alone for a sufficiently long time, I can conclude something about the past condition. The fact that it is "smooth" at the sides can only arise because it was much more satisfactorily separated in the past; because if it were not more satisfactorily separated in the past, in the time since then it would have become more mixed up than it is. It is therefore possible to tell, from the present, something about the past. [. . .]

Some people have proposed that the way the world became ordered is this. In the beginning the whole universe was just irregular motions, like the mixed water. We saw that if you waited long enough, with very few atoms, the water could have got separated accidentally. Some physicists (a century ago) suggested that all that has happened is that the world, this system that has been going on and going on, fluctuated. (That is the term used when it gets a little out of the ordinary uniform condition.) It fluctuated, and now we are watching the fluctuation undo itself again. [. . .] But I believe this theory to be incorrect. I think it is a ridiculous theory for the following reason. If the world were much bigger, and the atoms were all over the place starting from a completely mixed up condition, then if I happened to look only at the atoms in one place, and I found the atoms there separated, I would have no way to conclude that the atoms anywhere else would be separated. In fact if the thing were a fluctuation, and I noticed something odd, the most likely way that it got there would be that there was nothing odd anywhere else. That is, I would have to borrow odds, so to speak, to get the thing lopsided, and there is no use borrowing too much. In the experiment with the blue and white water, when eventually the few molecules in the box became separated, the most likely condition of the rest of the water would still be mixed up. And therefore, although when we look at the stars and we look at the world we see everything is ordered, if there were a fluctuation, the prediction would be that if we looked at a place where we have not looked before, it would be disordered and a mess. Although the separation of the matter into stars which are hot and space which is cold, which we have seen, could be a fluctuation, then in places where we have not looked we would expect to find that the stars are not separated from space. [. . .] Therefore I think it necessary to add to the physical laws the hypothesis that in the past the universe was more ordered, in the technical sense, than it is today—I think this is the additional statement that is needed to make sense, and to make an understanding of the irreversibility.

That statement itself is of course lopsided in time; it says that something about the past is different from the future. But it comes outside the province of what we ordinarily call physical laws, because we try today to distinguish between the statement of the physical laws which govern the rules by which the universe develops, and the law which states the condition that the world was in in the past. This is considered to be astronomical history—perhaps some day it will also be a part of physical law.

EDWARD HARRISON

The Golden Walls of Edgar Allan Poe, from
Darkness at Night: A Riddle of the Universe (1987)

E dward Harrison (1919–) is a British astronomer who lives in
the United States and has written a number of interesting
books about cosmology. The selection here is from an excellent
book that he wrote about Olbers's paradox. In 1826 Heinrich
Olbers, a German astronomer who worked out the method that is
still used for calculating the orbits of comets, asked why the sky
was dark. If the universe contains an infinite number of stars, why
don't we see one at every point in the night sky? Why are most
points dark instead of light? To this day, most answers to that
question are either wrong or hopelessly inarticulate, but Harrison's
book is a gem of clarity.

The chapter given here reports the astonishing information that
Edgar Allan Poe was the first to imagine the correct solution. Although perhaps it is not quite that astonishing. When science moves
this far out toward the edge of understanding, it needs imagination
more than anything else, and Poe had one of the freest imaginations
of his time. He was a master of approaching ideas from a new direction. It is well known that Poe anticipated two other figures who
would startle the early twentieth century: Arthur Conan Doyle and
Sigmund Freud. Now it seems that Einstein has joined the list.

The early Victorians were moderately tolerant of their amateur
scientists. Indeed, many investigators in the natural sciences, of independent means, whom we would nowadays regard as amateurs,
became esteemed members of learned societies. The gathering complexity of natural science and its specializations had yet to erect insur-
mountable barriers against nonprofessional contributors. The first
clear and correct solution to the riddle of darkness, though only qualitatively expressed, came from Edgar Allan Poe, the renowned poet,
essayist, critic, and amateur scientist.

In the June 1845 issue of the *United States Magazine and Demo-
cratic Review* Edgar Allan Poe published a moving essay, "The Power
of Words," in which he wrote,

Look down into the abysmal distances!—attempt to force the gaze down the multitudinous vistas of the stars, as we sweep slowly through them thus—and thus—and thus! Even the spiritual vision, is it not at all points arrested by the continuous golden walls of the universe?—the walls of the myriads of the shining bodies that mere number has appeared to blend into unity?

In February 1848, three years after referring to the golden walls of the universe, and only a year before he died at age forty, Edgar Allan Poe delivered a two-hour lecture "On the cosmogony of the universe" before a scant audience at the Society Library, New York, while the Rev. Dr. John Pringle Nichol, regius professor of astronomy at the University of Glasgow, also happened to be lecturing in New York. Nichol's *Views of the Architecture of the Heavens*, bringing to the attention of the public the astounding discoveries of contemporary astronomy, and written in a style of religious humility incapable of giving offense, had caused an immense stir in the English-speaking world. Undoubtedly, Edgar Allan Poe was greatly influenced by Nichol's popular work. Later in that year Poe amplified his lecture and published it as an essay entitled *Eureka: A Prose Poem*. In this imaginative masterpiece, dedicated to Alexander von Humboldt, he formulated his most daring cosmological ideas. He visualized a universe rhythmically expanding and collapsing with each pulse of the Heart Divine. In an apocalyptic vision he foresaw the collapse of the present cosmic era: "Then, indeed, amid unfathomable abysses, will be glaring unimaginable suns." Of this work he wrote in a letter, "What I have propounded will (in good time) revolutionize the world of Physical and Metaphysical Science."

Eureka failed to revolutionize the world of physics and metaphysics; its science was too metaphysical and its metaphysics too scientific for contemporary tastes. It constitutes, however, a most interesting and important contribution to cosmology, and only twenty-five years after [Heinrich] Olbers wrote his paper on the riddle of darkness, it contains the first anticipation of a formally correct solution. In *Eureka* Poe wrote,

Were the succession of stars endless, then the background of the sky would present us an uniform luminosity, like that displayed by the Galaxy—*since there could be absolutely no point, in all that background, at which would not exist a star.* The only mode, therefore, in which, under such a state of affairs, we could comprehend the *voids* which our telescopes find in innumerable directions, would be by supposing the distance of the invisible background so immense that no ray from it has yet been able to reach us at all.

The speed of light and the age of stars had at last come together to reveal a new aspect of an old problem.

In the twentieth century we have grown accustomed to the idea that our vision slices through space and time. When we gaze at the night sky, looking far out in space, we are fully aware that we see the apparitions of long ago. We find it difficult to understand why Descartes and other philosophers once viewed with alarm the prospect of splicing space and time together. Yet even to us, accustomed to the idea of looking out in space and back in time, the thought that at the horizon of the visible universe lies the creation, unveiled and open to inspection, comes as a shock.

Our vision extends a limited distance in a universe in which stars have been shining for a limited period of time. The starry universe may be infinite in space, or if not infinite then vast beyond measurement, and yet the part of it that we can see—the visible universe—is comparatively small, generally much too small to contain enough stars to cover the sky.

Poe hesitated: "That this *may* be so, who shall venture to deny?" He granted that the astronomical evidence favored a one-island universe of a single system of stars, but felt attracted to a many-island universe of multiple systems. May we not infer from the evidence, he said, that this

> perceptible Universe—that this cluster of clusters—is but one of a *series* of clusters of clusters, the rest of which are invisible through distance—through the diffusion of their light being so excessive, ere it reaches us, as not to produce upon our retinas a light-impression—or from there being no such emanation as light at all, in those unspeakably distant worlds—or, lastly, from the mere interval being so vast, that

the electric tidings of their presence in space, have not yet—through the lapsing myriads of years—been enabled to traverse that interval?

A few lines later: "Let me declare, only, that as an individual I myself feel impelled to the *fancy*—without daring to call it more—that there *does* exist a *limitless* succession of Universes, more or less similar to that of which we have cognizance." By universes I assume he meant galaxies. A few pages later, he added, "but the considerations through which, in this Essay, we have proceeded step by step, enable us clearly and immediately to perceive that *Space and Duration are one.*"

From the works of William and John Herschel, John Nichol, and Thomas Dick, and the recent reviews of the first volume of Alexander von Humboldt's *Kosmos*, which described the architecture of the universe and drew attention to the long periods of time taken by light to traverse interstellar space, Poe came to the conclusion that a possible solution to the riddle of darkness is that the light of the "golden walls" has yet to reach us. When we look far out in space we look far back to a time before the birth of stars.

>─+◊»·0·«+─<

FRED ALAN WOLF

Waves Without a Breeze, from
Taking the Quantum Leap (1986)

F red Alan Wolf (1934–) is an ambitious science writer who tries
to explain modern physics to the nonscientist. A fundamental
paradox of physics is the one that caused Isaac Newton and Robert
Hooke to argue. If, as Newton did, you test whether light is a particle, you will find that it is indeed a particle. If, as Hooke did, you
test whether it is a wave, you will find that it is indeed a wave, too.
Light appears to be both, and that duality is inherently difficult to
understand. Because light comes from the atoms themselves, something about the atom must allow it to send out this wave/particle.
What is that something? We could call it a "wavicle" and say it is
both a wave and a particle, but that is a name, not an explanation.
Although wavicles seem beyond our imagination, physics has developed a remarkably sophisticated set of equations for predicting and
describing events. These equations are reminiscent of the astronomical situation in 1492 when the math existed to predict eclipses, but
nobody could understand how such motions were possible. In this
selection, Wolf describes an aspect of quantum mechanics that grows
stranger and stranger as it is investigated. Wolf is skilled at presenting sensual analogies so that the reader can grasp just what it is they
do not understand.

Erwin Schroedinger, an Austrian physicist, found a mathematical
equation that explained the changing wave patterns inside an atom.

Schroedinger's equation provided a continuous mathematical
description. He viewed the atom as analogous to the vibrating violin
string. The movement of the electron from one orbit to another lower
energy orbit was a simple change of notes. As a violin string undergoes such a change, there is a moment when both harmonics can be
heard. This results in the well-known experience of harmony or, as
wave scientists call it, the phenomenon of *beats*. The beats between
two notes are what we hear as the harmony. These beats are perceived as a third sound. The vibrational pattern of the beats is determined by the difference in the frequencies of the two harmonics.

This was just what was needed to explain the observed frequency of the light waves or photons emitted when the electron in the atom undergoes a change from one orbit to the other. The light was a beat, a harmony, between the lower and upper harmonics of the Schroedinger-de Broglie waves. When we see atomic light, we are observing an atom singing harmony. With this explanation, Schroedinger hoped to save the continuity of physical processes.

However, physicists were not altogether comfortable with his wave equation. No one could imagine what the waves looked like. They had no medium to wave in, and they had no recognizable form in physical space. They didn't look like water waves or sound waves. Instead, they were abstract, mathematical waves described by mathematical functions.

Although a physical picture of a mathematical function is difficult to imagine, it is not impossible. If you have ever stepped into a shallow wading pool—one that has been recently visited by young children—you may have experienced a disconcerting physical manifestation of a mathematical function due to the children's unfortunate lack of bladder control. As you moved from one place to another in the pool, you undoubtedly noticed that there were warm spots and cold spots. The temperature of the water was not the same everywhere. Temperature was a mathematical function of location in the water. In time, the temperature could even change at a given point in the water. Temperature was also a function of the time of observation. In other words, the temperature was a mathematical function of space and time.

Similarly, Schroedinger's wave was a mathematical function of space and time. The only problem was that no one knew how to look for its "warm and cold spots"—in other words, its troughs and crests. Furthermore, as the atom became more complicated, the wave also became more complicated. For example, the wave describing one electron is a function of that electron's location in space and time. That's not too difficult. But if we are looking at a helium atom, there are two electrons present, but only one wave. The wave behavior, then, depends on the location of both electrons at the same time. And as the atomic number of an atom increases, the number of electrons contained within that atom increases. Uranium, which has an atomic number of 92, has 92 electrons and only one wave function describing it all. There was simply no convenient way to picture this wave.

But despite its unimaginability, Schroedinger's wave proved indispensable. For it explained a great many physical phenomena to which the classical model could no longer be applied. It was a successful mathematical way to explain light from any atom, molecular vibrations, and the ability of gases to absorb heat at extremely low temperatures. Physicists were excited and eager to apply Schroedinger's mathematics to anything they could get their hands on. They were like a gang of kids who had invaded the kitchen and, after many disappointing attempts to bake a cake, had suddenly discovered mother's cookbook. Schroedinger's formula gave the correct recipe in every physical application imaginable.

Everyone believed in Schroedinger's wave, even if no one knew how it waved in space and time. Somehow the wave had to exist. But even without a picture, the mathematics was enough—provided you knew how to read the mathematics cookbook. Could the wave make a particle? Was there a way to use the Schroedinger cookbook to bake a particle? Even that was not impossible for the master chef. But how could one use waves to make a particle? The answer lies in our concept of a particle. It is a tiny object distinguished from a wave by one outstanding characteristic: it is localized. It occupies a well-defined region of space. It moves from one region of space to another. You always know where it is. It exists at one place only at any given time.

Waves are different; they are not localized. They are spread over wide regions of space and can, in fact, occupy any region of space containing many locations at the same instant of time.

But waves can be added together. And when many waves are added, surprising results can be achieved. Schroedinger waves were no exception to the rule. Schroedinger waves could be added like ingredients to a recipe and produce a *Schroedinger pulse*.

A pulse is a special kind of wave. If you attach one end of a jump rope to a wall and take the other end in your hand, you can make a pulse by stretching the rope taut and giving it a sudden up-and-down movement. The pulse travels from your hand toward the wall and then reflects from the wall. This action resembles that of a ball thrown against the wall and bouncing from it. Perhaps that's all there was to an electron. It was a pulse on an invisible rope.

But there was something awfully embarrassing about the idea of a Schroedinger pulse: it got fatter as it got older. That is, it spread out and became wider each second it existed. The problem was it had nothing to hold it together. It was made up of other waves, and each of these waves had its own speed. With time, each wave would move apart from the others. The pulse would stay together only so long as the waves remained in harmony with each other.

Imagine, if you will, the pulse as a closely bunched herd of horses galloping around a racetrack bend. The horses can stay together for only a short time. Eventually, the group spreads out as each horse assumes its own pace. The slowest horses fall to the rear of the group, while the fastest ones move to the front. As time goes on, the distance between the slowest and fastest horses lengthens. In a similar manner, the pulse grows fatter as its slower waves fall out of synchronization with its faster waves.

Though big objects, like baseballs, also were made of waves, the larger the object was initially, the slower its waves spread. Thus a baseball maintained its shape because it was so big to begin with. The Schroedinger pulse describing the baseball was no embarrassment.

But an electron was a horse of a different color. While it was confined within an atom, the electrical forces of the atomic nucleus held its waves in rein. Its waves were only allowed to spread over a region the size of the atom, no further. But when an electron was no longer in such confinement, when it was set free, the waves making up its tiny pulse-particle size would begin to spread at an extremely rapid rate. In less than a millionth of a second, the electron pulse-particle would become as big as the nearest football stadium! But, of course, no one has ever seen an electron that big. All electrons appear, whenever they appear, as tiny spots.

This contradiction between our observations of electrons and Schroedinger's mathematical description of them uncovered a new problem: what prevented Schroedinger's pulses from growing so large? Little did anyone realize that question was to open the doors of paradox and mystery and lead us to a quite different picture of the universe. The answer to the question was: human observation kept them from growing so large.[2] We were on the verge of the discovery of a new discontinuity.

><+<>+O+<+>+<

---

[2] *This theme of the observer's role is picked up and developed further in the essay by Heinz R. Pagels.*

HEINZ R. PAGELS

Making the Observer Count, from
The Cosmic Code (1982)

Heinz R. Pagels (1939–1988) was another physicist with an impressive ability to explain modern physics even as it stands at the limits of intelligibility. One of the most famous doctrines of postclassical physics is that the observer affects the outcome. Many antiscience pundits take a cheerful attitude toward this idea, saying, "You see, people matter after all." But Pagels shows that this limitation is really a limitation, a barrier to understanding and imagining. Wolf's essay stressed just how puzzling physics has become in terms of its wave/particle duality. In this selection, Pagels goes a step further and shows how this confusing duality comes from the presence of the observer. Pagels makes the reader visualize the quantum world's un-visualiz-ability.

Once I tried to imagine what I would see if I could be shrunk down to the size of atoms. I would fly around the atomic nucleus and see what it was like to be an electron. But as I came to understand the meaning of the Copenhagen interpretation of [Niels] Bohr[3] and [Werner] Heisenberg I realized that the quantum theory, with its emphasis on super-realism, explicitly denies such a fantasy. I had been trying to form a mental picture of the atom based on the world of my ordinary visual experience, which obeys the laws of classical physics, and applied it precisely where quantum physics says such a picture cannot be maintained. Bohr would insist that if you want to indulge

[3] *During the 1920s, the Danish physicist Niels Bohr (1885–1962), based in Copenhagen, was the intellectual leader of the effort to understand the astonishing new evidence about the physics of things smaller than an atom. The result of that struggle was a new idea about the limits to scientific understanding. Bohr argued, "Evidence obtained under different experimental conditions cannot be comprehended within a single picture." Contrast this idea with the one expressed in the selection in which Newton talks about a decisive experiment (the Experimentum Crucis) to settle a matter. Thus, the "Copenhagen interpretation" is rather paradoxically named, because it cautions against making large interpretations of quantum evidence.*

this fantasy then you must precisely specify how the shrinking down to the size of atoms is to be accomplished. Suppose that instead of shrinking down myself I build a tiny little probe that will go down into the atom and tell me what it finds. But since the probe must also be built out of atoms and particles—there isn't anything else out of which to build it—the probe, too, becomes subject to the uncertainty relations, and then we can't even visualize the probe. You see we are stuck. All we can do is perform experiments on atoms and quantum particles resulting in measurements recorded on macroscopic-sized instruments. The quantum theory describes all possible such measurements; we cannot do better. The fantasy of the shrinking person is just that—a fantasy. [. . .]

Bohr wondered how we could even talk about the atomic world—it was so far removed from human experience. He struggled with this problem—how can we use ordinary language developed to cope with everyday events and objects to describe atomic events? Perhaps the logic inherent in our grammar was inadequate for this task. So Bohr focused on the problem of language in his interpretation of quantum mechanics. As he remarked, "It is wrong to think that the task of physics is to find out how Nature is. Physics concerns what we can say about Nature."

Bohr emphasized that when we are asking a question of nature we must also specify the experimental apparatus that we will use to determine the answer. For example, suppose we ask, "What is the position of the electron and what is its momentum?" In classical physics we do not have to take into account the fact that in answering the question—doing an experiment—we alter the state of the object. We can ignore the interaction of the apparatus and the object under investigation. For quantum objects like electrons this is no longer the case. The very act of observation changes the state of the electron.

The fact that observation can change what is being observed can be seen from examples drawn from ordinary life. The anthropologist who studies a small village isolated from modern life will by his mere presence alter village life. The object of his knowledge changes as a consequence of examination. The fact that people know they are being observed can alter their behavior.

Nature can be most passively accommodating to the quantum experimenter. If he wants to measure the position of an electron with arbitrarily great precision and sets up an apparatus to do this, no law of the quantum theory prevents a definite answer. By "position" I always mean a statistical averaging of many position measurements. The experimentalist would conclude that the electron is a particle, an object at a definite point in space. On the other hand, if he is interested in measuring the wavelength of the electron and sets up another apparatus to do this, he will also get a definite answer. Doing the experiment this way, he would conclude that the electron is a wave, not a particle. No conflict exists between the particle and the wave concept, because, as Bohr taught us, the outcome of the experiments depends on the experimental arrangement, and different experimental arrangements are needed to measure the position and the wavelength of the electron.

Now the experimenter gets persistent. He is fed up with this wave-particle, momentum-position duality nonsense and decides to settle the question once and for all by setting up an apparatus that will try to measure both the position and the momentum of the electron. Nature now becomes most stubborn, because the experimenter runs into the brick wall of the uncertainty relation. No amount of experimental technique seems to do any good, because it is a question of principle that he is up against. Why can't you measure the position and momentum simultaneously—what prevents it? [Max] Born describes it this way: "In order to measure space-coordinates and instants of time, rigid measuring rods and clocks are required. To measure momenta and energies, arrangements with movable parts are needed to take up and indicate the impact of the object. If quantum mechanics describes the interaction of the object and the measuring device, both arrangements are not possible." Born is describing that odd feature of the laws of quantum mechanics which imply that we cannot build an apparatus that measures both position and momentum simultaneously—the experimental arrangements for these two measurements are mutually exclusive. Trying simultaneously to measure both position and momentum precisely is like trying to look at the space both in front and behind your head without using a mirror. As soon as you turn to look behind you the space behind your head also turns. You cannot simultaneously see both the space in front of and behind you.

Particle and wave are what Bohr called complementary concepts, meaning they exclude one another. In the analogy between language and mathematics we gave previously, these complementary concepts are different representations of the same object. Physicists speak of the particle representation or the wave representation. Bohr's principle of complementarity asserts that there exist complementary properties of the same object of knowledge, one of which if known will exclude knowledge of the other. We may therefore describe an object like an electron in ways which are mutually exclusive—e.g., as wave or particle—without logical contradiction provided we also realize that the experimental arrangements that determine these descriptions are similarly mutually exclusive. Which experiment—and hence which description one chooses—is purely a matter of human choice. [. . .]

We thus come to the two crucial points about quantum reality that emerge from the work of Heisenberg and Bohr, the Copenhagen interpretation. The first point is that quantum reality is statistical, not certain. Even after the experimental arrangement has been specified for measuring some quantum property, it may be necessary to repeat the precise measurement again and again, because individual precise measurements are meaningless. The microworld is given only as a statistical distribution of measurements, and these distributions can be determined by physics. The attempt to form a mental picture of the position and momentum of a single electron consistent with a series of measurements results in the "fuzzy" electron. This is a human construct attempting to fit the quantum world into the limitations of everyday sense awareness. People who engage in such constructions or try to find objective meaning in individual events are really closet determinists.

The second main point is that it is meaningless to talk about the physical properties of quantum objects without precisely specifying the experimental arrangement by which you intend to measure them. Quantum reality is in part an observer-created reality. As the physicist John Wheeler says, "No phenomenon is a *real* phenomenon until it is an *observed* phenomenon." This is radically different from the orientation of classical physics. As Max Born put it, "The generation to which Einstein, Bohr, and I belong was taught that there exists an objective physical world, which unfolds itself according to immutable laws independent of us; we are watching this process as the audience watches a play in a theater. Einstein still believes that this should be the relation between the scientific observer and his subject." But with the quantum theory, human intention influences the structure of the physical world.

>–+–◊•–O–•◊–+–<

PAUL DAVIES
Schrödinger's Cats and Wigner's Friends, from
Other Worlds (1980)

B ritish astronomer Paul Davies (1946–) is a prolific writer, now
living in Australia, who enjoys depicting science at the furthest
edge of its confrontation with mystery. In this selection he recounts
some of the most peculiar paradoxes that arise from moving causality from event to observer. The piece itself is entertaining and
remarkably easy to follow if, as Piaget recommends, you simply go
with the flow of the other fellow's argument. Yet going with the
flow led me to laughter. Surely, any reader must insist, something is
wrong here. But what?

Taken as a whole, the various quantum essays in this book por-
tray the border between mystery and understanding. Oppenheimer
says that "classical physics just does not describe what happens in
the atom, so we need a new theory that does. We got one, and, lo,
classical physics is derivable from it." Heisenberg says that determinism does not hold at the atomic level. These last three essays by
Wolf, Pagels, and Davies show how mad people get when they try to
apply the language of quantum mathematics to everyday experience.
But then look at James Watson's piece on searching for the double
helix. The solution to the structure of DNA rests on quantum effects
in the chemical bond. So there is your border. On one side, effects.
We can talk about them. The other side, what it means. One word in
that direction and zap . . . we are off in mystery land.

Erwin Schrödinger, who invented the wave theory of quantum
mechanics, called attention to a curiosity that has become known as
the cat paradox. Suppose our microsystem consists of a radioactive
nucleus that may or may not decay after, say, one minute, according
to the laws of quantum probability. The decay is registered by a
Geiger counter, which is in turn attached to a hammer, in such a way
that if the nucleus decays and produces a response in the counter, it
releases a trigger which causes the hammer to fall and break a
cyanide capsule. The whole assembly is put into a sealed box along
with a cat. After one minute there is a fifty per cent chance that the
nucleus has decayed. The device is switched off automatically at this
stage. Is the cat alive or dead?

The answer would seem to be that there is a fifty-fifty chance of finding the cat alive when we look in the box. However, if we follow von Neumann and agree that the overlapping waves which represent the decayed and intact nucleus are correlated with overlapping waves describing the cat, then one cat-wave corresponds to 'live-cat,' the other to 'dead-cat.' But these waves are both present, and interfering (minutely) with one another. The state of the cat after one minute cannot be *either* 'alive' or 'dead' because of this overlap. On the other hand, what sense can we make of a 'live-dead' cat?

On the face of it, it seems that the cat's [. . .] fate is only determined when the experimenter opens the box and peers in to check on the cat's health. However, as he can choose to delay this final step as long as he pleases, the cat must continue to endure its suspended animation, until either finally dispatched from its purgatory, or resurrected to a full life by the obliging but whimsical curiosity of the experimenter.

The unsatisfactory aspect of this description is that the cat itself presumably knows whether it is alive or dead long before anyone looks in the box. It might be argued that a cat is not a proper observer, inasmuch as it does not possess the full awareness of its own existence that humans enjoy, so would be too dim-witted to know whether it was alive, dead, or alive-dead. To circumvent this objection we could replace the cat with a human volunteer, sometimes known to the physics fraternity as 'Wigner's friend,' after the physicist Eugene Wigner who has discussed this aspect of the paradox. [. . .] With such a capable accomplice installed in the box we can, if we find him alive at the end of the experiment, ask him what he felt during the period before the box was opened. There is no doubt that he would answer 'nothing,' in spite of the fact that his body was supposed to have been in a live-dead state for the duration of the experiment, after which it dramatically collapsed into a living condition once more. It is true that people sometimes complain of feeling half dead, but it is hard to imagine that quantum interference phenomena have much to do with this particular condition.

If we insist on adhering to quantum principles at all costs then we are driven into solipsism—the conclusion that the individual (in this case the reader) is the only one who really exists, all others being unconscious robots merely forming part of the scenery. If Wigner's friend is a robot, he cannot be relied upon to report truthfully his observations, for he does not really experience any. Now this is a big step, for it pitches the observer into the centre of reality in a way that is still more crucial than we have already come to accept. To escape from solipsism, Wigner himself has proposed that quantum theory cannot be correct under all circumstances; that when the conscious awareness of the observer is involved the theory breaks down and the description of the world as a set of overlapping waves is invalidated. Solipsism has attracted its share of adherents over the centuries, but most people find it unpalatable, including Wigner. In Wigner's interpretation of quantum theory, the minds of sentient beings occupy a central role in the laws of nature and in the organization of the universe, for it is precisely when the information about an observation enters the consciousness of an observer that the superposition of waves actually collapses into reality. Thus, in a sense, the whole cosmic panorama is generated by its own inhabitants! According to Wigner's theory, before there was intelligent life, the universe did not 'really' exist. This places a grave, indeed cosmic, responsibility on living things to sustain the existence of everything else, for if all life were to cease, all the other objects—from every distant star to the smallest subatomic particle—would no longer enjoy an independent reality, but would lapse into the limbo of superposition. The bonus gained from this awesome role is that Wigner's friend-in-the-box can now bring about the collapse into reality of the box contents—himself included—so that when Wigner eventually opens the box and asks him how he felt a few moments before, he can announce 'fine,' secure in the knowledge that he was one hundred per cent real already, without enlisting the help of Wigner's tardy observation of his condition to collapse his body and mind into reality.

Wigner's idea has, not unexpectedly, been widely criticized. Consciousness is normally regarded by scientists as at best ill-defined (is a cockroach conscious? a rat? a dog? . . .), and at worst non-existent physically. Yet it has to be conceded that all our observations, and through them all of science, are based ultimately on our consciousness of the surrounding world. [. . .]

A more serious objection to Wigner's ideas is exposed if two observers become involved in observing the same system, for then each has the power to collapse it into reality. To illustrate the sort of problems that can arise, suppose that we again consider a radioactive nucleus, the decay of which will trigger a Geiger counter, but this time there is no conscious observer immediately involved. It is arranged that after one minute, when the chance of a decay is fifty per cent, the experiment is terminated and the Geiger counter pointer is locked in whatever position it held, i.e. deflected if the nucleus decayed, undeflected if it had not, so that it can be read at any time thereafter. Rather than an experimenter looking directly at the pointer, the Geiger counter is photographed. When eventually the photograph is developed, the experimenter looks at it, without ever consulting the counter directly. According to Wigner, it is only at this final stage of the proceedings that reality appears, because reality owes its creation to the conscious act of observation by the experimenter, or anyone else. Thus we must conclude that before the photograph was scrutinized, the nucleus, the Geiger counter and the photograph were all in schizophrenic states consisting of overlapping alternative outcomes to the experiment even though the delay before the photograph is developed could be many years. This little corner of the universe has to hang around in unreality until the experimenter (or a curious onlooker) deigns to glance at the photograph.

The real problem arises if two successive photographs, call them A and B, are taken of the Geiger counter at the end of the experiment. As the pointer is locked in place we know that the image on A must be the same as the image on B. The snag arises if there are two experimenters also, call them Alan and Brian, and Brian looks at photograph B before Alan looks at A. Now B was taken after A, but scrutinized first. Wigner's theory requires Brian to be the conscious individual responsible for creating reality here, because he looks at his photographic record first. Suppose Brian sees a deflected pointer and pronounces that the nucleus had decayed. Naturally, when Alan looks at photograph A it will likewise show a deflected pointer. The difficulty is that when photograph A was taken, B did not even exist, so in some mysterious way Brian's glance at B causes A to become identical to B even though A was taken *before* B! It seems that we are forced to believe in backwards causation; Brian looking at a photograph, perhaps many years later, influences the operation of the camera for the preceding photograph.

"Every Intellect Which Strives After Generalization Must Feel the Temptation"

Albert Einstein

CHAPTER NINE

ANTOINE-LAURENT LAVOISIER

Preface, from
The Elements of Chemistry (1789)

Antoine-Laurent Lavoisier (1743–1794) is one of the heroes of science history. The essay in this volume by Herbert Butterfield explains the reasons for Lavoisier's importance, and Lavoisier's selection here shows the basis for it—his clear head. When scientists stand where chemists were in the late 1700s—full of facts and techniques but no way to make sense of them—they need more than a new idea about nature. They need a new idea about their own work. Lavoisier's novelty was to reform the language of chemists. He purged chemistry of its alchemical names and redefined central concepts such as "element." In this preface to his study of chemistry, he justifies giving chemicals new names. The system he introduced, still used by chemists today, forces clearer thought and aids in understanding. It also forced chemists to acknowledge what they knew and did not know, because if they did not know something, they could not talk about it. The importance of Lavoisier's book was immediately recognized. It was translated into English a year after appearing in French, and very quickly chemists began using its terminology. Lavoisier should have had many years ahead of him, but the French Revolution broke out in Paris the same year that his book appeared. During the height of the Terror, Lavoisier was marched to the guillotine for political reasons having nothing to do with his science. His fine head was cut off and tossed into a basket.

When I began the following Work, my only object was to extend
and explain more fully the Memoir which I read at the public meeting
of the Academy of Sciences in the month of April 1787, on the necessity of reforming and completing the Nomenclature of Chemistry.
While engaged in this employment, I perceived, better than I had ever
done before, the justice of the following maxims of the Abbé de Condillac, in his System of Logic, and some other of his works.

> We think only through the medium of words.—Languages
> are true analytical methods.—Algebra, which is adapted to its
> purpose in every species of expression, in the most simple,
> most exact, and best manner possible, is at the same time a
> language and an analytical method.—The art of reasoning is
> nothing more than a language well arranged.

Thus, while I thought myself employed only in forming a Nomenclature, and while I proposed to myself nothing more than to improve
the chemical language, my work transformed itself by degrees, without my being able to prevent it, into a treatise upon the Elements
of Chemistry.

The impossibility of separating the nomenclature of a science
from the science itself, is owing to this, that every branch of physical
science must consist of three things; the series of facts which are the
objects of the science, the ideas which represent these facts, and the
words by which these ideas are expressed. Like three impressions of
the same seal, the word ought to produce the idea, and the idea to be
a picture of the fact. And, as ideas are preserved and communicated
by means of words, it necessarily follows that we cannot improve the
language of any science without at the same time improving the science itself; neither can we, on the other hand, improve a science,
without improving the language or nomenclature which belongs to it.
However certain the facts of any science may be, and, however just
the ideas we may have formed of these facts, we can only communicate false impressions to others, while we want words by which these
may be properly expressed.

To those who will consider it with attention, the first part of this
treatise will afford frequent proofs of the truth of the above observations. But as, in the conduct of my work, I have been obliged to
observe an order of arrangement essentially differing from what has

been adopted in any other chemical work yet published, it is proper that I should explain the motives which have led me to do so.

It is a maxim universally admitted in geometry, and indeed in every branch of knowledge, that, in the progress of investigation, we should proceed from known facts to what is unknown. In early infancy, our ideas spring from our wants; the sensation of want excites the idea of the object by which it is to be gratified. In this manner, from a series of sensations, observations, and analyses, a successive train of ideas arises, so linked together, that an attentive observer may trace back to a certain point the order and connection of the whole sum of human knowledge.

When we begin the study of any science, we are in a situation, respecting that science, similar to that of children; and the course by which we have to advance is precisely the same which Nature follows in the formation of their ideas. In a child, the idea is merely an effect produced by a sensation; and, in the same manner, in commencing the study of a physical science, we ought to form no idea but what is a necessary consequence, and immediate effect, of an experiment or observation. Besides, he that enters upon the career of science, is in a less advantageous situation than a child who is acquiring his first ideas. To the child, Nature gives various means of rectifying any mistakes he may commit respecting the salutary or hurtful qualities of the objects which surround him. On every occasion his judgments are corrected by experience; want and pain are the necessary consequences arising from false judgment; gratification and pleasure are produced by judging aright. Under such masters, we cannot fail to become well informed; and we soon learn to reason justly, when want and pain are the necessary consequences of a contrary conduct.

In the study and practice of the sciences it is quite different; the false judgments we form neither affect our existence nor our welfare; and we are not forced by any physical necessity to correct them. Imagination, on the contrary, which is ever wandering beyond the bounds of truth, joined to self-love and that self-confidence we are so apt to indulge, prompt us to draw conclusions which are not immediately derived from facts; so that we become in some measure interested in deceiving ourselves. Hence it is by no means to be wondered, that, in the science of physics in general, men have often made suppositions, instead of forming conclusions. These suppositions, handed down from one age to another, acquire additional weight from the authorities by which they are supported, till at last they are received, even by men of genius, as fundamental truths.

The only method of preventing such errors from taking place, and of correcting them when formed, is to restrain and simplify our reasoning as much as possible. This depends entirely upon ourselves, and the neglect of it is the only source of our mistakes. We must trust to nothing but facts: These are presented to us by Nature, and cannot deceive. We ought, in every instance, to submit our reasoning to the test of experiment, and never to search for truth but by the natural road of experiment and observation. Thus mathematicians obtain the solution of a problem by the mere arrangement of data, and by reducing their reasoning to such simple steps, to conclusions so very obvious, as never to lose sight of the evidence which guides them.

Thoroughly convinced of these truths, I have imposed upon myself, as a law, never to advance but from what is known to what is unknown; never to form any conclusion which is not an immediate consequence necessarily flowing from observation and experiment; and always to arrange the facts, and the conclusions which are drawn from them, in such an order as shall render it most easy for beginners in the study of chemistry thoroughly to understand them. Hence I have been obliged to depart from the usual order of courses of lectures and of treatises upon chemistry, which always assume the first principles of the science, as known, when the pupil or the reader should never be supposed to know them till they have been explained in subsequent lessons. In almost every instance, these begin by treating of the elements of matter, and by explaining the table of affinities, without considering, that, in so doing, they must bring the principal phenomena of chemistry into view at the very outset: They make use of terms which have not been defined, and suppose the science to be understood by the very persons they are only beginning to teach. It ought likewise to be considered, that very little of chemistry can be learned in a first course, which is hardly sufficient to make the language of the science familiar to the ears, or the apparatus familiar to the eyes. It is almost impossible to become a chemist in less than three or four years of constant application.

These inconveniencies are occasioned not so much by the nature of the subject, as by the method of teaching it; and, to avoid them, I was chiefly induced to adopt a new arrangement of chemistry, which appeared to me more consonant to the order of Nature. I acknowledge, however, that in thus endeavouring to avoid difficulties of one kind, I have found myself involved in others of a different species, some of which I have not been able to remove; but I am persuaded, that such as remain do not arise from the nature of the order I have adopted, but are rather consequences of the imperfection under which chemistry still labours. This science still has many chasms, which interrupt the series of facts, and often render it extremely difficult to reconcile them with each other: It has not, like the elements of geometry, the advantage of being a complete science, the parts of which are all closely connected together: Its actual progress, however, is so rapid, and the facts, under the modern doctrine, have assumed so happy an arrangement, that we have ground to hope, even in our own times, to see it approach near to the highest state of perfection of which it is susceptible.

The rigorous law from which I have never deviated, of forming no conclusions which are not fully warranted by experiment, and of never supplying the absence of facts, has prevented me from comprehending in this work the branch of chemistry which treats of affinities,[1] although it is perhaps the best calculated of any part of chemistry for being reduced into a completely systematic body. Messrs Geoffroy, Gellert, Bergman, Scheele, De Morveau, Kirwan,[2] and many others, have collected a number of particular facts upon this subject, which only wait for a proper arrangement; but the principal data are still wanting, or, at least, those we have are either not sufficiently defined, or not sufficiently proved, to become the foundation upon which to build so very important a branch of chemistry. This science of affinities, or elective attractions, holds the same place with regard to the other branches of chemistry, as the higher or transcendental geometry does with respect to the simpler and elementary part; and I thought it improper to involve those simple and plain elements, which

---

[1] *Chemical affinities are chemical attractions. Two chemicals that combine easily are said to have an affinity for each other. Chemists naturally wondered how to account for those affinities.*

[2] *In this list the most important chemist is the Swedish Carl Wilhelm Scheele (1742–1786), who made great contributions to Lavoisier's own work, but the most important chemist for working on the nature of chemical affinity was another Swede, Tobern Olof Bergman (1735–1784), who compiled extensive tables of affinities among acids and bases.*

I flatter myself the greatest part of my readers will easily understand, in the obscurities and difficulties which still attend that other very useful and necessary branch of chemical science.

Perhaps a sentiment of self-love may, without my perceiving it, have given additional force to these reflections. Mr de Morveau is at present engaged in publishing the article *Affinity* in the Methodical Encyclopaedia; and I had more reasons than one to decline entering upon a work in which he is employed.

It will, no doubt, be a matter of surprise, that in a treatise upon the elements of chemistry, there should be no chapter on the constituent and elementary parts of matter; but I shall take occasion, in this place, to remark, that the fondness for reducing all the bodies in nature to three or four elements, proceeds from a prejudice which has descended to us from the Greek Philosophers. The notion of four elements, which, by the variety of their proportions, compose all the known substances in nature, is a mere hypothesis, assumed long before the first principles of experimental philosophy or of chemistry had any existence. In those days, without possessing facts, they framed systems; while we, who have collected facts, seem determined to reject them when they do not agree with our prejudices. The authority of these fathers of human philosophy still carries great weight, and there is reason to fear that it will even bear hard upon generations yet to come.

It is very remarkable, that, notwithstanding of the number of philosophical chemists who have supported the doctrine of the four elements, there is not one who has not been led by the evidence of facts to admit a greater number of elements into their theory. The first chemists that wrote after the revival of letters, considered sulphur and salt as elementary substances entering into the composition of a great number of substances; hence, instead of four, they admitted the existence of six elements. Beccher assumes the existence of three kinds of earth, from the combination of which, in different proportions, he supposed all the varieties of metallic substances to be produced. Stahl gave a new modification to this system; and succeeding chemists have taken the liberty to make or to imagine changes and additions of a similar nature. All these chemists were carried along by the influence of the genius of the age in which they lived, which contented itself with assertions without proofs; or, at least, often admitted as proofs the slightest degrees of probability, unsupported by that strictly rigorous analysis required by modern philosophy.

All that can be said upon the number and nature of elements is, in my opinion, confined to discussions entirely of a metaphysical nature. The subject only furnishes us with indefinite problems, which may be solved in a thousand different ways, not one of which, in all probability, is consistent with nature. I shall therefore only add upon this subject, that if, by the term *elements,* we mean to express those simple and indivisible atoms of which matter is composed, it is extremely probable we know nothing at all about them; but, if we apply the term *elements,* or *principles of bodies,* to express our idea of the last point which analysis is capable of reaching, we must admit, as elements, all the substances into which we are capable, by any means, to reduce bodies by decomposition. Not that we are entitled to affirm, that these substances we consider as simple may not be compounded of two, or even of a greater number of principles; but, since these principles cannot be separated, or rather since we have not hitherto discovered the means of separating them, they act with regard to us as simple substances, and we ought never to suppose them compounded until experiment and observation has proved them to be so.

The foregoing reflections upon the progress of chemical ideas naturally apply to the words by which these ideas are to be expressed. Guided by the work which, in the year 1787, Messrs. de Morveau, Berthollet, de Fourcroy, and I composed upon the Nomenclature of Chemistry, I have endeavoured, as much as possible, to denominate simple bodies by simple terms, and I was naturally led to name these first. It will be recollected, that we were obliged to retain that name of any substance by which it had been long known in the world, and that in two cases only we took the liberty of making alterations; first, in the case of those which were but newly discovered, and had not yet obtained names, or at least which had been known but for a short time, and the names of which had not yet received the sanction of the public; and, secondly, when the names which had been adopted, whether by the ancients or the moderns, appeared to us to express evidently false ideas, when they confounded the substances, to which they were applied, with others possessed of different, or perhaps opposite qualities. We made no scruple, in this case, of substituting other names in their room, and the greatest number of these were borrowed from the Greek language. We endeavoured to frame them in such a manner as to express the most general and the most characteristic quality of the substances; and this was attended with the additional advantage both of assisting the memory of beginners, who find it difficult to remember a new word which has no meaning, and of accustoming them early to admit no word without connecting with it some determinate idea.

To those bodies which are formed by the union of several simple substances we gave new names, compounded in such a manner as the nature of the substances directed; but, as the number of double combinations is already very considerable, the only method by which we could avoid confusion, was to divide them into classes. In the natural order of ideas, the name of the class or genus is that which expresses a quality common to a great number of individuals: The name of the species, on the contrary, expresses a quality peculiar to certain individuals only.

These definitions are not, as some may imagine, merely metaphysical, but are established by Nature. "A child," says the Abbé de Condillac, "is taught to give the name *tree* to the first one which is pointed out to him. The next one he sees presents the same idea, and he gives it the same name. This he does likewise to a third and a fourth, till at last the word *tree*, which he first applied to an individual, comes to be employed by him as the name of a class or a genus, an abstract idea, which comprehends all trees in general. But, when he learns that all trees serve not the same purpose, that they do not all produce the same kind of fruit, he will soon learn to distinguish them by specific and particular names." This is the logic of all the sciences, and is naturally applied to chemistry.

The acids, for example, are compounded of two substances, of the order of those which we consider as simple; the one constitutes acidity, and is common to all acids, and, from this substance, the name of the class or the genus ought to be taken; the other is peculiar to each acid, and distinguishes it from the rest, and from this substance is to be taken the name of the species. But, in the greatest number of acids, the two constituent elements, the acidifying principle, and that which it acidifies, may exist in different proportions, constituting all the possible points of equilibrium or of saturation. This is the case in the sulphuric and the sulphurous acids; and these two states of the same acid we have marked by varying the termination of the specific name.

Metallic substances which have been exposed to the joint action of the air and of fire, lose their metallic lustre, increase in weight, and assume an earthy appearance. In this state, like the acids, they are compounded of a principle which is common to all, and one which is peculiar to each. In the same way, therefore, we have thought proper to class them under a generic name, derived from the common principle; for which purpose, we adopted the term *oxide*; and we distinguish them from each other by the particular name of the metal to which each belongs.

Combustible substances, which in acids and metallic oxides are a specific and particular principle, are capable of becoming, in their turn, common principles of a great number of substances. The sulphurous combinations have been long the only known ones in this kind. Now, however, we know, from the experiments of Messrs. Vandermonde, Monge, and Berthollet, that charcoal may be combined with iron, and perhaps with several other metals; and that, from this combination, according to the proportions, may be produced steel, plumbago, &c. We know likewise, from the experiments of M. Pelletier, that phosphorus may be combined with a great number of metallic substances. These different combinations we have classed under generic names taken from the common substance, with a termination which marks this analogy, specifying them by another name taken from that substance which is proper to each.

The nomenclature of bodies compounded of three simple substances was attended with still greater difficulty, not only on account of their number, but, particularly, because we cannot express the nature of their constituent principles without employing more compound names. In the bodies which form this class, such as the neutral salts, for instance, we had to consider, 1st, The acidifying principle, which is common to them all; 2d, The acidifiable principle which constitutes their peculiar acid; 3d, The saline, earthy, or metallic basis, which determines the particular species of salt. Here we derived the name of each class of salts from the name of the acidifiable principle common to all the individuals of that class; and distinguished each species by the name of the saline, earthy, or metallic basis, which is peculiar to it.

A salt, though compounded of the same three principles, may, nevertheless, by the mere difference of their proportion, be in three different states. The nomenclature we have adopted would have been defective, had it not expressed these different states; and this we attained chiefly by changes of termination uniformly applied to the same state of the different salts.

In short, we have advanced so far, that from the name alone may be instantly found what the combustible substance is which enters into any combination; whether that combustible substance be combined with the acidifying principle, and in what proportion; what is the state of the acid; with what basis it is united; whether the saturation be exact, or whether the acid or the basis be in excess.

It may be easily supposed that it was not possible to attain all these different objects without departing, in some instances, from established custom, and adopting terms which at first sight will appear uncouth and barbarous. But we considered that the ear is soon habituated to new words, especially when they are connected with a general and rational system. The names, besides, which were formerly employed, such as *powder of algaroth, salt of alembroth, pompholix, phagadenic water, turbith mineral, colcothar,* and many others, were neither less barbarous nor less uncommon. It required a great deal of practice, and no small degree of memory, to recollect the substances to which they were applied, much more to recollect the genus of combination to which they belonged. The names of *oil of tartar per deliquium, oil of vitriol, butter of arsenic and of antimony, flowers of zinc,* &c. were still more improper, because they suggested false ideas: For, in the whole mineral kingdom, and particularly in the metallic class, there exists no such thing as butters, oils, or flowers; and, in short, the substances to which they give these fallacious names, are nothing less than rank poisons.

When we published our essay on the nomenclature of chemistry, we were reproached for having changed the language which was spoken by our masters, which they distinguished by their authority, and handed down to us. But those who reproach us on this account, have forgotten that it was Bergman and Macquer themselves who urged us to make this reformation. In a letter which the learned Professor of Upsal, M. Bergman, wrote, a short time before he died, to M. de Morveau, he bids him *spare no improper names; those who are learned, will always be learned, and those who are ignorant will thus learn sooner.*

ALFRED WALLACE

On the Tendency of Varieties to Depart Indefinitely from the Original Type (1859)

Alfred Russel Wallace (1823–1913) was a surveyor with a passion for investigating nature. Despite his lack of formal training, he traveled to the upper reaches of the Amazon in search of clues to the "origin of species" (a term he used in 1847). He supported himself by providing museums with wildlife specimens. Then he sailed to the East Indies where he continued collecting biological specimens. By 1858 he was second to no one in his detailed understanding of the relation between species and environment.

While he was in Ternate, one of the spice islands in what is now Indonesia, Wallace suffered from malaria. During an attack of fever, he suddenly understood the relationship between the struggle to survive and the evolution of species. He quickly wrote the following essay and mailed it to England. Three months later it reached its addressee, Charles Darwin, who was thunderstruck. Here, in excellent prose, was the theory he had struggled over for decades. Fortunately for Darwin, Wallace was all curiosity and no personal ambition. Wallace agreed that Darwin should be credited with the idea, and when he published his own book about evolution, he even titled it *Darwinism* (1889). Darwin is now famous around the world, whereas Alfred Wallace is remembered only by specialists. Yet Wallace's essay still shines with its clear outline of the process of natural selection.

One of the strongest arguments which have been adduced to prove the original and permanent distinctness of species is, that *varieties* produced in a state of domesticity are more or less unstable, and often have a tendency, if left to themselves, to return to the normal form of the parent species; and this instability is considered to be a distinctive peculiarity of all varieties, even of those occurring among wild animals in a state of nature, and to constitute a provision for preserving unchanged the originally created distinct species.

In the absence of scarcity of[3] facts and observations as to *varieties*
occurring among wild animals, this argument has had great weight
with naturalists, and has led to a very general and somewhat preju-
diced belief in the stability of species. Equally general, however, is the
belief in what are called "permanent or true varieties,"—races of animals which continually propagate their like, but which differ so
slightly (although constantly) from some other race, that the one is
considered to be a *variety* of the other. Which is the *variety* and which
the original *species*, there is generally no means of determining, except
in those rare cases in which the one race has been known to produce
an offspring unlike itself and resembling the other. This, however,
would seem quite incompatible with the "permanent invariability of
species," but the difficulty is overcome by assuming that such varieties
have strict limits, and can never again vary further from the original
type, although they may return to it, which, from the analogy of the
domesticated animals, is considered to be highly probably, if not certainly proved.

It will be observed that this argument rests entirely on the assumption, that *varieties* occurring in a state of nature are in all
respects analogous to or even identical with those of domestic animals, and are governed by the same laws as regards their permanence
or further variation. But it is the object of the present paper to show
that this assumption is altogether false, that there is a general principle in nature which will cause many *varieties* to survive the parent
species, and to give rise to successive variations departing further and
further from the original type, and which also produces in domesticated animals, the tendency of varieties to return to the parent form.

The Struggle for Existence

The life of wild animals is a struggle for existence. The full exertion
of all their faculties and all their energies is required to preserve their
own existence and provide for that of their infant offspring. The possibility of procuring food during the least favourable seasons, and of
escaping the attacks of their most dangerous enemies, are the primary
conditions which determine the existence both of individuals and of
entire species. These conditions will also determine the population of

---

[3] *Surely Wallace meant to cross out either "absence of" or "scarcity of."*

a species; and by a careful consideration of all the circumstances we may be enabled to comprehend, and in some degree to explain, what at first sight appears so inexplicable—the excessive abundance of some species, while others closely allied to them are very rare.

The Law of Population of Species

The general proportion that must obtain between certain groups of animals is readily seen. Large animals cannot be so abundant as small ones; the carnivora must be less numerous than the herbivora; eagles and lions can never be so plentiful as pigeons and antelopes; the wild asses of the Tartarian deserts cannot equal in numbers the horses of the more luxuriant prairies and pampas of America. The greater or less fecundity of an animal is often considered to be one of the chief causes of its abundance or scarcity; but a consideration of the facts will show us that it really has little or nothing to do with the matter. Even the least prolific of animals would increase rapidly if unchecked, whereas it is evident that the animal population of the globe must be stationary, or perhaps, through the influence of man, decreasing. Fluctuations there may be; but permanent increase, except in restricted localities, is almost impossible. For example, our own observation must convince us that birds do not go on increasing every year in a geometrical ratio, as they would do, were there not some powerful check to their natural increase. Very few birds produce less than two young ones each year, while many have six, eight, or ten; four will certainly be below the average; and if we suppose that each pair produce young only four times in their life, that will also be below the average, supposing them not to die either by violence or want of food. Yet at this rate how tremendous would be the increase in a few years from a single pair! A simple calculation will show that in fifteen years each pair of birds would have increased to nearly ten millions! whereas we have no reason to believe that the number of the birds of any country increases at all in fifteen or in one hundred and fifty years. With such powers of increase the population must have reached its limits, and have become stationary, in a very few years after the origin of each species. It is evident, therefore, that each year an immense number of birds must perish—as many in fact as are born; and as on the lowest calculation the progeny are each year twice as numerous as their parents, it follows that, whatever be the

average number of individuals existing in any given country, *twice that
number must perish annually,*—a striking result, but one which seems
at least highly probable, and is perhaps under rather than over the
truth. It would therefore appear that, as far as the continuance of the
species and the keeping up the average number of individuals are concerned, large broods are superfluous. On the average all above *one*
become food for hawks and kites, wild cats and weasels, or perish of
cold and hunger as winter comes on. This is strikingly proved by the
case of particular species; for we find that their abundance in individuals bears no relation whatever to their fertility in producing offspring.
Perhaps the most remarkable instance of an immense bird population
is that of the passenger pigeon of the United States, which lays only
one, or at most two eggs, and is said to rear generally but one young
one. Why is this bird so extraordinarily abundant, while others producing two or three times as many young are much less plentiful? The
explanation is not difficult. The food most congenial to this species,
and on which it thrives best, is abundantly distributed over a very
extensive region, offering such difference of soil and climate, that in
one part or another of the area the supply never fails. The bird is capable of a very rapid and long-continued flight, so that it can pass without fatigue over the whole of the district it inhabits, and as soon as the
supply of food begins to fail in one place is able to discover a fresh
feeding-ground.[4] This example strikingly shows us that the procuring
a constant supply of wholesome food is almost the sole condition requisite for ensuring the rapid increase of a given species, since neither
the limited fecundity, nor the unrestrained attacks of birds of prey and
of man are here sufficient to check it. In no other birds are these peculiar circumstances so strikingly combined. Either their food is more
liable to failure, or they have not sufficient power of wing to search
for it over an extensive area, or during some season of the year it
becomes very scarce, and less wholesome substitutes have to be found;
and thus, though more fertile in offspring, they can never increase be-

---

[4] *Wallace's account offers an insight into what happened to the passenger pigeon. It became
extinct early in the twentieth century, in part because of extensive hunting, but also because
of the widespread loss of habitat to a settled, modern human population and because, as
Wallace notes, it was not very good at reproducing itself, even when, very late in the game,
people began to say that something should be done to ensure the bird's survival.*

yond the supply of food in the least favourable seasons. Many birds can only exist by migrating, when their food becomes scarce, to regions possessing a milder, or at least a different climate, though, as these migrating birds are seldom excessively abundant, it is evident that the countries they visit are still deficient in a constant and abundant supply of wholesome food. Those whose organization does not permit them to migrate when their food becomes periodically scarce, can never attain a large population. This is probably the reason why woodpeckers are scarce with us, while in the tropics they are among the most abundant of solitary birds. Thus, the house sparrow is more abundant than the redbreast, because its food is more constant and plentiful,—seeds of grasses being preserved during the winter, and our farm-yards and stubble-fields furnishing an almost inexhaustible supply. Why, as a general rule, are aquatic, and especially sea birds, very numerous in individuals? Not because they are more prolific than others, generally the contrary; but because their food never fails, the sea-shores and river-banks daily swarming with a fresh supply of small mollusca and crustacea. Exactly the same laws will apply to mammals. Wild cats are prolific and have few enemies; why then are they never as abundant as rabbits? The only intelligible answer is, that their supply of food is more precarious. It appears evident, therefore, that so long as a country remains physically unchanged, the numbers of its animal population cannot materially increase. If one species does so, some other requiring the same kind of food much diminish in proportion. The numbers that die annually must be immense; and as the individual existence of each animal depends upon itself, those that die must be the weakest—the very young, the aged, and the diseased,— while those that prolong their existence can only be the most perfect in health and vigour—those who are best able to obtain food regularly, and avoid their numerous enemies. It is, as we commenced by remarking, "a struggle for existence," in which the weakest and least perfectly organized must always succumb.

The Abundance or Rarity of a Species Dependent Upon Its More or Less Perfect Adaptation to the Conditions of Existence

It seems evident that what takes place among the individuals of a species must also occur among the several allied species of a group,— viz., that those which are best adapted to obtain a regular supply of food, and to defend themselves against the attacks of their enemies and the vicissitudes of the seasons, must necessarily obtain and preserve a superiority in population; while those species which from some defect of power or organization are the least capable of counteracting the vicissitudes of food, supply, &c., must diminish in numbers, and, in extreme cases, become altogether extinct. Between these extremes the species will present various degrees of capacity for ensuring the means of preserving life; and it is thus we account for the abundance or rarity of species. Our ignorance will generally prevent us from accurately tracing the effects to their causes; but could we become perfectly acquainted with the organization and habits of the various species of animals, and could we measure the capacity of each for performing the different acts necessary to its safety and existence under all the varying circumstances by which it is surrounded, we might be able even to calculate the proportionate abundance of individuals which is the necessary result.

If now we have succeeded in establishing these two points—1st, *that the animal population of a country is generally stationary, being kept down by a periodical deficiency of food, and other checks*; and, 2nd, *that the comparative abundance or scarcity of the individuals of the several species is entirely due to their organization and resulting habits, which, rendering it more difficult to procure a regular supply of food and to provide for their personal safety in some cases than in others, can only be balanced by a difference in the population which have to exist in a given area*—we shall be in a condition to proceed to the consideration of *varieties*, to which the preceding remarks have a direct and very important application.

Useful Variations Will Tend to Increase; Useless or Hurtful Variations to Diminish

Most or perhaps all the variations from the typical form of a species must have some definite effect, however slight, on the habits or capacities of the individuals. Even a change of colour might, by rendering them more or less distinguishable, affect their safety; a greater or less development of hair might modify their habits. More important changes, such as an increase in the power or dimensions of the limbs or any of the external organs, would more or less affect their mode of procuring food or the range of country which they inhabit. It is also evident that most changes would affect, either favourably or adversely, the powers of prolonging existence. An antelope with shorter or weaker legs must necessarily suffer more from the attacks of the feline carnivora; the passenger pigeon with less powerful wings would sooner or later be affected in its powers of procuring a regular supply of food; and in both cases the result must necessarily be a diminution of the population of the modified species. If, on the other hand, any species should produce a variety having slightly increased powers of preserving existence, that variety must inevitably in time acquire a superiority in numbers. These results must follow as surely as old age, intemperance, or scarcity of food produce an increased mortality. In both cases there may be many individual exceptions; but on the average the rule will invariably be found to hold good. All varieties will therefore fall into two classes—those which under the same conditions would never reach the population of the parent species, and those which would in time obtain and keep a numerical superiority. Now, let some alteration of physical conditions occur in the district—a long period of drought, a destruction of vegetation by locusts, the irruption of some new carnivorous animal seeking "pastures new"—any change in fact tending to render existence more difficult to the species in question, and tasking its utmost powers to avoid complete extermination; it is evident that, of all the individuals composing the species, those forming the least numerous and most feebly organized variety would suffer first, and, were the pressure severe, must soon become extinct. The same causes continuing in action, the parent species would next suffer, would gradually diminish in numbers, and with a recurrence of similar unfavourable conditions might also become extinct. The superior variety would then alone remain, and on a return to favourable circumstances would rapidly increase in numbers and occupy the place of the extinct species and variety.

Superior Varieties Will Ultimately Extirpate the Original Species

The *variety* would now have replaced the *species*, of which it would be a more perfectly developed and more highly organized form. It would be in all respects better adapted to secure its safety, and to prolong its individual existence and that of the race. Such a variety *could not* return to the original form; for that form is an inferior one, and could never compete with it for existence. Granted, therefore, a

"tendency" to reproduce the original type of the species, still the variety must ever remain preponderant in numbers, and under adverse
physical conditions *again alone survive*. But this new, improved, and
populous race might itself, in course of time, give rise to new varieties,
exhibiting several diverging modifications of form, any of which, tending to increase the facilities for preserving existence, must by the same
general law, in their turn become predominant. Here, then, we have
progression and continued divergence deduced from the general laws
which regulate the existence of animals in a state of nature, and from
the undisputed fact that varieties do frequently occur. It is not, however, contended that this result would be invariable; a change of physical conditions in the district might at times materially modify it, rendering the race which had been the most capable of supporting
existence under the former conditions now the least so, and even causing the extinction of the newer and, for a time, superior race, while
the old or parent species and its first inferior varieties continued to
flourish. Variations in unimportant parts might also occur, having no
perceptible effect on the life-preserving powers; and the varieties so
furnished might run a course parallel with the parent species, either
giving rise to further variations or returning to the former type. All we
argue for is, that certain varieties have a tendency to maintain their
existence longer than the original species, and this tendency must
make itself felt; for though the doctrine of chances or averages can
never be trusted to on a limited scale, yet, if applied to high numbers,
the results come nearer to what theory demands, and, as we approach
to an infinity of examples, become strictly accurate. Now the scale on
which nature works is so vast—the numbers of individuals and periods of time with which she deals approach so near to infinity, that any
cause, however slight, and however liable to be veiled and counter-
acted by accidental circumstances, must in the end produce its full
legitimate results.

The Partial Reversion of Domesticated Varieties Explained

Let us now turn to domesticated animals, and inquire how varieties
produced among them are affected by the principles here enunciated.
The essential difference in the condition of wild and domestic animals
is this,—that among the former, their well-being and very existence
depend upon the full exercise and healthy condition of all their senses

and physical powers, whereas, among the latter, these are only partially exercised, and in some cases are absolutely unused. A wild animal has to search, and often to labour, for every mouthful of food—to exercise sight, hearing, and smell in seeking it, and in avoiding dangers, in procuring shelter from the inclemency of the seasons, and in providing for the subsistence and safety of its offspring. There is no muscle of its body that is not called into daily and hourly activity; there is no sense or faculty that is not strengthened by continual exercise. The domestic animal, on the other hand, has food provided for it, is sheltered, and often confined, to guard it against the vicissitudes of the seasons, is carefully secured from the attacks of its natural enemies, and seldom even rears its young without human assistance. Half of its senses and faculties are quite useless; and the other half are but occasionally called into feeble exercise, while even its muscular system is only irregularly called into action.

Now when a variety of such an animal occurs, having increased power or capacity in any organ or sense, such increase is totally useless, is never called into action, and may even exist without the animal ever becoming aware of it. In the wild animal, on the contrary, all its faculties and powers being brought into full action for the necessities of existence, any increase becomes immediately available, is strengthened by exercise, and must even slightly modify the food, the habits, and the whole economy of the race. It creates as it were a new animal, one of superior powers, and which will necessarily increase in numbers and outlive those inferior to it.

Again, in the domesticated animal all variations have an equal chance of continuance; and those which would decidedly render a wild animal unable to compete with its fellows and continue its existence are no disadvantage whatever in a state of domesticity. Our quickly fattening pigs, short-legged sheep, pouter pigeons, and poodle dogs could never have come into existence in a state of nature, because the very first step towards such inferior forms would have led to the rapid extinction of the race; still less could they now exist in competition with their wild allies. The great speed but slight endurance of the race horse, the unwielding strength of the ploughman's team, would both be useless in a state of nature. If turned wild on the pampas, such animals would probably soon become extinct, or under favorable circumstances might each lose those extreme qualities which would never be called into action, and in a few generations would revert to a common type, which must be that in which the various powers and faculties are so proportioned to each other as to be best adapted to procure food and secure safety,—that in which by the full exercise of every part of his organization the animal can alone continue to live. Domestic varieties, when turned wild, *must* return to something near the type of the original wild stock, *or become altogether extinct.*

Lamarck's Hypothesis Very Different from That Now Advanced

We see, then, that no inference as to varieties in a state of nature can be deduced from the observation of those occurring among domestic animals. The two are so much opposed to each other in every circumstance of their existence, that what applies to the one is almost sure not to apply to the other. Domestic animals are abnormal, irregular, artificial; they are subject to varieties which never occur and never can occur in a state of nature; their very existence depends altogether on human care: so far are many of them removed from that just proportion of faculties, that true balance of organization, by means of which alone an animal left to its own resources can preserve its existence and continue its race.

The hypothesis of Lamarck—that progressive changes in species have been produced by attempts of animals to increase the development of their own organs, and thus modify their structure and habits—has been repeatedly and easily refuted by all writers on the subject of varieties and species, and it seems to have been considered that when this was done the whole question has been finally settled; but the view here developed renders such an hypothesis quite unnecessary, by showing that similar results must be produced by the action of principles constantly at work in nature. The powerful retractile talons of the falcon- and the cat-tribes have not been produced or increased by the volition of those animals; but among the different varieties which occurred in the earlier and less highly organized forms of these groups, *those always survived longest which had the greatest facilities for seizing their prey.* Neither did the giraffe acquire its long neck by desiring to reach the foliage of the more lofty shrubs, and constantly stretching its neck for the purpose, but because any varieties which occurred among its antitypes with a longer neck than usual *at once secured a fresh range of pasture over the same ground as their shorter-necked companions, and on the first scarcity of food were thereby enabled to outlive them.* Even the peculiar colours of many animals, especially insects, so closely resembling the soil or the leaves or the trunks on which they habitually reside, are explained on the same principle; for though in the course of ages varieties of many tints have occurred, *yet those races having colours best adapted to concealment from their enemies would inevitably survive the longest.* We have also here an acting cause to account for that balance so often observed in nature,—a deficiency in one set of organs always being compensated by an increased development of some others— powerful wings accompanying weak feet, or great velocity making up for the absence of defensive weapons; for it has been shown that all varieties in which an unbalanced deficiency occurred could not long continue their existence. The action of this principle is exactly like that of the centrifugal governor of the steam engine, which checks and corrects any irregularities almost before they become evident; and in like manner no unbalanced deficiency in the animal kingdom can ever reach any conspicuous magnitude, because it would make itself felt at the very first step, by rendering existence difficult and extinction almost sure to follow. An origin such as is here advocated will also agree with the peculiar character of the modifications of form and structure which obtain in organized beings—the many lines of divergence from a central type, the increasing efficiency and power of a particular organ through a succession of allied species, and the remarkable persistence of unimportant parts such as colour, texture of plumage and hair, forms of horns or crests, through a series of species differing considerably in more essential characters. It also furnishes us with a reason for that "more specialized structure" which Professor Owen states to be a characteristic of recent compared with extinct forms, and which would evidently be the result of the progressive modification of any organ applied to a special purpose in the animal economy.

⊷⊶⊙⊷⊶

HERMANN VON HELMHOLTZ
The Conservation of Energy, from a lecture (1863)

Hermann von Helmholtz (1821–1894) was the ideal nineteenth-century scientist. He was curious about everything and found something valuable to add to physics, psychology, chemistry, and geology. He also had a specific philosophical goal. As a university student he had sworn an oath of loyalty to the idea that everything in the universe could be reduced to chemical and physical principles. He then began a lifetime crusade to find the principles that would cover everything from musical sensations to moving glaciers.

If we had to choose one of his feats to stand for all the others, it would probably be his contribution to the doctrine of the conservation of energy. Energy was the great nineteenth-century generalization, just as information has been the twentieth century's. Once the people of the 1800s grasped the idea of energy, the technical issues of running heavy machinery became much easier to contemplate. In the lecture given here, Helmholtz explains to a popular audience his notion that like matter, energy could be neither created nor destroyed. Before Helmholtz began giving these lectures, German professors almost never spoke to ordinary people about their studies. But Helmholtz did it as part of his missionary work for science.

Every great deed of which history tells us, every mighty passion which art can represent, every picture of manners, of civic arrangements, of the culture of peoples of distant lands or of remote times, seizes and interests us, even if there is no exact scientific connection among them. We continually find points of contact and comparison in our conceptions and feelings; we get to know the hidden capacities and desires of the mind, which in the ordinary peaceful course of civilised life remain unawakened.

It is not to be denied that, in the natural sciences, this kind of interest is wanting. Each individual fact, taken by itself, can indeed arouse our curiosity or our astonishment, or be useful to us in its practical applications. But intellectual satisfaction we obtain only from a connection of the whole, just from its conformity with law. [. . .] There is a kind, I might almost say, of artistic satisfaction, when we are able to survey the enormous wealth of Nature as a regularly-ordered whole—a kosmos, an image of the logical thought of our own mind.

The last decades of scientific development have led us to the recognition of a new universal law of all natural phenomena, which, from its extraordinarily extended range, and from the connection which it constitutes between natural phenomena of all kinds, even of the remotest times and the most distant places, is especially fitted to give us an idea of what I have described as the character of the natural sciences. [. . .] This law is *the Law of the Conservation of Force* [. . . and it] asserts, that the *quantity of force which can be brought into action in the whole of Nature is unchangeable*, and can neither be increased nor diminished. [. . .]

The idea of work for machines, or natural processes, is taken from comparison with the working power of man; and we can therefore best illustrate from human labour the most important features of the question with which we are concerned. [. . .] Both the arm of the blacksmith who delivers his powerful blows with the heavy hammer, and that of the violinist who produces the most delicate variations in sound, and the hand of the lace-maker who works with threads so fine that they are on the verge of the invisible, all these acquire the force which moves them in the same manner and by the same organs, namely, the muscles of the arms. [. . .]

Just so it is with machines: they are used for the most diversified arrangements. We produce by their agency an infinite variety of movements, with the most various degrees of force and rapidity, from powerful steam hammers and rolling mills, where gigantic masses of iron are cut and shaped like butter, to spinning and weaving frames, the work of which rivals that of the spider. [. . .] But one thing is common to all these differences; they all need a *moving force*, which sets and keeps them in motion, just as the works of the human hand all need the moving force of the muscles. [. . .]

We have nothing to do here with the manifold character of the actions and arrangements which the machines produce; we are only concerned with an expenditure of force.

This very expression which we use so fluently, 'expenditure of force,' which indicates that the force applied has been expended and lost, leads us to a further characteristic analogy between the effects of the human arm and those of machines. The greater the exertion, and the longer it lasts, the more is the arm *tired*, and the more *is the store of its moving force for the time exhausted*. We shall see that this peculiarity of becoming exhausted by work is also met with in the moving forces of inorganic nature; indeed, that this capacity of the human arm of being tired is only one of the consequences of the law with which we are now concerned. When fatigue sets in, recovery is needed, and this can only be effected by rest and nourishment. We shall find that also in the inorganic moving forces, when their capacity for work is spent, there is a possibility of reproduction, although in general other means must be used to this end than in the case of the human arm. [. . .]

Let us now consider that moving force which we know best, and which is simplest—gravity. It acts, for example, as such in those clocks which are driven by a weight. This weight, fastened to a string, which is wound round a pulley connected with the first toothed wheel of the clock, cannot obey the pull of gravity without setting the whole clockwork in motion. Now I must beg you to pay special attention to the following points: the weight cannot put the clock in motion without itself sinking; did the weight not move, it could not move the clock, and its motion can only be such a one as obeys the action of gravity. Hence, if the clock is to go, the weight must continually sink lower and lower, and must at length sink so far that the string which supports it is run out. The clock then stops. [. . .]

But we can wind up the clock by the power of the arm, by which the weight is again raised. When this has been done, it has regained its former capacity, and can again set the clock in motion. [. . .]

The case is somewhat different when Nature herself raises the weight, which then works for us. She does not do this with solid bodies, at least not with such regularity as to be utilised; but she does it abundantly with water, which, being raised to the tops of mountains by meteorological processes, returns in streams from them. The gravity of water we use as moving force, the most direct application being in what are called *overshot* wheels. [. . .] It is [. . .] the weight of the falling water which turns the wheel, and furnishes the motive power. But you will at once see that the mass of water which turns the wheel must necessarily fall in order to do so, and that though, when it has reached the bottom, it has lost none of its gravity, it is no longer in a position to drive the wheel, if it is not restored to its original position, either by the power of the human arm or by means of some other natural force. If it can flow from the mill-stream to still lower levels, it may be used to work other wheels. But when it has reached its lowest level, the sea, the last remainder of the moving force is used up, which is due to gravity—that is, to the attraction of the earth, and it cannot act by its weight until it has been again raised to a high level. As this is actually effected by meteorological processes, you will at once observe that these are to be considered as sources of moving force.

Water power was the first inorganic force which man learnt to use instead of his own labour or of that of domestic animals. According to Strabo, it was known to King Mithridates of Pontus, who was also otherwise celebrated for his knowledge of Nature; near his palace there was a water wheel. Its use was first introduced among the Romans in the time of the first Emperors. Even now we find water mills in all mountains, valleys, or wherever there are rapidly-flowing regularly-filled brooks and streams. [. . .]

You all know how powerful and varied are the effects of which steam engines are capable; with them has really begun the great development of industry which has characterised our century before all others. Its most essential superiority over motive powers formerly known is that it is not restricted to a particular place. The store of coal and the small quantity of water which are the sources of its power can be brought everywhere, and steam engines can even be made movable, as is the case with steam ships and locomotives. [. . .]

We see, then, that heat can produce mechanical power; but in the cases which we have discussed we have seen that the quantity of force which can be produced by a given measure of a physical process is always accurately defined, and that the further capacity for work of the natural forces is either diminished or exhausted by the work which has been performed. How is it now with *Heat* in this respect?

This question was of decisive importance in the endeavour to extend the law of the Conservation of Force to all natural processes. In the answer lay the chief difference between the older and newer views in these respects. Hence it is that many physicists designate that view of Nature corresponding to the law of the conservation of force with the name of *Mechanical Theory of Heat*.

The older view of the nature of heat was that it is a substance, very fine and imponderable indeed, but indestructible, and unchangeable in quantity, which is an essential fundamental property of all matter. And, in fact, in a large number of natural processes, the quantity of heat which can be demonstrated by the thermometer is unchangeable. [. . .]

But one relation of heat—namely, that to mechanical work—had not been accurately investigated. [. . .] It was already known that whenever two bodies in motion rubbed against each other, heat was developed anew, and it could not be said whence it came.

The fact is universally recognised; the axle of a carriage which is badly greased and where the friction is great, becomes hot—so hot, indeed, that it may take fire; machine wheels with iron axles going at a great rate may become so hot that they weld to their sockets.

A powerful degree of friction is not, indeed, necessary to disengage an appreciable degree of heat; thus, a lucifer match, which by rubbing is so heated that the phosphoric mass ignites, teaches this fact. Nay, it is enough to rub the dry hands together to feel the heat produced by friction, and which is far greater than the heating which takes place when the hands lie gently on each other. [. . .]

So long as it was only a question of the friction of solids, in which particles from the surface become detached and compressed, it might be supposed that some changes in structure of the bodies rubbed might here liberate latent heat, which would thus appear as heat of friction. [. . .]

Heat can also be produced by the impact of imperfectly elastic bodies as well as by friction. This is the case, for instance, when we produce fire by striking flint against steel, or when an iron bar is worked for some time by powerful blows of the hammer.

If we inquire into the mechanical effects of friction and of inelastic impact, we find at once that these are the processes by which all terrestrial movements are brought to rest. A moving body whose motion was not retarded by any resisting force would continue to move to all eternity. The motions of the planets are an instance of this. This is apparently never the case with the motion of the terrestrial bodies, for they are always in contact with other bodies which are at rest, and rub against them. We can, indeed, very much diminish their friction, but never completely annul it. [. . .]

If we review the results of all these instances, which each of you could easily add to from your own daily experience, we shall see that friction and inelastic impact are processes in which mechanical work is destroyed, and heat produced in its place. [. . .]

Joule [. . .] measured [. . .] the amount of work which is destroyed by the friction of solids and by the friction of liquids; and, on the other hand, he has determined the quantity of heat which is thereby produced, and has established a definite relation between the two. His experiments show that when heat is produced by the consumption of work, a definite quantity of work is required to produce that amount of heat which is known to physicists as the *unit of heat;* the heat, that is to say, which is necessary to raise one gramme of water through one degree centigrade. The quantity of work necessary for this is, according to Joule's best experiments, equal to the work which a gramme would perform in falling through a height of 425 metres. [. . .]

Exactly the same relations between heat and work were also found in the reverse process—that is, when work was produced by heat. [. . .] A gas which is allowed to expand with moderate velocity becomes cooled. Joule was the first to show the reason of this cooling. For the gas has, in expanding, to overcome the resistance which the pressure of the atmosphere and the slowly yielding side of the vessel oppose to it. [. . .] Gas thus performs work, and this work is produced at the cost of its heat. Hence the cooling. If, on the contrary, the gas is suddenly allowed to issue into a perfectly exhausted space where it finds no resistance, it does not become cool, as Joule has shown; or if individual parts of it become cool, others become warm; and, after the temperature has become equalised, this is exactly as much as before the sudden expansion of the gaseous mass. [. . .]

Thus then: a certain quantity of heat may be changed into a definite quantity of work; this quantity of work can also be retransformed into heat, and, indeed, into exactly the same quantity of heat as that from which it originated; in a mechanical point of view, they are exactly equivalent. Heat is a new form in which a quantity of work may appear.

These facts no longer permit us to regard heat as a substance, for its quantity is not unchangeable. It can be produced anew from the *vis viva* of motion destroyed; it can be destroyed, and then produces motion. We must rather conclude from this that heat itself is a motion, an internal invisible motion of the smallest elementary particles of bodies. If, therefore, motion seems lost in friction and impact, it is not actually lost, but only passes from the great visible masses to their smallest particles; while in steam engines the internal motion of the heated gaseous particles is transferred to the piston of the machine, accumulated in it, and combined in a resultant whole. [. . .]

We turn now to another kind of natural forces which can produce work—I mean the chemical. We have to-day already come across them. They are the ultimate cause of the work which gunpowder and the steam engine produce; for the heat which is consumed in the latter, for example, originates in the combustion of carbon—that is to say, in a chemical process. The burning of coal is the chemical union of carbon with the oxygen of the air, taking place under the influence of the chemical affinity of the two substances. [. . .]

The application of electrical currents opens out a large number of relations between the various natural forces. We have decomposed water into its elements by such currents, and should be able to decompose a large number of other chemical compounds. On the other hand, in ordinary galvanic batteries electrical currents are produced by chemical forces. [. . .]

Let us review these examples once more, and recognise in them the law which is common to all.

A raised weight can produce work, but in doing so it must necessarily sink from its height, and, when it has fallen as deep as it can fall, its gravity remains as before, but it can no longer do work. [. . .]

Heat can perform work; it is destroyed in the operation. Chemical forces can perform work, but they exhaust themselves in the effort.

Electrical currents can perform work, but to keep them up we must consume either chemical or mechanical forces, or heat.

We may express this generally. *It is a universal character of all known natural forces that their capacity for work is exhausted in the degree in which they actually perform work.*

We have seen, further, that when a weight fell without performing any work, it *either* acquired velocity or produced heat. We might also drive a magneto-electrical machine by a falling weight; it would then furnish electrical currents.

We have seen that chemical forces, when they come into play, produce either heat or electrical currents or mechanical work.

We have seen that heat may be changed into work; there are apparatus (thermo-electric batteries) in which electrical currents are produced by it. Heat can directly separate chemical compounds; thus, when we burn limestone, it separates carbonic acid from lime.

Thus, whenever the capacity for work of one natural force is destroyed, it is transformed into another kind of activity. Even within the circuit of inorganic natural forces, we can transform each of them into an active condition by the aid of any other natural force which is capable of work. [. . .]

It follows thence *that the total quantity of all the forces capable of work in the whole universe remains eternal and unchanged throughout all their changes.* All change in nature amounts to this, that force can change its form and locality without its quantity being changed. The universe possesses, once for all, a store of force which is not altered by any change of phenomena, can neither be increased nor diminished, and which maintains any change which takes place on it.

You see how, starting from considerations based on the immediate practical interests of technical work, we have been led up to a universal natural law, which, as far as all previous experience extends, rules and embraces all natural processes; which is no longer restricted to the practical objects of human utility, but expresses a perfectly general and particularly characteristic property of all natural forces, and which, as regards generality, is to be placed by the side of the laws of the unalterability of mass, and the unalterability of the chemical elements.

At the same time, it also decides a great practical question which has been much discussed in the last two centuries, to the decision of which an infinity of experiments has been made and an infinity of apparatus constructed—that is, the question of the possibility of a perpetual motion. By this was understood a machine which was to work continuously without the aid of any external driving force. The solution of this problem promised enormous gains. Such a machine would have had all the advantages of steam without requiring the expenditure of fuel. Work is wealth. A machine which could produce work from nothing was as good as one which made gold. This problem had thus for a long time occupied the place of gold making, and had confused many a pondering brain. That a perpetual motion could not be produced by the aid of the then known mechanical forces could be demonstrated in the last century by the aid of the mathematical mechanics which had at that time been developed. But to show also that it is not possible even if heat, chemical forces, electricity, and magnetism were made to co-operate, could not be done without a knowledge of our law in all its generality. The possibility of a perpetual motion was first finally negatived by the law of the conservation of force, and this law might also be expressed in the practical form that no perpetual motion is possible, that force cannot be produced from nothing; something must be consumed.

>─┼─◆>─○─<◆─┼─◄

ALBERT EINSTEIN

Two Theories of Relativity, from *Relativity: The Special and General Theory, a Popular Exposition* (1916)

Along with Mozart and Shakespeare, Albert Einstein (1879–1955) is one of those special creators whose achievement was so unusual that it is nearly impossible to see how a normal, human source yielded so unprecedented and immortal a result. How did he do it? Einstein did not like to reveal himself, but he did make a strong effort to make his theories of relativity clear to interested amateurs. Shortly after publishing his general theory of relativity, he wrote a popular account of relativity. Werner Heisenberg, the author of physics's uncertainty principle, later recalled that he read the book as a schoolboy and that it moved him from thinking about a career in mathematics to one in physics. Most of Einstein's book gave no hint of how he managed to know the things he asserted, but there is a moment in the selection here when he gives us a peek at his hand. He writes about a problem that had blocked his progress. It seemed that there was a kind of motion that was not relative. For example, if a car stops abruptly, the passengers fly forward, but an observer watching the car stop like that is not sent flying. This difference suggests that the motion in the car is confined to the car and is not relative to the observer. To solve this problem, Einstein had to redefine gravity.

In this selection he does not go that far, but he does give us a rare look at the working of his own mind. He tells us he wanted to solve the problem, describing himself as an intellect striving for generalization—about as fine a definition of Einstein at work as has ever been written—and being tempted toward a solution. Why tempted? Because he had no ready explanation for why passengers fly forward in cars but outside observers do not. Here was a kind of motion that Galileo had not considered but that was part of everyday life in the modern world. In that scene, drawn as sparsely as haiku, Einstein showed his nature: the striving intellect, resisting the temptation to cheat, boldly facing up to what Galileo left out.

Every motion must be considered only as a relative motion. [. . .] [Imagine an] embankment and [a] railway carriage. We can express the fact of the motion here taking place in the following two forms, both of which are equally justifiable:

a. The carriage is in motion relative to the embankment.

b. The embankment is in motion relative to the carriage.

In (a) the embankment, in (b) the carriage, serves as the body of reference in our statement of the motion taking place. If it is simply a question of detecting or describing the motion involved, it is in principle immaterial to what reference-body we refer the motion. [. . .] This is self-evident, but it must not be confused with the much more comprehensive statement called "the principle of relativity," which we have taken as the basis of our investigations.

The principle we have made use of not only maintains that we may equally well choose the carriage or the embankment as our reference-body for the description of any event (for this, too, is self-evident). Our principle rather asserts what follows: If we formulate the general laws of nature as they are obtained from experience, by making use of

a. the embankment as reference-body,

b. the railway carriage as reference-body,

then these general laws of nature (*e.g.* the laws of mechanics or the law of the propagation of light *in vacuo*) have exactly the same form in both cases. This can also be expressed as follows: For the *physical* description of natural processes, neither of the reference-bodies K, K' is unique (lit.[5] "specially marked out") as compared with the other. Unlike the first, this latter statement need not of necessity hold *a priori*; it is not contained in the conceptions of "motion" and "reference-body" and derivable from them; only *experience* can decide as to its correctness or incorrectness.

Up to the present, however, we have by no means maintained the equivalence of *all* bodies of reference K in connection with the formulation of natural laws. Our course was more on the following lines. In the first place, we started out from the assumption that there exists a reference-body K, whose condition of motion is such that the Galilean law holds with respect to it: A particle left to itself and sufficiently far removed from all other particles moves uniformly in a straight line. With reference to K (Galilean reference-body) the laws of nature were to be as simple as possible. But in addition to K, all bodies of reference

---

[5] *Translator's aside.*

K' should be given preference in this sense, and they should be exactly equivalent to K for the formulation of natural laws, provided that they are in a state of *uniform rectilinear and non-rotary motion* with respect to K; all these bodies of reference are to be regarded as Galileian reference-bodies. The validity of the principle of relativity was assumed only for these reference-bodies, but not for others (*e.g.*, those possessing motion of a different kind). In this sense we speak of the *special* principle of relativity, or special theory of relativity.

In contrast to this we wish to understand by the "general principle of relativity" the following statement: All bodies of reference K, K', etc., are equivalent for the description of natural phenomena (formulation of the general laws of nature), whatever may be their state of motion. But before proceeding farther, it ought to be pointed out that this formulation must be replaced later by a more abstract one. [. . .]

Since the introduction of the special principle of relativity has been justified, every intellect which strives after generalisation must feel the temptation to venture the step towards the general principle of relativity. But a simple and apparently quite reliable consideration seems to suggest that, for the present at any rate, there is little hope of success in such an attempt. Let us imagine ourselves transferred to our old friend the railway carriage, which is travelling at a uniform rate. As long as it is moving uniformly, the occupant of the carriage is not sensible of its motion, and it is for this reason that he can without reluctance interpret the facts of the case as indicating that the carriage is at rest, but the embankment in motion. Moreover, according to the special principle of relativity, this interpretation is quite justified also from a physical point of view.

If the motion of the carriage is now changed into a non-uniform motion, as for instance by a powerful application of the brakes, then the occupant of the carriage experiences a correspondingly powerful jerk forwards. The retarded motion is manifested in the mechanical behaviour of bodies relative to the person in the railway carriage. The mechanical behaviour is different from that of the case previously considered, and for this reason it would appear to be impossible that the same mechanical laws hold relatively to the non-uniformly moving carriage, as hold with reference to the carriage when at rest or in uniform motion. At all events it is clear that the Galileian law does not hold with respect to the non-uniformly moving carriage.

Eine einfache theoretis…

…, macht die Annahme …

…ichtstrahlen in einem G…

…ne Deviation erfahren.

Grav. Feld

→ Lichtstrahl

…neurande müsste diese …

…tragen und wie $\frac{1}{R}$ ab…

…indig vom Sonnen- ~~oberradius~~) — Mittelpunk…

$\uparrow 0,84''$

Sonne

…wäre deshalb von grö…

…, bis zu wie grossen So…

Style in the Scientific Imagination

The union of literary skill with scientific understanding has yielded a small stream of classics that are read and remembered years after they first appear. Yet professional scientists are often uneasy about these works. They, too, read and remember them. Indeed, in autobiographies they commonly refer to some such work as having helped steer them toward science when they were young. But they also worry that these works are not real science.

Some scientists would even say that, for example, Primo Levi is simply popularizing, taking from science rather than contributing to it. But that view misses his achievement. Levi and the other contributors to this part ordered many small understandings into a large whole. Chemists and biochemists produced the facts that Levi used. (It does not matter that Levi, too, was a chemist, for he is not reporting the results of his own research.) Then Levi organized those facts into a larger context so even a nonscientist who never once wondered about anything chemical could be stirred and think, "Oh, I see."

Chemists reading Levi's story may realize that although they know the facts that Levi cites, somehow they never imagined them neatly arranged in this way. The unity of their knowledge had eluded them. Sometimes professionals feel uneasy about this sense of broader understanding. Shouldn't they have realized all this before? Their discomfort underestimates what the writers in this section did. Literary skills are not, like gargoyles on cathedral walls, adornments that contribute nothing to keeping the building up. Instead, these writers self-consciously create style and form because without them, their pieces would not come together.

Naive readers often dismiss literary skills by saying that they are interested only in the substance of things. They have half a point: Truth sits to one side while, from a distance, the lens of language hides and distorts what it claims to reveal. But the half they miss is that in regard to understanding, there is no escaping language. Every writer must arrange words into a form, and if they do not work to find the best word, the readers will be left to hunt it out on their own. If writers cannot find a way to lead readers through the information, the readers will become lost, then confused, and finally angry. Creating these words and forms often demands that the writers have a broader understanding of the data than do the specialists who provided them.

The accounts in this part fall into two overlapping groups. Some, like the essays by J. B. S. Haldane and Julian Huxley, are stories about science. Although the writers present many facts, the organizing principle has been the science that understands them. Other essays, like the selections by Primo Levi and Louise Young, are stories of phenomena. Scientists may appear, but the stories concern natural processes that are worth knowing in themselves. Rachel Carson's vivid account of the endless drift of debris to the ocean floor is a nearly perfect example of this genre's ability to dramatize a process without falsifying it.

"It Is in Some Fashion a History"

Primo Levi

⊷⊷○⊷⊷

GALILEO GALILEI

The Speed of Falling Bodies, from
Two New Sciences (1638)

In 1638 Galileo Galilei (1564–1642) published a book in Holland, even though he was under house arrest in Italy. Its appearance caused the same sort of sensation produced centuries later when forbidden works of Soviet writers appeared suddenly in the West. Like Galileo's work on the Two Systems, this new book was written as a dialogue, although it was no longer a debate. Simplicio is still the representative of Aristotle, but he is much more amenable to what Salviati (that is, Galileo) has to say. Sagredo is again the reader's representative, the intelligent onlooker to be convinced.

Galileo explored the physics of motion all his adult life. In 1582, when he was 18 years old, he noticed that the swing of a pendulum took the same amount of time no matter what the size of the swing was. If a pendulum needed 2 seconds to swing through an arc of 4 inches, it would also need 2 seconds to swing through 8 inches. The observation gave him a way of timing events. In about 1604 he proved the thesis of the excerpt here, that bodies fall at the same rate no matter what their weight. By the time of this book, Galileo had been thinking about and teaching its contents for more than 30 years.

It was in this work that Galileo moved scientific understanding beyond that of the Greeks. The ancients worked out the static mechanics of geometric space and of physical weights, but motion defeated them. They had no clear way of thinking about it. But thanks to Galileo and the ideas in this selection, we can make sense of it.

CHAPTER

TEN

SALVIATI: I greatly doubt that Aristotle ever tested by experiment whether it be true that two stones, one weighing ten times as much as the other, if allowed to fall, at the same instant, from a height of, say, 100 cubits, would so differ in speed that when the heavier had reached the ground, the other would not have fallen more than 10 cubits.

SIMPLICIO: His language would seem to indicate that he had tried the experiment, because he says: *We see the heavier;* now the word *see* shows that he had made the experiment.

SAGREDO. But I, Simplicio, who have made the test can assure you that a cannon ball weighing one or two hundred pounds, or even more, will not reach the ground by as much as a span ahead of a musket ball weighing only half a pound, provided both are dropped from a height of 200 cubits.

SALVIATI: But, even without further experiment, it is possible to prove clearly, by means of a short and conclusive argument, that a heavier body does not move more rapidly than a lighter one provided both bodies are of the same material and in short such as those mentioned by Aristotle. But tell me, Simplicio, whether you admit that each falling body acquires a definite speed fixed by nature, a velocity which cannot be increased or diminished except by the use of force or resistance.

SIMPLICIO: There can be no doubt but that one and the same body moving in a single medium has a fixed velocity which is determined by nature and which cannot be increased except by the addition of momentum or diminished except by some resistance which retards it.

SALVIATI: If then we take two bodies whose natural speeds are different, it is clear that on uniting the two, the more rapid one will be partly retarded by the slower, and the slower will be somewhat hastened by the swifter. Do you not agree with me in this opinion?

SIMPLICIO: You are unquestionably right.

SALVIATI: But if this is true, and if a large stone moves with a speed of, say, eight while a smaller moves with a speed of four, then when they are united, the system will move with a speed less than eight; but the two stones when tied together make a stone larger than that which before moved with a speed of eight. Hence the heavier body moves with less speed than the lighter; an effect which is contrary to your supposition. Thus you see how, from your assumption that the heavier body moves more rapidly than the lighter one, I infer that the heavier body moves more slowly.

SIMPLICIO: I am all at sea because it appears to me that the smaller stone when added to the larger increases its weight and by adding weight I do not see how it can fail to increase its speed or, at least, not to diminish it.

SALVIATI: Here again you are in error, Simplicio, because it is not true that the smaller stone adds weight to the larger.

SIMPLICIO: This is, indeed, quite beyond my comprehension.

SALVIATI: It will not be beyond you when I have once shown you the mistake under which you are laboring. Note that it is necessary to distinguish between heavy bodies in motion and the same bodies at rest. A large stone placed in a balance not only acquires additional weight by having another stone placed upon it, but even by the addition of a handful of hemp its weight is augmented six to ten ounces according to the quantity of hemp. But if you tie the hemp to the stone and allow them to fall freely from some height, do you believe that the hemp will press down upon the stone and thus accelerate its motion or do you think the motion will be retarded by a partial upward pressure? One always feels the pressure upon his shoulders when he prevents the motion of a load resting upon him; but if one descends just as rapidly as the load would fall how can it gravitate or press upon him? Do you not see that this would be the same as trying to strike a man with a lance when he is running away from you with a speed which is equal to, or even greater, than that with which you are following him? You must therefore conclude that, during free and natural fall, the small stone does not press upon the larger and consequently does not increase its weight as it does when at rest.

SIMPLICIO: But what if we should place the larger stone upon the smaller?

SALVIATI: Its weight would be increased if the larger stone moved more rapidly; but we have already concluded that when the small stone moves more slowly it retards to some extent the speed of the larger, so that the combination of the two, which is a heavier body than the larger of the two stones, would move less rapidly, a

conclusion which is contrary to your hypothesis. We infer therefore that large and small bodies move with the same speed provided they are of the same specific gravity.

SIMPLICIO: Your discussion is really admirable; yet I do not find it
easy to believe that a bird-shot falls as swiftly as a cannon ball.

SALVIATI: Why not say a grain of sand as rapidly as a grindstone?
But, Simplicio, I trust you will not follow the example of many
others who divert the discussion from its main intent and fasten
upon some statement of mine which lacks a hair's-breadth of the
truth and, under this hair, hide the fault of another which is as
big as a ship's cable. Aristotle says that "an iron ball of one hundred pounds falling from a height of one hundred cubits reaches
the ground before a one-pound ball has fallen a single cubit." I
say that they arrive at the same time. You find, on making the
experiment, that the larger outstrips the smaller by two finger-
breadths, that is, when the larger has reached the ground, the
other is short of it by two finger-breadths; now you would not
hide behind these two fingers the ninety-nine cubits of Aristotle,
nor would you mention my small error and at the same time pass
over in silence his very large one. Aristotle declares that bodies of
different weights, in the same medium, travel (in so far as their
motion depends upon gravity) with speeds which are proportional to their weights; this he illustrates by use of bodies in
which it is possible to perceive the pure and unadulterated effect
of gravity, eliminating other considerations, for example, figure
as being of small importance, influences which are greatly dependent upon the medium which modifies the single effect of gravity
alone. Thus we observe that gold, the densest of all substances,
when beaten out into a very thin leaf, goes floating through the
air; the same thing happens with stone when ground into a very
fine powder. But if you wish to maintain the general proposition
you will have to show that the same ratio of speeds is preserved
in the case of all heavy bodies, and that a stone of twenty pounds
moves ten times as rapidly as one of two; but I claim that this is
false and that, if they fall from a height of fifty or a hundred
cubits, they will reach the earth at the same moment.

SIMPLICIO: Perhaps the result would be different if the fall took place not from a few cubits but from some thousands of cubits.

SALVIATI: If this were what Aristotle meant you would burden him with another error which would amount to a falsehood; because, since there is no such sheer height available on earth, it is clear that Aristotle could not have made the experiment; yet he wishes to give us the impression of his having performed it when he speaks of such an effect as one which we see.

LUCRETIUS

The Persistence of Atoms, from
On the Nature of Things (ca. 60 B.C.)

Titus Lucretius Carus (ca. 94–ca. 50 B.C.) wrote an epic poem presenting the atomic theory of Epicurus, a Greek philosopher who had lived 250 years earlier. The poem was a remarkable feat. No one before or since has succeeded so well at turning reasoning into poetic assertion. Adding to the achievement was Lucretius' Latin. In those days, Greek was the language of learning. But Lucretius's generation of Latin writers struggled to make their language subtle and rich enough to express Greek ideas. They managed so well that they made Latin the language of philosophy for the next 1600 years.

Belief in the atomic world survived almost unchanged into the 1900s. It held that all matter resolved itself into "indivisibles," tiny particles that could not be made smaller. This idea was the principal materialist doctrine before Einstein. In the following passage, Lucretius asserts the prime doctrine of classical physics that matter can be neither created nor destroyed. In 1905, as part of a footnote in his paper on the special theory of relativity, Einstein indicated that oh, by the way, matter could be destroyed. Since then, the doctrine that Lucretius expressed so ably has been reworked to assert that although matter can be transformed into energy, and vice versa, the sum total of existing matter and energy can neither grow nor shrink. This modern translation is by Frank O. Copley.

This fright, this night of the mind, must be dispelled
not by the rays of the sun, nor day's bright spears,
but by the face of nature and her laws.
And this is her first, from which we take our start:
nothing was ever by miracle made from nothing.
You see, all mortal men are gripped by fear
because they see so many things on earth
and in the sky, yet can't discern their causes
and hence believe that they are acts of god.

But in all this, when we have learned that nothing
can come from nothing, then we shall see straight through
to what we seek: whence each thing is created
and in what manner made, without god's help.
    If things were made from nothing, then all kinds
could spring from any source: they'd need no seed.
Man could have burst from ocean, from dry land
the bearers of scales, and from thin air the birds;
cows, horses, sheep, and the rest, and all wild beasts
would breed untrue, infesting farm and forest.
Nor would one tree produce one kind of fruit;
no, they would change, and all could bear all kinds.
For if there were no factors governing birth,
how could we tell who anyone's mother was?
But things are formed, now, from specific seeds,
hence each at birth comes to the coasts of light
from a thing possessed of its essential atoms.
Thus everything cannot spring from anything,
for things are unique; their traits are theirs alone.
And why in spring do we see roses, grain
in summer, vines produce at autumn's call,
if not because right atoms in right season
have streamed together to build each thing we see,
while weather favors and life-giving earth
brings delicate seedling safe to land and light?
But if they came from nothing, they'd spring up
all helter-skelter in seasons not their own;
for there would be no atoms to be kept
from fertile union at untimely hours.
Nor would things when they grown have need of time
for seeds to combine, if they could grow from nothing.
Why! Babes in arms would turn into men forthwith,
and forest would leap from sprouts new-sprung of earth.
Yet clearly such things never occur: all growth

is gradual, regular, from specific seed,
and with identity kept. Hence learn that things
can grow only when proper substance feeds them.
To this we add: without her seasonal rains
Earth could not send up offspring rich in joy,
nor, further, could living creatures without food
beget their kind or keep their hold on life.
Better conceive of many atoms shared
by many things, as letters are by words,
than of a single thing not made of atoms.
To continue: why could nature not produce
men of such size that they could cross the seas
on foot, and with bare hands pull hills apart,
and live the lifetime of ten thousand men,
if not because each thing has but one substance
marked and designed to bring it into being?
Admit then: nothing can be made of nothing
since things that are created must have seed
from which to come forth to the gentle breezes.
Finally, since we see tilled fields excel
untilled, and pay more profit on our toil
surely prime bodies must exist in soil.
Plowing the fertile furrow, turning up
the earth, we bring these bodies to the surface.
But if there were none such, everything would grow
spontaneously, and better, without our labors.
And now add this: nature breaks up all things
into their atoms; no thing dies off to nothing.
For if a thing were mortal in all its parts,
it would be whisked away, just drop from sight,
since there would be no need of force to wrench
one part from another, or to dissolve their bonds.
But things are made of atoms; they are stable.
Until some force comes, hits them hard, and splits them,

or seeps to their inner parts and makes them burst,
nature brings no destruction to our sight.
Besides, take things that time removes through aging:
if when they died their matter were all consumed,
whence does Venus bring animals forth to life
kind after kind, and earth, the magic-maker,
nourish, increase, and feed them, kind by kind?
Whence could native fountains and far-flung rivers
supply the sea? Whence ether feed the stars?
For everything of mortal mass long since
has been used up as boundless time passed by.
But if the stuff of which this sum of things
is built has lasted down through empty ages,
surely it is endowed with deathless nature;
no thing, therefore, can be reduced to nothing.

▻─┼─◆─○─◆─┼─◅

J. B. S. HALDANE

Food Control in Insect Societies, from
Possible Worlds (1928)

John Burden Sanderson Haldane (1892–1964) spent much of the 1920s and early 1930s developing a mathematical link between genetics and evolution. His work, known as the "modern synthesis," puts him among the finalists for this century's title of leading biologist. Haldane also wrote extensively for the general public. He had a knack for stringing facts together in a way that kept them lively and interesting, and many of today's older scientists still remember how much they were intrigued in their youth by Haldane's writing. In the following essay, readers can spot his secret: Haldane knew how to connect natural facts to human concerns. In this selection, for example, he contrasts modern humanity's poor dietary habits with a series of astonishing facts about the way insects manage their food. Although insects are Haldane's subject, human interest is the hook he uses to snare his readers.

Man's habits change more rapidly than his instincts. To-day we are born with instincts appropriate to our palaeolithic ancestors, and when we follow our instincts alone we behave in a palaeolithic manner. It is probable that primitive man, like a wild animal, 'knew' pretty well what was good for him in the way of food. Modern man does not, and when he does he cannot get it. Sedentary workers consume meals appropriate to hunters. Women of fashion attempt to supply the energy needed for dancing by the ingestion of large amounts of chocolate. Man, in fact, must use his reason to arrive at an appropriate diet. But the members of insect societies have solved a similar problem on instinctive and physiological lines. They have brought about the best possible division of a communal food supply by methods which, if strange and often disgusting to human minds, are as effective as any system of food control invented by man.

Let us see what are the prerequisites of a rational distribution. Apart from water, salts, and vitamins which are only required in tiny quantities, foodstuffs may be classified as carbohydrates, fats, and proteins. Carbohydrates include sugar, starch, and the like, fats embrace the chemically similar oils and waxes. Neither contain nitrogen or sulphur, and they are mainly useful as fuel; that is to say in order by combining with oxygen to give up energy which can be used by the animal for heating itself, or working its muscles and other organs. The proteins, on the other hand, are required to build up the living tissues during growth, and repair it after injury or the wear and tear of everyday life. If we compare the requirements of an animal and a motor vehicle, water serves the same function in both, of cooling and carrying away unwanted substances, carbohydrates and fats correspond to petrol, proteins to spare parts, and probably vitamins to lubricating oil. As a matter of fact proteins can act as a source of energy, just as spare woodwork for a train could be used as fuel, but most animals find them unsatisfactory as the sole source.

It is clear that a growing animal needs relatively more proteins than an adult. A baby lives on milk which an adult would instinctively supplement with starchy foods. But the baby requires some fat and carbohydrate as fuel to keep itself warm and work its tiny muscles, and these exist in milk. The wasp grub is cold-blooded and sluggish. It requires very little but proteins. And the adult worker with its short life of intense exertion needs little protein but plenty of fuel. Hence, even though the food which the workers give to the grubs consists very largely of chewed-up flies, it contains more carbohydrate than necessary. When a worker comes to feed a grub by regurgitation from its crop the grub thanks it by secreting a drop of fluid containing sugar for which it has no use, but which is valuable fuel for an active insect.

The bees have taken things a stage further. Their sources of food are the nectar of flowers, a nearly pure solution of sugar in water, and the pollen, which consists largely of proteins. Even from the same flower one bee never collects both nectar and pollen. And in the hive the nectar is stored as honey, and the pollen separately as 'bee-bread.' The honey is used primarily as a source of energy and heat during the winter, the bee-bread along with some honey as food for the grubs. What is more remarkable is the fact that a grub gets a different mixture according to its future career. Queens and workers come from fertilized eggs, drones from unfertilized, but the difference between queens and workers seems to be determined by the type of food given to the larva. So that in the hive food control is also birth control.

The most bizarre system of all is found among termites. These insects live almost entirely on wood, which most animals cannot digest. Strictly speaking the termites cannot do so either, but their intestines contain protozoa which can, just as the horse and cow digest their hay with the help of bacteria. There is evidence that these or other organisms in their bodies can even fix atmospheric nitrogen like the bacteria found in the roots of leguminous plants, thus dispensing with the need for proteins in the diet of their hosts. But this digestion is too slow a process to come to completion in the body of a single insect, so the partially digested excreta of one are eaten by another until the process is complete, and the final indigestible residue is also so incapable of putrefaction that it can be used for nest-building. This apparently repulsive process only corresponds to the passing on of half-digested food by one segment of our intestines to the next. A single termite has not a long enough intestine for the whole process. But it is only certain of the termites that can play their part in this communal digestion. Besides queens and males the termite nest usually contains several different castes of workers and soldiers with large jaws. These jaws are too clumsy to allow of wood-chewing, so the soldiers are fed by the workers with so-called saliva, as is the queen.

Termite societies therefore rest on a basis of physiological functions and of instincts, each one as complex and highly organized as those which form the basis of the relationship between a mammalian mother and her children. But alas, insect societies are no more perfect than human, and parasites can as easily find a place in an economic system determined by instinct, as in the products of intelligence, enlightened self-interest, or whatever else is at the basis of human economies. Whether the correct form of demand for food in an ant's or termite's nest is a gentle stroking of the donor, an offer of a drop of some sweet secretion, or what not, some unprincipled insect will generally be found to make it. Students of human society will compare these parasites with brewers, burglars, bolsheviks, bankers, bishops, or bookmakers according to their tastes. Occasionally they are of some value to the community, for instance by joining in its defence, generally they are useless, so far as we can see; and often they devour not only food, but larval ants.

Humanity is engaged in the awkward passage from an instinctive to a rational choice of food. 'A little of w'ot yer fancy does yer good' is no longer a sufficient guide for us, as it is for the insects, and we do not yet know quite enough to rush to the opposite extreme, though the experience of the war showed that a fairly strict rationing on scientific lines is already a possibility. But every month we are approaching nearer to a knowledge of the dietaries best suited for any individual case, a knowledge which will be as efficient as the instinct of the insect, and infinitely more elastic.

>-+◆>-○-<+-+-<

JULIAN HUXLEY

Animals Courting, from
The Yale Review (1943)

Julian Huxley (1887–1975) was the grandson of "Darwin's bull-dog," T. H. Huxley, and was himself an important biologist and theorist of evolution. He was also a splendid and generous teacher who could inspire the best in his pupils. He persuaded his student Charles Elton to write *Animal Ecology* (1927), the book generally credited with making the study of animal environments a discipline in its own right. Huxley urged another student, Edmund Ford, to write *Mendelism and Evolution* (1931), the book that first synthesized Darwin's ideas with modern genetics.

During the middle part of the century Huxley was one of the most visible scientists, writing and appearing on broadcasts. His book *Evolution: The Modern Synthesis* was a standard on paper-back racks for decades. It completed the work done by Ford, Haldane, and others that linked genetic and evolutionary theory.

In interests and abilities, Huxley was very much like his contemporary, Haldane. Both men had a talent for stringing facts like beads on a chain and, by connecting facts to human issues, making them seem fascinating and worth knowing. In this essay, Huxley builds on his own early projects observing birds in the wild to link evolution and what today is called ethology, the study of animal behavior.

The trouble about acquiring knowledge is that it reveals fresh ignorance: the more facts, the more questions. Thanks to observant hunters, gamekeepers, amateur naturalists, and professional biologists, we are now in possession of an immense body of facts about the peculiarities of behavior and structure connected with the securing of a mate in animals. But the facts immediately turn themselves into a series of insistent questions—why, why, why?

Why do stags not have beautiful tails (or rather, to be accurate, trains) like peacocks, or, vice versa, why do not peacocks have bony excresecences on their heads like stags? Why are male dogs or horses not provided with bright-colored adornments like cock pheasants or sage grouse, and why don't they sing like nightingales or mocking birds? Why are male elephant seals or bustards much bigger than females, while in most creatures the sexes are approximately equal in size? Why are most mammals restricted to blacks, whites, and grays, browns, russets, and yellows, while monkeys, like birds (and fish), run the gamut of color?

Within the one group of birds, why are some conspicuous in both sexes (like crows or loons), some inconspicuous in both sexes (like sparrows or skylarks), some again conspicuous in the male sex but inconspicuous in the female (like most ducks and some finches)? When there are special plumes or ornaments used for display, why are these sometimes restricted to the males (as in pheasants or birds of paradise), but in other species found in both sexes (as in herons or grebes)? Why is song almost entirely confined to one group of birds? And why do birds sing anyhow? Why do skylarks and pipits sing in the air, while most birds sing from a perch?

Why do so many birds, such as prairie chicken, ruff, or birds of paradise, have special communal courting grounds, or at least perform their courtship while in flocks or groups? How is it possible for evolution to have produced adornments which are actually a hindrance to their possessor in the ordinary affairs of life, like the train of the peacock? [. . .]

That is a small selection of the questions which the facts insist on asking. Curiously enough, no one bothered very much about answering this kind of question until after the middle of the last century. This was mainly because the theory of special creation still held the field in biology: the thousands of different species of animals were supposed to have been created, once and for all, with all their peculiarities of construction and behavior as we observe them today. Furthermore, even biologists had scarcely begun then to interpret animal behavior except in human terms. In the field we are here considering, this anthropomorphic tendency often showed itself by interpretation of animal display in terms of human courtships—bird song was some sort of serenade, bright plumage was the equivalent of putting on your best suit, fighting was a duel for possession of a bride, and so on.

With the acceptance of the idea of evolution, however, all this was changed. Animals had not always been what we see today. They had gradually become their modern selves, and all their characteristics demanded an explanation, for nothing could have evolved without some biological reason. Darwin himself was the first to tackle the kind of question I have been posing, and propounded a special theory to account for special masculine weapons, such as deer antlers, and for special masculine adornments, such as the peacock's train. According to him, such characteristics owed their evolution to what he called "sexual selection"—selection based on a struggle or competition between rival males for the possession of a mate. The successful males would reproduce themselves, the others not; and so the characters making for success in this sexual struggle would be inherited and progressively developed generation by generation.

This was a reasonable hypothesis at the time it was propounded, in 1871; but gradually new facts came to light which it did not meet or cover: There were the numerous cases, apparently unknown to Darwin, of special adornments developed by both sexes and used in mutual display. There was the fact that many kinds of bright masculine colors are not used in display towards females at all, but only in threat against rival males. And there was the difficulty that in most song birds, the males are not only monogamous but do not begin their courtship display until *after* they are mated for the season, so that the display could have nothing to do with the choice of mates.

For these and other reasons, Darwin's theory of sexual selection fell into disrepute in the early years of the present century. However, the facts remained, and continued to pose their questions. Gradually, as the result of a great deal of patient observation and a certain amount of experiment, the answers began to shape themselves, until today we can at least make general sense of the situation. Masculine weapons and bright display characters do owe their evolution to selection, thought it is not sexual selection in the diagrammatic sense in which Darwin employed the term.

With regard to obvious weapons, the position is much as Darwin stated it. They have been evolved to fight for the possession of mates. Furthermore, as is to be expected, they are more striking when the battle is for a whole harem of females, for then the selective advantage to the victor, in the shape of having more descendants, is multiplied many times. The antlers of stags, the mane and heavy forequarters of the male bison, the huge size of the male elephant seal are cases in point. But there was a further subtlety that has only come to light since Darwin's day. Fighting is exhausting and dangerous: so, while it will pay, biologically speaking, to fight for mates if fighting is necessary, it will pay still better if the object can be secured without a fight. Thus characters are evolved which serve for threat, as symbols of fighting strength, or even for bluff, in lieu of fighting strength. [. . .]

There is another kind of threat, which is very common in birds, being specially developed where the males stake out a large territory, as is the case in most song birds. Round this territory the whole business of reproduction later centres. The females will only mate with males that are in possession of a territory; the nest is built there, and in the territory is the area in which the parents find food for their young. Males will fight each other viciously for the possession of a territory; but they prefer not to have to fight, and so a male without a territory will often prefer to go on in search of an unclaimed area rather than run the risks of battle. Accordingly, it is biologically important to advertise the fact of being a territorial owner: it may keep a number of rivals from trespassing, and so obviate the need for fighting. At the same time, the advertisement has to be directed at the other sex also—here is a territory complete with cock, at the service of the first hen bird to take up her quarters there. As the advertisement has to serve two purposes at once—attraction to mates as well as warning to rivals—it must [. . .] be a mere symbol of possession, and concentrate on conspicuousness.

This appears to indicate the origin of bird song, and also of some of the conspicuous colors of male song birds. The yearly business of reproduction begins with the males staking out their territories. In migratory species, the males migrate first, and may be in possession of a territory for several days before there are any hen birds in the country at all. Once in possession, the males spend a considerable part of each day singing, which they generally do from one or other of a few conspicuous perches—or on the wing, if they are birds of open country, like the lark. The male's song is usually at its best before he has been joined by a female. Thus full song is a sign of a male in occupation of his territory. Another male will normally not wish to risk a serious fight by invading the territory, but will pass on, warned by the song, until he finds an unoccupied area. Conversely, the hen birds will be attracted by the song, for it advertises a potential nest. If there is another hen already in possession, a newcomer will have to fight; if not, she simply settles in, and the pair is then normally mated for the season.

Of course, everyone knows that birds may sing at other times of the year (as after the autumn moult) or in other circumstances (for instance, as a result of anger); it is also doubtless true that the individual bird usually sings because it "feels good." But this does not contradict the idea that song owes its evolutionary origin to the need for male conspicuousness on the territory; and this indeed is the only view that will fit the facts.

The conspicuous colors of so many male song birds, like goldfinch and cardinal and blackbirds, serve the same sort of purpose. They advertise the presence of the cock bird as prominently as possible.

Such is the broad general theory of the biological meaning of song and male conspicuousness in song birds. When, however, we explore the detailed differences between different species, we run up against a great many other interesting points and principles. One of the most striking things about song birds (though it is one which naturalists do not seem to have bothered about until quite recently) is the degree of difference which we often find between the males of closely related species. In the thrush family, for instance, the primitive coloration is brown with spotted breast and underparts. But the American robin has a chestnut-red breast; the European blackbird is jet black all over with golden bill, and so on. Or again, almost every species of finch and bunting is distinctively colored in the male.

It is the word *distinctive* which gives us the key. The cock bird on its territory needs to be conspicuous; but he needs also to be different from the cock birds of other species, especially those of closely related species, so that there shall be as little biological wastage as possible through hybridization, through hens of another species presenting themselves as potential mates, or through battles with other mates which are not really rivals.

Distinctiveness to the ear is demanded as much as to the eye; and thus we find that, on the average, related species differ as much in their songs as in their plumage. However, when we go into details, we are confronted with some cases where related species look very alike but have highly distinct songs. Why is this? Such birds generally look alike because they need to escape detection, and have developed a coloration which matches their background—mottled brown for the pipits of the moors; pale green for the leaf-warblers of our deciduous trees. The need for visual protection has overridden the need for visual distinctiveness. All the more reason, then, for accentuating the distinctiveness of what is left—namely, song. In some cases, most remarkable results have been effected by evolution. Thus the two European leaf-warblers, the chiffchaff and the willow-warbler, are almost exactly alike to look at, and often overlap in the same area. But while the latter has a typical warble song, the chiffchaff has a song of two repeated notes (from which it takes its name). Sometimes, however, it gives a brief and faint warble at the close of its chiffchaffing, revealing that its song has been evolved from one like the willow-warbler's—and evolved as it has just in order to be as different as possible. [. . .]

However, all these elaborate characters of song and distinctive coloration, combining warning to rivals with attraction to mates, are territorial advertisements, and not display or courtship in the usual sense. Territorial birds do have a real display, but this does not begin until *after* a female has settled in an occupied territory, and the pair is therefore mated for the season or the brood. So obviously Darwin's original idea that courtship had something to do with the selection of mates will not work here. Yet the male's displays may be elaborate and often repeated, with drooping of wings, fanning of tail, raising of crest, strange posturings and antics; and when bright plumage is present such displays actually show it off to special advantage. They could not have been evolved unless they had some biological meaning and advantage. For a long time, this puzzle remained unsolved. In the last few decades, however, we seem to have reached the solution. For one thing, they stimulate the female's readiness to mate; and for another, they have a physiological effect on the reproductive organs and help to ripen and activate them.

Readiness to mate may be determined mainly by purely physiological agencies, such as the discharge into the bloodstream of hormones from the gland part of the reproductive organs: this is what happens to female cats or dogs or other lower mammals "on heat." In this state they are not only ready but eager for mating; accordingly, no display by the male is needed—and none is to be found.

In birds, however, this does not happen; there is no such special discharge of "mating-readiness" hormones over a brief period, and readiness to mate is largely a psychological and emotional matter. Male display in birds is thus a device for stimulating the hen's emotions to a pitch at which she is ready to mate. Sometimes, as when one sees a peacock displaying all his magnificence before an apparently quite indifferent hen, this may seem a trifle far-fetched. However, we can remind ourselves that male display can be looked on as an advertisement, and in human affairs manufacturers are willing to pay large amounts of hard cash for advertisements, even though these may have no effect on 99 people out of 100, or on 99 occasions out of 100. The only question that matters, both for the human advertiser and the male bird, is whether the expenditure produces results.

That is one point; but there is another. It is now well established that in birds emotional stimuli, apparently acting through the nervous system and the ductless glands, may help in the ripening of the eggs. In doves, an isolated hen may even ripen and lay eggs (though these of course are sterile) as the result of seeing a male in another cage display at her. So a great deal of the apparently fruitless display of male birds (as of cock sparrows before the hens) is really serving to ripen the eggs in the female's ovaries.

This would be of particular biological advantage in bad seasons. It is well known that inclement weather may discourage birds from breeding. When the weather has been cold and wet, the number of eggs in a clutch is on the average lower than in good seasons. So any stimulating effect of display on reproduction would be of biological advantage both to the individual male through his leaving more descendants, and to the species. [. . .]

Quite recently, a further interesting fact has come to light: this effect of display and general emotional excitement on the reproductive organs can extend from the individual to the group. In other words, if a number of birds have closely congregated together in the breeding season, the effect of a male courting his mate may have a stimulating effect on other females, and in general, the excitement caused by display is shared by the group as a whole. In gulls, this has the effect of making egg-laying a little earlier in large than in small colonies, and in concentrating it into a shorter time, which reduces the total toll of eggs and young birds taken by enemies.

This group-stimulating effect at once accounts for the fact that so many birds conduct their display publicly, in crowded groups—either at the nest in various gregarious breeders, or at special places used for nothing else. Communal display grounds are found with all kinds of birds, from grouse to birds of paradise, from waders like the ruff to song birds like manakins.

There are two other quite different kinds of bird courtship that must be mentioned, since they show so clearly the connection of type of display with mode of life. One is the exaggeration of purely masculine display, which we find in birds which are polygamous or promiscuous in their mating habits, instead of monogamous (at any rate, for each breeding season) like those so far described. Pheasants, peacocks, various grouse, birds of paradise, ruffs—in all these the females alone brood the eggs and look after the young, and therefore must be dull and inconspicuous, while the function of the males in reproduction is confined to securing as many mates as possible, and they are accordingly very conspicuous, with an elaborate display. [. . .]

At the other end of the scale are those numerous kinds of birds where both cock and hen play equal or almost equal parts in all reproductive activities, both helping in building the nest, in brooding the eggs, and in feeding and looking after the young birds. In most of these birds (unlike the bird of paradise or the barnyard rooster), this equality is reflected in appearance and in display—both sexes having special bright plumage, and using it to show off to each other in similar displays. [. . .]

The biological meaning of mutual display is in the main the same as that of one-sided display in territorial birds—to stimulate readiness to mate and to ripen the reproductive organs. But it appears to have another function, too. The mutual ceremonies seem to give great emotional satisfaction to the pair, and to act as a bond between them, helping to keep them together throughout the breeding season. This is biologically important, since the co-operation of the parent birds is necessary if the brood is to be hatched and reared. In confirmation of this, we find that mutual displays, instead of coming to an end when the eggs are laid, are often continued right through the breeding season, until the young can fend for themselves.

I have already said something of the need for escaping notice (by enemies) against the need for attracting notice (from rivals and mates). This has led to a number of very interesting results. The two needs make contradictory biological demands—the one for inconspicuousness, the other for conspicuousness. The contradiction is reconciled in various quite different ways, according to the circumstances of the case. When the species is defenseless and lives in open surroundings, where it is in urgent need of concealment from enemies, then, as with skylarks, the coloration may be entirely protective in both sexes, and masculine display may be reduced to posturing, with drooping of wings, and spreading of tail, without any brilliance to set it off. Or species that need protection may have some bright sexual plumage, and yet keep it hidden most of the time, flashing it out only for display, as with the yellow crest of the goldcrest, or the white of the bustard, or the orange pouches of the prairie chicken.

In most territorial birds, however, the need for concealment does not seem so great, and the males are generally brightly or strikingly colored, as with finches, buntings, or blackbirds. As the hen has to brood the eggs, she is in greater need of concealment, and her plumage is correspondingly duller. Exceptions to this are found in birds which nest in holes, where the brooding female cannot be seen; here, as in nuthatches, the force of heredity can work unopposed by selection, and the bright plumage of the males is much more completely transferred to the females.

In ducks, the males neither brood the eggs nor help in looking after the young, and are, therefore, of much less value to the next generation than the females. Hence, if enemies are going to take a certain toll of the species, it is better that the males should suffer, and so, while the ducks must be made as inconspicuous as possible, it will be an advantage to the species (though not to the individual males!) for the drakes to be conspicuous. This is probably one reason for the bright colors of so many kinds of male ducks, which make their possessors conspicuous all the time, and not only for the purposes of threat or courtship.

Finally, we reach a point where we can give quantitative expression to the forces working for conspicuousness. In monogamous territorial birds, courtship display characters will have what we may call a fractional reproductive advantage—for instance, they may, as has been said, counteract the depressing effects of bad weather and stimulate the hen to lay rather more eggs than she otherwise would have done. But the advertisement characters which warn rivals off an occupied territory and attract a mate to it—these may have a whole unit of reproductive advantage: either a bird secures a territory and a mate, or it does not, while the display characters come into play only when a mate has already been secured. So, as their advantage is larger, it will override the need for concealment to a greater extent; and in point of fact we find that the most conspicuous male characters, both of plumage and voice, are concerned with territorial advertisement, while those used exclusively in display are less conspicuous, or are revealed only during display itself.

Finally, when the species is polygamous or promiscuous, the unit reproductive advantage may be multiplied a number of times, if the male secures several mates. Nature here is playing for very high stakes, and we find that display characters (which are those that count in such cases) may be developed to such an extent that not merely do they leave little room for concealment, but they may actually hinder the cock bird in its day-to-day struggle for existence. The train of the peacock is a hindrance; and the wings of the male argus pheasant have become so entirely devoted to purposes of display that they are almost useless for flight. [. . .]

Whereas the colors of the lower mammals are entirely restricted to shades between black and white, and between yellow, russet, and brown, in primates we find blues, violets, greens, and true reds as well. This is without question to be correlated with the fact that all lower mammals (if we may judge from all those which have been investigated) are color blind, whereas primates can see in color. If birds were color blind, all their display plumage would be in pigments which made striking tone contrasts with each other; we should never have witnessed such gorgeous productions of nature as the peacock's train or the glories of the birds of paradise, which depend mainly on color contrast. [. . .]

All these posturings, these songs, this gorgeous plumage, this distinctiveness of voice or appearance, these expressions of hostile excitement—all these are devices for projecting into the mind of another individual the fact that a sexual situation exists. At the lowest level, the devices are the simplest and crudest symbols—an exciting odor, a huge claw brandished aloft. But as sense organs become more elaborate and emotional life grows more complex, the display devices follow suit, until they come to constitute a large fraction of the beauty and the exciting strangeness of life. All the most elaborate beauty of birds, all their songs; the antlers of deer, with their recurrent growth and shedding, which if they were not familiar would be among the most astonishing facts of nature; the masculine grandeur of lion or bison; the cheerful music of frogs and toads, of grasshoppers and crickets; the bizarre faces of our monkey cousins; *haute couture* and the huge cosmetics industry—these are all in their various ways the product of this trend of evolution to project this or that aspect of a sexual situation into the consciousness of potential or actual mate or rival.

The type of display varies with the kind of sexual situation and the animal's general way of life. The peacock or the bird of paradise reminds us that in certain conditions beauty is of more avail than brutality; the mutual courtship of grebe or egret, with pleasures shared by both sexes, shows that one-sided masculinity is not an evolutionary necessity, and that sex equality can be in some conditions the correct biological solution. It is at least entertaining to speculate as to what possible new beauties and shared delights sex display, if sublimated and consciously guided and rooted in economic prosperity and equality, may have in store for the human species, which is still in the very early stages of its evolutionary career.

‹—•◦•—›

RACHEL CARSON

The Long Snowfall, from
The Sea Around Us (1951)

Rachel Carson (1907–1964) won the National Book Award for her science classic *The Sea Around Us*. She had worked for many years as a biologist with the U.S. Bureau of Fisheries (now the U.S. Fish and Wildlife Service), and her book is firsthand evidence that bureaucrats sometimes do have a soul.

When speaking to people about what they thought should be included in this anthology, I was told repeatedly that I should include Rachel Carson's *Silent Spring* (1962), the book that inspired the modern environmental movement. I knew from the outset, however, that I would turn to the book that established Carson's authority to write about environmental destruction. *The Sea Around Us* established Carson as one of the great poets of science and nature. So, when she wrote about ecological damage, she already had a world of admirers who were ready to take her seriously. Carson was the one who, when I was a schoolboy, first showed me that science writing could be humane and moving.

An ironic aside here is that those studying the ocean floor that Carson celebrates soon discovered that the floor was being created from the bottom up. Deep fissures in the middle of the oceans are pushing up new seafloor and spreading continents apart. Carson's portrait of the ocean floor's being built by debris from above is thus incomplete, and a writer today would feel obliged to focus on the moving floor. Carson does show us, however, that the modern emphasis also is incomplete. The seafloor is not just spreading; it is being continuously dusted with the refuse of time.

Every part of earth or air or sea has an atmosphere peculiarly its own, a quality or characteristic that sets it apart from all others. When I think of the floor of the deep sea, the single, overwhelming fact that possesses my imagination is the accumulation of sediments. I see always the steady, unremitting, downward drift of materials from above, flake upon flake, layer upon layer—a drift that has continued for hundreds of millions of years, that will go on as long as there are seas and continents.

For the sediments are the materials of the most stupendous 'snowfall' the earth has ever seen. It began when the first rains fell on the barren rocks and set in motion the forces of erosion. It was accelerated when living creatures developed in the surface waters and the discarded little shells of lime or silica that had encased them in life began to drift downward to the bottom. Silently, endlessly, with the deliberation of earth processes that can afford to be slow because they have so much time for completion, the accumulation of the sediments has proceeded. So little in a year, or in a human lifetime, but so enormous an amount in the life of earth and sea.

The rains, the eroding away of the earth, the rush of sediment-laden waters have continued, with varying pulse and tempo, throughout all of geologic time. In addition to the silt load of every river that finds its way to the sea, there are other materials that compose the sediments. Volcanic dust, blown perhaps half way around the earth in the upper atmosphere, comes eventually to rest on the ocean, drifts in the currents, becomes waterlogged, and sinks. Sands from coastal deserts are carried seaward on off-shore winds, fall to the sea, and sink. Gravel, pebbles, small boulders, and shells are carried by icebergs and drift ice, to be released to the water when the ice melts. Fragments of iron, nickel, and other meteoric debris that enter the earth's atmosphere over the sea—these, too, become flakes of the great snowfall. But most widely distributed of all are the billions upon billions of tiny shells and skeletons, the limy or silicious remains of all the minute creatures that once lived in the upper waters.

The sediments are a sort of epic poem of the earth. When we are wise enough, perhaps we can read in them all of past history. For all is written here. In the nature of the materials that compose them and in the arrangement of their successive layers the sediments reflect all that has happened in the waters above them and on the surrounding lands. The dramatic and the catastrophic in earth history have left their trace in the sediments—the outpourings of volcanoes, the advance and retreat of the ice, the searing aridity of desert lands, the sweeping destruction of floods.

The book of the sediments has been opened only within the lifetime of the present generation of scientists, with the most exciting progress in collecting and deciphering samples made since 1945. Early oceanographers could scrape up surface layers of sediment from the sea bottom with dredges. But what was needed was an instrument, operated on the principle of an apple corer, that could be driven vertically into the bottom to remove a long sample or 'core' in which the order of the different layers was undisturbed. Such an instrument was invented by Dr. C.S. Piggot in 1935, and with the aid of this 'gun' he obtained a series of cores across the deep Atlantic from Newfoundland to Ireland. These cores averaged about 10 feet long. [. . .] A piston core sampler, developed by the Swedish oceanographer Kullenberg about 10 years later, now takes undisturbed cores 70 feet long. The rate of sedimentation in the different parts of the ocean is not definitely known, but it is very slow; certainly such a sample represents millions of years of geologic history.

Another ingenious method for studying the sediments has been used by Professor W. Maurice Ewing of Columbia University and the Woods Hole Oceanographic Institution. Professor Ewing found that he could measure the thickness of the carpeting layer of sediments that overlies the rock of the ocean floor by exploding depth charges and recording their echoes; one echo is received from the top of the sediment layer (the apparent bottom of the sea), another from the 'bottom below the bottom' or the true rock floor. The carrying and use of explosives at sea is hazardous and cannot be attempted by all vessels, but this method was used by the Swedish *Albatross* as well as by the *Atlantis* in its exploration of the Atlantic Ridge. Ewing on the *Atlantis* also used a seismic refraction technique by which sound waves are made to travel horizontally through the rock layers of the ocean floor, providing information about the nature of the rock.

Before these techniques were developed, we could only guess at the thickness of the sediment blanket over the floor of the sea. We might have expected the amount to be vast, if we thought back through the ages of gentle, unending fall—one sand grain at a time, one fragile shell after another, here a shark's tooth, there a meteorite fragment— but the whole continuing persistently, relentlessly, endlessly. It is, of course, a process similar to that which has built up the layers of rock that help to make our mountains, for they, too, were once soft sediments under the shallow seas that have overflowed the continents from time to time. The sediments eventually became consolidated and cemented and, as the seas retreated again, gave the continents their thick, covering layers of sedimentary rocks—layers which we can see uplifted, tilted, compressed, and broken by the vast earth movements. And we know that in places the sedimentary rocks are many thousands of feet thick. Yet most people felt a shock of surprise and wonder when Hans Pettersson, leader of the Swedish Deep Sea Expedition, announced that the *Albatross* measurements taken in the Atlantic basin showed sediment layers as much as 12,000 feet thick. [. . .]

If more than two miles of sediments have been deposited on the floor of the Atlantic, an interesting question arises: has the rocky floor sagged a corresponding distance under the terrific weight of the sediments? Geologists hold conflicting opinions. The recently discovered Pacific sea mounts may offer one piece of evidence that it has. If they are, as their discoverer called them, 'drowned ancient islands,' then they may have reached their present stand a mile or so below sea level through the sinking of the ocean floor. Hess believed the islands had been formed so long ago that coral animals had not yet evolved; otherwise the corals would presumably have settled on the flat, planed surfaces of the sea mounts and built them up as fast as their bases sank. In any event, it is hard to see how they could have been worn down so far below 'wave base' unless the crust of the earth sagged under its load.

One thing seems probable—the sediments have been unevenly distributed both in place and time. In contrast to the 12,000-foot thickness found in parts of the Atlantic, the Swedish oceanographers never found sediments thicker than 1000 feet in the Pacific or in the Indian Ocean. Perhaps a deep layer of lava, from ancient submarine eruptions on a stupendous scale, underlies the upper layers of the sediments in these places and intercepts the sound waves.

Interesting variations in the thickness of the sediment layer of the Atlantic Ridge and the approaches to the Ridge from the American side were reported by Ewing. As the bottom contours became less even and began to slope up into the foothills of the Ridge, the sediments thickened, as though piling up into mammoth drifts 1000 to 2000 feet deep against the slopes of the hills. Farther up in the mountains of the Ridge, where there are many level terraces from a few to a score of miles wide, the sediments were even deeper, measuring up to 3000 feet. But along the backbone of the Ridge, on the steep slopes and peaks and pinnacles, the bare rock emerged, swept clean of sediments.

Reflecting on these differences in thickness and distribution, our minds return inevitably to the simile of the long snowfall. We may think of the abyssal snowstorm in terms of a bleak and blizzard-ridden arctic tundra. Long days of storm visit this place, when driving snow fills the air; then a lull comes in the blizzard, and the snowfall is light. In the snowfall of the sediments, also, there is an alteration of light and heavy falls. The heavy falls correspond to the periods of mountain building on the continents, when the lands are lifted high and the rain rushes down their slopes, carrying mud and rock fragments to the sea; the light falls mark the lulls between the mountain-building periods, when the continents are flat and erosion is slowed. And again, on our imaginary tundra, the winds blow the snow into deep drifts, filling in all the valleys between the ridges, piling the snow up and up until the contours of the land are obliterated, but scouring the ridges clear. In the drifting sediments on the floor of the ocean we see the work of the 'winds,' which may be the deep ocean currents, distributing the sediments according to laws of their own, not as yet grasped by human minds.

We have known the general pattern of the sediment carpet, however, for a good many years. Around the foundations of the continents, in the deep waters off the borders of the continental slopes, are the muds of terrestrial origin. There are muds of many colors—blue, green, red, black, and white—apparently varying with climatic changes as well as with the dominant soils and rocks of the lands of their origin. Farther at sea are the oozes of predominantly marine origin—the remains of the trillions of tiny sea creatures. Over great areas of the temperate oceans the sea floor is largely covered with the remains of unicellular creatures known as foraminifera, of which the most abundant genus is Globigerina. The shells of Globigerina may be recognized in very ancient sediments as well as in modern ones, but over the ages the species have varied. Knowing this, we can date approximately the deposits in which they occur. But always they have been simple animals, living in an intricately sculptured shell of carbonate of lime, the whole so small you would need a microscope to see its details. After the fashion of unicellular beings, the individual Globigerina normally did not die, but by the division of its substance became two. At each division, the old shell was abandoned, and two new ones were formed. In warm, lime-rich seas these tiny creatures have always multiplied prodigiously, and so, although each is so minute, their innumerable shells blanket millions of square miles of ocean bottom, and to a depth of thousands of feet.

In the great depths of the ocean, however, the immense pressures and the high carbon-dioxide content of deep water dissolve much of the lime long before it reaches the bottom and return it to the great chemical reservoir of the sea. Silica is more resistant to solution. It is one of the curious paradoxes of the ocean that the bulk of the organic remains that reach the great depths intact belong to unicellular creatures seemingly of the most delicate construction. The radiolarians remind us irresistibly of snow flakes, as infinitely varied in pattern, as lacy, and as intricately made. Yet because their shells are fashioned of silica instead of carbonate of lime, they can descend unchanged into the abyssal depths. So there are broad bands of radiolarian ooze in the deep tropical waters of the North Pacific, underlying the surface zones where the living radiolarians occur most numerously.

Two other kinds of organic sediments are named for the creatures whose remains compose them. Diatoms, the microscopic plant life of the sea, flourish most abundantly in cold waters. There is a broad belt of diatom ooze on the floor of the Antarctic Ocean, outside the zone of glacial debris dropped by the ice pack. There is another across the North Pacific, along the chain of great deeps that run from Alaska to Japan. Both are zones where nutrient-laden water wells up from the depths, sustaining a rich growth of plants. The diatoms, like the radiolarians, are encased in silicious coverings—small, boxlike cases of varied shape and meticulously etched design.

Then, in relatively shallow parts of the open Atlantic, there are patches of ooze composed of the remains of delicate swimming snails, called pteropods. These winged mollusks, possessing transparent shells of great beauty, are here and there incredibly abundant. Pteropod ooze is the characteristic bottom deposit in the vicinity of Bermuda, and a large patch occurs in the South Atlantic.

Mysterious and eerie are the immense areas, especially in the North Pacific, carpeted with a soft, red sediment in which there are no organic remains except sharks' teeth and the ear bones of whales. This red clay occurs at great depths. Perhaps all the materials of the other sediments are dissolved before they can reach this zone of immense pressures and glacial cold.

The reading of the story contained in the sediments has only begun. When more cores are collected and examined we shall certainly decipher many exciting chapters. Geologists have pointed out that a series of cores from the Mediterranean might settle several controversial problems concerning the history of the ocean and of the lands around the Mediterranean basin. For example, somewhere in the layers of sediment under this sea there must be evidence, in a sharply defined layer of sand, of the time when the deserts of the Sahara were formed and the hot, dry winds began to skim off the shifting surface layers and carry them seaward. Long cores recently obtained in the western Mediterranean off Algeria have given a record of volcanic activity extending back through thousands of years, and including great prehistoric eruptions of which we know nothing.

The Atlantic cores taken more than a decade ago by Piggot from the cable ship *Lord Kelvin* have been thoroughly studied by geologists. From their analysis it is possible to look back into the past 10,000 years or so and to sense the pulse of the earth's climatic rhythms; for the cores were composed of layers of cold-water Globigerina faunas (and hence glacial stage sediments), alternating with Globigerina ooze characteristic of warmer waters. From the clues furnished by these cores we can visualize interglacial stages when there were periods of mild climates, with warm water overlying the sea bottom and warmth-loving creatures living in the ocean. Between these periods the sea grew chill. Clouds gathered, the snows fell, and on the North American continent the great ice sheets grew and the ice mountains moved out to the coast. The glaciers reached the sea along a wide front; there they produced icebergs by the thousand. The slow-moving, majestic processions of the bergs passed out to sea, and because of the coldness of much of the earth they penetrated farther south than any but stray bergs do today. When finally they melted, they relinquished their loads of silt and sand and gravel and rock fragments that had become frozen into their under surfaces as they made their grinding way over the land. And so a layer of glacial sediment came to overlie the normal Globigerina ooze, and the record of an Ice Age was inscribed.

Then the sea grew warmer again, the glaciers melted and retreated, and once more the warmer-water species of Globigerina lived in the sea—lived and died and drifted down to build another layer of

Globigerina ooze, this time over the clays and gravels from the glaciers. And the record of warmth and mildness was again written in the sediments. From the Piggot cores it has been possible to reconstruct four different periods of the advance of the ice, separated by periods of warm climate.

It is interesting to think that even now, in our lifetime, the flakes of a new snow storm are falling, falling, one by one, out there on the ocean floor. The billions of Globigerina are drifting down, writing their unequivocal record that this, our present world, is on the whole a world of mild and temperate climate. Who will read their record, ten thousand years from now?

LOUISE B. YOUNG

How Ice Changed the World, from
The Blue World (1983)

L ouise Young (1919–) has been an environmentalist and naturalist since long before green became a political color. She was
trained as a physicist and, during World War II, helped develop radar antennas for the navy. After the war she turned to science writing, specializing in pieces on the environment.

In this selection, Young's science and environmentalism come
together into an account of the ice ages and what effect they had.
Young's environmentalism gives meaning to the facts of ice, snow,
glaciers, and the traces they leave behind. These facts in turn support
her enthusiasm for nature. Professional scientists are discouraged
from voicing their own feelings in their reports, and most poets are
discouraged from delving too deeply into data. But Young shows
how much can be accomplished when a writer brings them together.
We readers feel glad to be alive in an age of knowledge.

The afternoon may be so clear that you dare not make a sound, lest it fall
in pieces. And on such a day I have seen the sky shatter like a broken
goblet, and dissolve into iridescent tipsy fragments—ice crystals falling
across the face of the sun.

—RICHARD E. BYRD, *ALONE*

The long sensuous days of summer in the temperate zone of the
Northern Hemisphere are past as the planet Earth swings relentlessly
in its orbit. The faces of the lands we know turn away from the sun a
little farther each day. The shadows of the night lengthen week by
week and the winds spin out of the north. One day the first snowflakes fall softly from the darkening sky.

On the lakes and ponds a thin film of ice crystals makes a
translucent skin, hardly visible to the naked eye but changing the way
the surface turns back the light. The wake of the moon is diffused

and softened. The water's restless movement is quieted; its body rises and falls almost imperceptibly as though it had fallen asleep.

The days pass and the ice thickens, clearly visible now in thin sheets and disks that nudge against each other. As they move and grow, they acquire rounded shapes with little upturned edges— "pancakes," these ice platelets are called, but they are more like shallow saucers. Along the shore they pile up and slowly hour by hour build fantastic castles with high walls and moats and turrets and battlements and deep wells down through the ice. Into these holes waves send sudden jets of water that spout fountains high into the air. The drops freeze as they fall in a sparkling shower of crystals, diamond-bright.

Along the drifted country roads, beside the shining seashores, through the softened spaces of forests and cities dressed in white, for a brief period of time frozen water has redesigned our world. But in much more important and fundamental ways ice has been creatively at work since the crust of the earth first began to form four billion years ago.

The contours, the color, and texture of all the land surfaces have been affected by ice, because without ice we would not have rain, which distributes the moisture across the continents, erodes and transforms the hard rocks of the continents into friable soil, provides a universal solvent, and distributes the most essential ingredients for living things. Raindrops do not usually form directly from clouds. Water passes first through the frozen state; raindrops are melted crystals of ice.

When water evaporates from the surface of the seas or lakes, it becomes a gas, composed of single molecules free to move independently among the other molecules of the atmosphere. Warm air can hold more water in the vapor form than can air at lower temperatures. When the atmosphere cools, the water molecules tend to condense back again into tiny cloud droplets many times smaller than raindrops—a billion cloud droplets would not fill the hollow of your hand. These tiny drops of water are the substance of all the clouds that veil the sky, the mists that rise from marshlands on a summer night, the dew that collects on meadow grass at dawn. But they are at least a million times smaller than raindrops. They are so light that they remain airborne, suspended by the rapidly dancing molecules of the gases in the atmosphere, and they do not tend to come together to form larger drops. So for many decades meteorologists were baffled in their attempt to explain how raindrops form. Then it was discovered that a frozen cloud droplet is the seed that makes raindrops. A growing ice crystal attracts water-vapor molecules, persuading them to give up the individuality they had preserved so carefully before the crystal appeared. As each molecule is added to the crystal it draws in others. Like an elaborate circle dance, an intricate, symmetrical pattern is built up in ever-widening spheres. A unique, original masterpiece of form suddenly materializes from what appeared to be empty space.

Soon the ice crystal becomes heavy enough to fall through the cloud and as it collides with cloud droplets, little splinters of ice break off, leaving a trail of tiny fragments that act as nuclei for new crystals. These also grow and fall and splinter. In the space of just a minute or two the cloud becomes a flurry of snowflakes. As they descend into warmer air they melt and descend as rain upon the land—rain to fill the lake basins and swell the rushing rivers, to carve out canyons, and soften the harsh rock surfaces, to tear down the mountain ranges stone by stone and return their substance to the sea.

In wintertime, of course, the snow crystals do not melt. They swirl gently downward, collecting in a downy blanket on the ground, covering every twig and rooftop. Freshly fallen snow is incredibly airy and light. A typical powder snow has twenty parts of air to just one of water. But as it remains on the ground, it gradually becomes compacted into a solid mass. Packed snow melts very slowly even when the summer sun shines on it. If the season is short and cold, the snow does not all melt away. It accumulates year after year. Thus a glacier is formed.

Many glaciers exist today at high altitudes, creating breathtaking mountain scenery. The ice collects in bowl-shaped depressions near the mountain peaks where the weight and movement of the growing ice mass carve steep-walled basins called cirques. Eventually the ice overflows the basin and moves downhill, becoming a mighty river of ice, and great ice falls are formed where it cascades over precipitous cliffs. As the glacier moves, it scoops out depressions that later, after the ice has melted, may become filled with water, making glacial lakes, or "tarns." Along mountainous seacoasts in the high latitudes glaciers have cut troughs deep down into the continental crust below present sea level, creating the beautiful fjord landscapes where the sea runs in shining threads far inland between towering rock walls.

In the polar regions extensive ice sheets cover the land and much of the water surface. Greenland and the Antarctic continent are almost entirely buried beneath glaciers measuring as much as two miles in thickness. The Antarctic ice sheet alone covers one and a half times the area of the forty-eight contiguous United States. In some places it extends beyond the borders of the continent, attached to the land but riding out over the sea.

Travelers to these frozen parts of the world have been surprised to find that the great ice masses are not all white or shades of gray. Even in the frozen state water retains its characteristic color, diluted to a faint gleaming aqua in the vertical walls of the great glaciers but intensified in the shadowy crevasses to a midnight blue. When seen from above, the icebergs that float free are a deep opaque tone like lapis lazuli. Admiral Byrd described his first impression of the remarkable ice sheet that he called "the Barrier" on his pioneering trip to Antarctica in 1928: "Near the water's edge, the Barrier in places was honeycombed with caves, of bewildering shapes and sizes which, when the sun struck them at just the right angle, blazed with a rich blue coloration."

Several months later Byrd flew from Little America over Marie Byrd Land, enjoying a really sweeping view of this tremendous continental glacier:

> There was great beauty here, in a way that things which are also terrible can be beautiful. Glancing to the right, one had the feeling of observing the twilight of an eternity. Over the water and submerged land crept huge tongues of solid ice and snow, ploughing into the outer fringe of shelf ice and accomplishing wide destruction. There were cliffs that must rise hundreds of feet. Once I caught sight of a cliff as it fell into the sea. From the great height of the plane it was just a small pellet falling from a toy wall. Not a sound penetrated through the noise of the engines. Yet thousands of tons must have collapsed in one frightful convulsion. To the right were the mountains, cold and gray, and from them fell, in places, ice falls which were perhaps 500 feet in height. . . .

Here was the ice age in its chill flood tide. Here was a continent throttled and overwhelmed. Here was the lifeless waste born of one of the greatest periods of refrigeration that the earth has ever known. Seeing it, one could scarcely believe that the Antarctic was once a warm and fertile climate, with its own plants and trees of respectable size.

This mighty field of ice is only a small remnant of the last glaciation. Ice sheets have covered much larger portions of the planet at many times in the past. Just 18,000 years ago the lands that now support the streets and tall buildings of New York, Berlin, Stockholm, and Warsaw were bent under a heavy burden of ice. Frequent storms raged as they do now in Antarctica and winds lashed with a force and fury unknown on earth today. The story of these great ice ages can be reconstructed from the marks they left on the land held so long in their frozen grip.

As ice sheets recede they reveal characteristic impressions. There are grooves and striations on bedrock, making roughly parallel scars, like clawmarks on a treetrunk. These are sure signs that a glacier has pulled sharp rocks across this surface as it moved slowly forward. Low, rounded hillocks of assorted rock and stone occurring in gently curving arcs mark the places where the glacier's advance was halted and some of the rock debris which it carried with it from faraway places was deposited. These glacial moraines border outwash plains where finer material, like sand and gravel, also foreign in origin, was spread by meltwater streams in sheets before the leading edge of the ice. The plains are often pockmarked with small depressions known as kettles where buried ice blocks melted. Now these holes form little lakes and ponds. Here and there long, narrow ridges of coarse sands and gravel can be seen extending in sinuous courses for miles. This material was carried by streams that emerged from tunnels at the ice margins.

The amount of clay, soil, sand, and rock moved by continental glaciers staggers the imagination. Across hundreds of thousands of square miles the glacial drift was deposited and accumulated to great thicknesses as the ice melted. It covered the valleys, the canyons, and lowlands more deeply than the hilltops and plateaus, so the broad-scale effect of retreating ice sheets was to smooth out the landforms and create great plains. In the United States land extending from east-ern Ohio through South Dakota was already relatively flat because it had been laid down in shallow seas, but it was given an additional smoothing by ice sheets several times during the last two million years. Glacial drift lies 50 to 200 feet thick over Iowa and Illinois. On top of the glacial drift is a layer of wonderfully fertile soil, which was also a gift of the ice age.

During the major glaciations winds were accelerated by extreme temperature differences on the planet. They blew across the ice sheets, picking up sand and dust from the alluvial valleys and outwash plains on the melting edge of the ice and distributing it across adjacent lands. Gradually this loess was built up into thick blankets extending over large portions of the continents. Loess is a fine-grained sediment, eas-ily cultivated, and has a varied mineral composition, providing the im-portant nutrients for plant growth. It is especially favorable for grain crops like corn and wheat, and it forms the parent soil for some of the richest agricultural lands on earth. Soft brown soil evolved from loess lies in a deep layer across the plain states of America and European Russia. In northern China loess accumulations sometimes reach thick-nesses as great as 300 feet.

While the ice sheets were putting the finishing touches on the most fertile breadbaskets of the world, they were also busy carving out another important geographical feature—the Great Lakes. Before the last ice epoch this part of North America was a wide lowland area that had been severely eroded by streams. The heavy mass of the great advancing ice sheets scoured this lowland, deepened the basins, and altered the drainage patterns. As successive glaciations advanced and retreated, the places underlaid by the weakest strata subsided, creating enormous shallow bowls that soon became filled with water. The courses of the rivers and the shapes of the lakes changed many times during this period, finally culminating in the five Great Lakes that we know today.

Ice has left its mark almost everywhere on the land surfaces, even in places that now lie on the equator and in the scorching sands of the deserts. In the early 1960s French geologists exploring for oil in the central Sahara came upon rocks scarred with glacial striations and grooves. They were working in the Hoggar region of southern

Algeria, one of the most forbidding places on earth. A wilderness of jagged rocks and sands, it is blistering hot on summer days: temperatures of 137° F in the shade have been recorded. At these times no ant or snake or even lizard stirs until dusk. And the Arabs travel across the desert only after the burning sun has left the sky.

The age of the grooved rocks discovered by the French geologists could be estimated by distinctive fossils found embedded in the layers of rock directly above and below the strata bearing the glacial marks. They proved to be latest Ordovician, approximately 440 million years old.

Throughout the same northwestern region of Africa geophysicists had recently been taking measurements of magnetism in ancient rocks, and the results were quite surprising. Rocks dated between 400 and 500 million years old bore evidence that the South Pole had been located between the Cape Verde Islands and Morocco—a region that now lies on the Tropic of Cancer 23 degrees north of the equator. The equator at that time seems to have run diagonally across northern Europe, eastern North America, and Mexico.

This information suggested that a polar ice sheet had covered what is now the Sahara Desert 440 million years ago, but the French geologists were cautious about drawing this conclusion. Striations can be made by mountain glaciers as well as by polar sheets. Perhaps a great mountain range had occupied this part of North Africa. It was necessary to obtain more information before the question could be settled. [. . .]

The most striking evidence was found in the eastern part of the Hoggar in the valley of Wadi Tafassasset. Here mile after mile of parallel striations had scarred the rocks as though a great comb had been dragged across the land from southeast to northwest. From the air these scars can be followed for hundreds of miles. The conclusion was inescapable, even for the most cautious: a polar ice sheet had lain across the land that is now one of the hottest and driest places on earth 440 million years ago.

Five glacial epochs (major ice intervals during which briefer ice ages may wax and wane) have been identified in the geologic record. Evidence of the earliest one is widespread in eastern Canada, where rocks 2.2 billion years old are marked with glacial striations and contain an unsorted jumble of large boulders, pebbles, and sandlike till dropped from melting ice. This period of extreme cold occurred long before life began to burgeon on earth. The next epoch of which we have any knowledge took place between 700 and 600 million years ago, when only soft-bodied organisms were present in the sea. The history of this Precambrian glaciation is fragmentary, but it seems to have involved Australia, South Africa, China, Europe, and North America. In fact, the entire world may have been covered with a seal of ice. The glaciers were not just polar phenomena; they even affected areas that lay on the equator at that time.

After this ice epoch the world passed through a long period when there were no glaciers or polar caps. About 200 million years after the Precambrian glaciation the Ordovician ice sheet formed in Africa near the South Pole of that period. By Silurian times, a few million years later, the ice had disappeared and tropical conditions had returned to the earth. Again there was a long warm interval—perhaps 150 million years—before the next ice epoch descended. (There is some evidence of glaciers in South America about 350 million B.P., but it has not been proved that a continental ice mass was responsible. They may have been just mountain glaciers.) In late Carboniferous time—about 290 million B.P.—ice sheets formed in Antarctica, which was near the South Pole at that time also. As the plates drifted around the world, other continents moved into the polar region—South America, Africa, India, and finally Australia. Glaciations appeared in each of these areas successively and did not entirely disappear from the planet until about 270 million years ago. Once more the earth was without large ice masses for a very long time until 2 to 3 million B.P., when the last ice epoch began, the glaciation that has had a profound effect on man and is still a dominant feature of the earth's climate.

Evidence of the most recent ice epoch is relatively fresh, so it is not surprising that these glacial remains were the first to be identified. They told a dramatic story of successive advances and retreats across the continents. For many years, however, the number and complexity of these Pleistocene glaciations were grossly underestimated. Until the late 1950s it was universally assumed that four advances to moderate latitudes had occurred in the last two million years. Only after the development of several sophisticated techniques for studying ancient climate and temperature conditions was the more complex history revealed.

Cores obtained by deep-sea drilling contain shells of small marine animals. Under a microscope paleontologists can count the numbers of fossil species whose range is markedly affected by climate. Using these counts, they can estimate to within two or three degrees the temperature of the water in which these organisms lived. Another measure utilizes the relative abundances of two isotopes of oxygen in the fossil shells: ordinary oxygen 16 and the heavier isotope, oxygen 18. When water evaporates from the ocean, water molecules containing the lighter isotope leave the surface more easily. So water with a higher proportion of oxygen 16 to oxygen 18 becomes cloud droplets and then the snow that creates the great glaciers. A larger proportion of oxygen 18 is left behind in the ocean water. This changing ratio is recorded in the shells of marine organisms.

Using these techniques for analyzing seafloor cores, geologists found a much more rapid succession of recent ice ages than had been identified in the glacial markings on terrestrial rocks. It appears that there may have been as many as twenty major glaciations in the last two million years. The cold periods have lasted much longer than the warm ones. Glacial periods averaging about 100,000 years in length have actually dominated the earth's climate for the whole period of time that human beings have occupied the planet. The warm interglacial periods have been only brief respites from the cold—typically 10,000 years in duration—and the one we have been enjoying now for about 9,000 years may have nearly run its course. [. . .]

Spring is the supreme gift of the Ice King. Every year it reenacts for us the drama of life transforming the barren earth. A few weeks—just a little fraction—out of any lifetime can be well spent observing this extraordinary unfolding of life in all its exquisite detail.

This year I watched spring come to a little plot of woodland near my home in Illinois. It was a small, protected glen where the southern sun shone strong through the still-barren branches of beech and hawthorne trees and a bank on the north held back the coldest winds. A tiny rill whispered over shining stones before it sank into a deep pool where the flow was held back by the decaying log of a giant oak tree. Along the edges of this pool the grass was beginning to turn a fresher green. A mat of periwinkle, its old leaves bronzed here and there with winter frost, was putting forth tiny pointed new leaf buds along its stems. A few early midges danced in a shaft of sunshine.

In this sheltered place I sat down on a cushion of moss. Carefully pushing aside the thick mat of oak leaves and pine needles, I uncovered a whole miniature world of resurgent life. In this little space a dozen different plants had pushed firm tips of dark curled leaves through the soil that was still cool to the touch and crumbly in my fingers. I could not yet guess what shape would grow from each of the thrusting tips. The identifying bulbs and roots lay hidden beneath the spongy soil where I could see and feel the stirrings of life as bacteria and earthworms wove their blind patterns of activity in the darkness.

The days passed and the earth warmed. I returned each morning to watch the individual shape of each plant unfurl. First came the distinctive heart-shaped leaves of the violet and then its sweet-faced lavender blossoms. The windflower shook out its feathery foliage in the cool breeze. The dark red tip of the bloodroot pushed upward through the moist leafmold, taut and purposeful as a phallus bearing its carefully shielded seed of life. It rose high above the tangled mat of vegetation. Then gradually the deeply lobed leaf unfolded, resigning its precious charge, a single bud that slowly swelled into a waxy flower with a golden center.

On slender, tremulous stems the twin-flower released pairs of nodding pink blossoms high above its satin leaves and filled the soft woodland air with a delicious fragrance.

The dark flags of large-flowered trillium slowly unfurled to reveal the precise mathematical perfection of its form. Three symmetrical leaves opened out in unison, forming a shallow cup that bore a tiny green bud in its center. The bud grew and the green calyx split into three pointed sepals that curled back, releasing the blossom. And the blossom unfolded its triangular petals in unison, culminating in the perfect three-lobed flower. Three upon three upon three, an equation written in time.

Fiddles of the cinnamon fern pressed upward toward the sunlight, raising their tightly packed whorls on delicate green stalks. These intricately involuted scrolls looked like the shapes of the great spiral galaxies scaled down to the size of my thumb. I watched the coils of organic matter expand and tiny tentacles reach out, groping toward the sunshine. If I were as big to a galaxy as I am to a fern, I thought, I could watch it unfurl and reach pale fingers out into space to catch the cosmic-shine.

Overhead, the birch trees began to put forth a fretwork of silvery buds that seemed to hold the light, giving it back slowly as though they each had a radiant source of their own. Against the deep shadows of the woods these buds of light looked innumerable and yet as distinct as star clusters in the Milky Way. The flowering of spring here in this little glen was a tiny universe compacted in space and time.

All too soon the season ripened into summer. The petals fell from the wood anemone and the foam flower. The fringed maple flowers were replaced by dark green leaves; the delicate tender green of spring gave way to a heavy canopy of leaves. Now the rays of light could not penetrate the dark shadows, and the ground was wet and soggy underfoot. No light breezes stirred the humid air. So might the world have seemed when jungles dominated the earth and no glaciers floated in the polar seas.

Ice gave us the fragile beauty of spring, the returning robin, the snow goose, and the crocus, and perhaps something more. Mankind grew up during the Pleistocene ice epoch, and the rigor of the glacial climate may have accelerated his evolution. Although the theory is speculative and can probably never be proved, it has been suggested that the challenge of a deteriorating environment was a factor leading to the rapid development of the brain. As the birds invented seasonal migrations, man developed unique human responses. In order to survive the severe winters he learned to plan ahead, to join in cooperation with others of his kind, to build shelters, to cover his body with animal skins, and he discovered the use of fire. In a world of eternal summertime where the living was easy the mysterious potential of mind might have remained dormant. The symphonies of Beethoven, the poetry of Keats, the space probes to Saturn, and messages to the stars—seeds of these wonders and others still unknown have been dropped one by one from the retreating glaciers, released from the shining splendor of ice.

❧

RICHARD PRESTON

Dark Time, from
First Light (1987)

R ichard Preston (1954–) specializes in the classic *New Yorker* magazine–style article that builds a mountain of facts in prose so fine that the readers don't care how much they have to read about frogs if they can only keep reading this hypnotic stuff. Preston succeeds by evoking pictures of what he describes. I think it is impossible to read this selection, for example, and come away with the same image of the solar system we began with. Preston knows the image we first were given. It is that clean chart we saw in school, with the sun here, the planets there, and the orbits they follow marked like interstates through the heavens. Preston replaces this with images of debris moving through the solar system. In this selection, every paragraph seems to report something new about the heavens, about astronomy, and about the people who study the sky.

On a mesa near Flagstaff, Arizona, a low house made of concrete blocks sits in a forest of ponderosa pines. It resembles a bomb shelter. A second house rests on top of it—a soaring structure with walls made of volcanic boulders and glass. One evening in October of 1985, I was sitting at the dining-room table inside the house made of boulders and glass, with the planetary scientist Eugene M. Shoemaker. We had just finished a dinner of homemade enchiladas, and he was examining a newspaper. He read aloud, "Astronomers Locate Possible Distant Galaxy." Then he put on a pair of half glasses and peered at the story. "What in the heck is this? What's this mean?" he said. "There's only about a hundred billion visible galaxies."

"I wonder which one they found," Carolyn Shoemaker remarked dryly as she cleared the dishes from the table. Night had fallen, and a steady rain poured down.

Gene dropped the newspaper on the table. He had a clipped mustache, and a robust face tanned by years of prospecting for the remains of giant craters left by asteroids and comets that had struck the earth. He wore a bolo tie with a clasp of Hopi silver in the shape of an eagle. "Astronomers have essentially abandoned the solar system," he said. "In the nineteenth century, the solar system was the object of central interest in astronomy. As their tools improved, astronomers focused their attention on what they called the larger questions." His face creased into a grin, and he added, "So the geophysicists and geologists came along and adopted this orphan—the solar system."

A drumroll of rain hit the house and reverberated along ponderosa rafters. Gene and Carolyn and their three children had lived in the lower house until it started to feel cramped, whereupon the roof seemed a natural place to begin another house. Eventually, the children had grown up and moved out, and now Gene's mother lived in the lower house.

"We're going to go after something new at Palomar this month," Gene said. He picked up a sheet of computer paper and unfolded it. It was headed "Known Trojans," and it contained a list of heroes from the Trojan War—Achilles, Patroclus, Hektor, Nestor, Priam. Each name was that of a minor planet in orbit around the sun and each name was followed by a long string of numbers describing that planet's orbit. A minor planet is the same thing as an asteroid. Gene said that lately he had been thinking about these Trojans, and he had begun to wonder if there might be a lot of *unknown* Trojans out there. He ran his finger across blocks of text composed of numbers. "Look at these orbital elements," he said "You can just see that the Trojan clouds are really enormous." I couldn't tell anything at all from looking at the computer paper, but when Gene looked at the strings of numbers, evidently he could see in his mind's eye two immense, uncharted clouds of asteroids out by Jupiter. "These clouds cover a heck of a lot of sky," he said. "The kind of sky we can explore with a little wide-field telescope."

"If the rain stops, Gene," Carolyn said from the kitchen, where she was washing the dishes.

A lugubrious noise hammered the roof. He looked up. "This is kind of discouraging," he said.

Carolyn came out of the kitchen. She said gently, "If it's raining in Flagstaff, Gene, then it's probably raining on Palomar Mountain."

What had sparked Gene's interest in the Trojan asteroids was the fact that Carolyn had recently discovered a new one while she was searching through some negatives. She had been looking for earth-crossing asteroids—stray asteroids whose orbits cross the earth's orbit, putting the earth at some risk of being hit—but instead she had found this Trojan. It was big, as minor planets go—a sooty ball about eighty miles around, and by far the largest thing the Shoemakers had ever found. Having discovered it, they were entitled to give it a name. By long-standing tradition, this type of planet is named after a hero from the Trojan War. They studied a copy of the Iliad. "The big names were all taken," Gene said "We thought we were going to have to scrape the barrel. Get into the minor troops." Then they came across the name Paris. "For some reason, Paris had never been used. I don't know why. Paris was the guy who started the war."

There were two clouds of Trojans, one on either side of Jupiter and sharing its orbit. Trojans are distant, slow-moving asteroids, darker than anthracite coal, and for these reasons only forty had been found, whereas in the main asteroid belt, which lies between Jupiter and Mars, thousands of minor planets had been found. The earth was about to swing by one of the two clouds of Trojan asteroids. The moonless time of the month, which astronomers call dark time, had arrived, and the Shoemakers had decided to devote some of their allotted dark time on the eighteen-inch Schmidt telescope at the Palomar Observatory, in California, to a search for Trojan asteroids.

The Trojan cloud had never been completely explored. Scattered pinpoints of light, barely resolvable on a photographic emulsion, they fanned out for half a billion miles on either side of Jupiter. Nobody knew for sure how they had got there or what they were made of. In 1906, the German astronomer Max Wolf discovered an asteroid in Jupiter's orbit, wobbling sixty degrees head of Jupiter, as if Jupiter were pushing it along. Wolf named it Achilles. Achilles had somehow wandered into a region of space where the gravitational fields of Jupiter and the sun formed a stable pocket of gravity that left the asteroid bobbing up and down in limbo. The French mathematician J.-L. Lagrange had predicted such a peculiarity in orbital systems more than a century before, calculating that there would be dimples in gravity sixty degrees on either side of any body in orbit around another body. A stray object that happened to fall into one of these dimples would oscillate inside the dimple and never leave it without a push.

Achilles was the first object found trapped in Jupiter's leading Lagrangian point. Then Patroclus was found in Jupiter's trailing

Lagrangian point, travelling sixty degrees in Jupiter's wake, as if Jupiter were pulling it along on a track. It soon became apparent that two hosts of minor planets bracket Jupiter. A style of naming them became established: with a few exceptions (Patroclus among them), asteroids that travel ahead of Jupiter are named for heroes from the Greek side of the war, while asteroids that trail Jupiter are named for heroes from the Trojan side. The two clouds are known formally as the Greeks and the Trojans, but astronomers usually refer to both clouds simply as Trojans.

The core of the Greek cloud was now at the top of the sky around midnight. "We are looking at an opportunity to find a heck of a lot more of these guys," Gene said. "I think there could be two hundred thousand Trojan planets bigger than a kilometre across, total, for both clouds. I would add that this is not the received wisdom about the Trojan planets." He suspected that the Trojan cloud might extend far above and below the plane of the solar system, where nobody had ever looked systematically for Trojans before. "That's where we are hoping to hit pay dirt," he said.

Asteroid clouds contain a range of debris—everything from dust and sand through boulders to small worlds. If Carolyn, searching photographic negatives after the run, could discover a handful of big Trojans, that would imply the existence of many small objects. Big Trojans, lurking in parts of the sky where they were not supposed to, would betray a haze of Trojans swarming like no-see-ums among them. Carolyn would also be scanning the negatives for earth-crossing asteroids booming past, since one of those could turn up at any time, on any film. [. . .]

In 1932, Karl Reinmuth, an asteroid hunter in Heidelberg, Germany, discovered the first recognized earth-crosser. It appeared on a glass photographic plate as a bright line among the stars—a near-earth object a mile in diameter, travelling quickly. He named it Apollo, after the god who drove the chariot of the sun, because the asteroid's orbit took it near the sun. (Apollo, as it happens, was also the god who fired invisible arrows at mortals to kill them instantly.) Apollo was in an unstable orbit and vanished, only to be unexpectedly rediscovered in 1973 by Harvard astronomers while it was making a swift apparition across a moonless sky. Five years after he sighted Apollo, Reinmuth found the second recognized earth-crosser,

Hermes. Hermes missed us by about five hundred thousand miles—
or twice the distance from the earth to the moon—and disappeared.
Hermes is still lost. Nobody knows when it will be back. [. . .]

In his mind's eye, Gene holds a peculiar vision of the solar system, and it is not any solar system that I had ever heard of. In school-
books, the solar system is pictured as a series of flat concentric circles
centered on the sun, each circle representing the orbit of a planet. In
Gene's mind, the solar system is a spheroid: a dynamic, evolving
cloud of debris, filigreed with bands and shells of shrapnel, full of
bits and pieces of material likely to be pumped into long ellipses and
tangles and wobbling orbits, which carry the drifting projectiles all
over the place—minor planets that every once in a while take a hook
into a major planet, causing a major explosion. "There's just a zoo of
beasts out there, roaming the solar system," he once remarked. Now
he began to talk about comets. One of the more interesting shells of
debris, to him, was the solar system's reservoir of comets. Comets are
lumps of crumbly material, some of them miles across, containing
slilica dust, carbonaceous compounds, and various kinds of ice—
trash left over after the planets formed. Beyond Pluto, the outermost
known planet, there is a spherical shell of comets known as the Oort
cloud—named after the Dutch astronomer Jan Oort, who demonstrated its existence. The Oort cloud contains a prodigious number of
comets—perhaps a trillion (nobody can say how many) travelling
in nearly circular orbits that may average about a light-year from
the sun. If Pluto's orbit were the size of a dime, a typical Oort comet
would circle about ten yards out. From the Oort cloud, the sun
would look like a bright star. Out there, a typical comet moves
around the sun slowly, at about three hundred miles per hour; it can
feel the gravity of stars other than the sun. Since all the stars in the
galaxy are in motion around the galactic center, a star passes near the
Oort cloud now and then. A comet subjected to a gravitational pull
from a passing star can in some cases be slowed down nearly to a
halt—to an orbital speed of around five or ten miles per hour. Then it
does what any object does if it is suspended all but motionless over its
star: it falls toward the sun. By the time it reaches the inner solar system, the comet is falling at outrageous velocity. It takes a hairpin turn
around the sun and heads back for the Oort cloud. Some comets
actually hit the sun. Thinking about comets dropping through the

solar system, Gene wondered out loud how often comets got trapped in the zone of planets. He wondered, for example, if the Trojan planets were dead, black, burned-out comets that had been trapped in Jupiter's Lagrangian points during some early epoch in the history of the solar system. [. . .]

How these pieces of rock came together is an interesting story. It happens that one of matter's more common habits on an astronomical scale is to collect into a rotating pancake of gas and dust particles known as an accretion disk. The solar system began as an accretion disk. When the pressure and density at the center of the disk rose beyond a critical level, thermonuclear ignition occurred and the sun was born. The disk, thinner now, formed into ice-and-rock balls. Called planetesimals, these were the ancestors of planets. The planetesimals collided and stuck together under mutual gravity, growing into planets. As the planetesimals orbited the sun, they separated into rings. Jupiter probably accreted first, from a very dense ring. As the planets fattened, their accretion rates slowed. They ate up the available planetesimals until only a few were left.

Some planetesimals took their sweet time coming home. Gene Shoemaker's studies of the cratering rate of the moon show that even about a billion years after the formation of the earth and the moon, late-arriving planetesimals were occasionally pounding into the moon, their impacts creating the lunar maria—seas of lava that welled up from wounds on the moon's face. The earth suffered the same late heavy bombardment, but weather erased the scars long ago. The bombardment has dwindled to almost nothing today. Almost. The planets have never quite left off growing. The earth is now gaining about twenty tons a day through perpetual rain of dust from space. Every once in a while, it gains two billion tons in one second.

Astronomers used to think that the main asteroid belt might be the rubble of an exploded planet. Now they think that it is material left over from a planet that never formed. Jupiter disturbed a ring of planetesimals in the region now occupied by the belt, preventing the ring from accreting into a planet. Jupiter's gravity raked these planetesimals, mixed them up, tossed them around. They could not stick together. Every time two planetesimals collided, they broke into fragments, and Jupiter pulled the fragments this way and that, causing more collisions and the production of more fragments. Jupiter is still churning the main belt: accidents still happen out there. Hammered by repeated impacts, the asteroids are covered with a layer of dust and rubble, and some may even be piles of bashed fragments barely clinging together under their own gravity. Jupiter has already thrown most of the mass of the asteroid belt off into deep space. "If you took all the asteroids in the main belt and wadded them up into a ball, you would get something about a tenth of the mass of the moon," Gene says.

Earth-crossing asteroids are steadily disappearing: they either collide with the earth or are whiplashed away by a close encounter with it. Fresh earth-crossers must be coming from somewhere, and the main asteroid belt is the largest known reservoir of rock and metal fragments in the solar system. While most main-belt asteroids are in stable orbits that do not come near the earth, the belt appears to be pumping asteroids into earth-crossing orbits. The belt is itself gathered into rings, separated by clear lanes called the Kirkwood gaps. Jupiter sweeps these lanes clean. Any fragment that falls by chance into a Kirkwood gap enters a resonating dance with Jupiter, which can flip the asteroid away. Orbital specialists believe that the Kirkwood gaps, along with other unstable areas in and near the main belt, are a source of many earth-crossers. For example, two asteroids can collide in the belt; a fragment can drift into a Kirkwood gap; Jupiter can pull the fragment from the Kirkwood gap and throw it into an orbit near Mars; then Mars can throw the object inward toward the earth. Saturn can also pull an asteroid from a Kirkwood gap and throw it directly at the earth. [. . .]

On the morning of June 30, 1908, a fireball passed over Siberia, making "even the light of the sun appear dark," according to a witness. In the minutes that followed, a brilliant mushroom cloud boiled up into the stratosphere above the valley of the Middle Tunguska River. Fifty miles from ground zero, a shock wave picked up a tent full of Evenkian nomads and tossed it like a purse through the air. A man sitting on a porch at the Vanovara trading station, seventy miles from ground zero, experienced a bath of radiant heat, followed by a shock wave that plucked him from the porch and threw him several yards through the air, knocking him senseless. The blast incinerated and flattened forests over hundreds of square miles, and the roar broke windows and crockery up to six hundred miles away. A pressure wave travelled twice around the earth. The following night, the sky glowed so brightly over Europe that a person in London could read a newspaper outdoors at midnight; and the cause of it all was the impact of a comet or an Apollo asteroid about two hundred feet in diameter.

One afternoon in 1912, in Holbrook, Arizona, the section boss of the Santa Fe railroad and his family were eating dinner. They heard a "terrific crash." One of the boys ran outside. He said, "It's raining rocks out here!" His father went out. As they looked, the entire plain for a mile eastward filled with puffs that reminded them of bullets "kicking up dust." Fourteen thousand meteorites landed.

In Johnstown, Colorado, on July 6, 1924, a funeral was taking place in the graveyard behind the little Elwell church, when there came a sound like machine-gun fire, and then a *thwock* as a football-size meteorite hit a nearby road. Mr. Clingenpeel, the undertaker, dug it up.

On April 28, 1927, in Aba, Japan, Mrs. Kuriyama's five-year-old daughter was playing in the garden when she cried out. She had been struck on the head by a meteorite the size of a mung bean, which her mother found resting on her neckband. This stone now sits in a museum in Japan. It is called the Aba.

An Air Force pilot flying at high altitude over Alabama on November 30, 1954, saw a bright light, like a falling star, heading in the general direction of the town of Sylacauga. Meanwhile, in Sylacauga, across the street from the Comet Drive-In Theatre, Mrs. E. Hulitt Hodges had just fallen asleep on her couch, when a loud noise woke her up. She jumped to her feet. At first, she thought that the gas heater had exploded. Then she felt a pain in her side. The falling star had punctured her roof, bounced off her radio, and bruised her viciously on the hip. Now lying insolently on her rug was the Sylacauga—eight and a half pounds of hypersthene stone, fresh from out beyond Mars.

On August 10, 1972, something came in from space over Utah. For two minutes, it moved northward over Idaho and Montana, going at least Mach 20. It may have skipped off the atmosphere over Canada or, as Gene Shoemaker suspects, it may have coasted to a relatively gentle impact somewhere in the Canadian forest. Mr. James Baker, vacationing at Jackson Lake, Wyoming, took an extraordinary photograph of it. In the picture, his wife is standing on a dock. She is obviously startled. She is looking toward the Grand Tetons, where, high over the peaks, a fireball is leaving a ruler-straight trail through the upper atmosphere. This was almost certainly an Apollo object, overtaking the earth from behind. It was probably made of metal and somewhat bigger than a diesel locomotive. Had it come in at a steeper angle, Mr. Baker might have photographed a two-kiloton mushroom cloud boiling abaft the Tetons.

Gene and Carolyn Shoemaker are cosmic weather forecasters. The chance of rocks, they report, is a hundred per cent. The smaller objects tend to vaporize in the upper atmosphere, but in doing so they can sometimes release the heat and shock of an atomic bomb. Gene estimates that about once a year a meteoroid explodes in the upper atmosphere, in a Hiroshima-force "event," as he likes to call such an occurrence. Since water covers two-thirds of the earth, many airbursts happen unwitnessed over the sea. Every twenty-five years, perchance, there might be an event that nudged a megaton—the power of a hydrogen bomb. An event like the Tunguska blast will occur roughly once every three centuries. Gene gives between five- and twenty-percent odds that an impact twice as powerful as Tunguska will occur during the next seventy-five years.

What if Carolyn actually found an object heading straight for the earth?

"If it were heading right *at* us, it wouldn't appear to move," Carolyn said, from the control desk. "So I might not notice it in the films. It would look like a star—until it was on top of us, when it would suddenly appear to move very fast. But by then it might be too late to tell anyone."

Gene dragged a stool under the telescope and contorted himself in order to fit into the tight space. "Naw, this isn't going to work," he said. He kicked the stool aside and sat on the floor. "This is a terrible angle," he said, peering into the guide scope. "Suppose that a twelve-megaton air-burst went off over a politically unstable region," he said. "Suppose it happened over Pakistan, and suppose Pakistan had the bomb. The heat, the light, getting knocked on your fanny by the shock wave—a large number of people would swear they had been nuked. The political leadership might say, 'Oh, those S.O.B.s! They've nuked us!' And respond with a real nuclear attack on someone else."

Over geologic time, sooner or later, there will came to pass what Gene likes to call a "major event," to distinguish it from an event. Major events happen roughly once every hundred thousand years. Two-thirds of the asteroids land in the sea. If the Tunguska horror had been made by an object the size of a small office building, then a responsible scientist must consider what might happen if a projectile half a mile wide clobbered the earth. Gene knew what would happen: a mountain coming in vertically would punch through the atmosphere in one second. "A bow wave in front of it opens a hole in the atmosphere, and the atmosphere burns, making nitrogen oxides," he said. He pointed a flashlight at the base of the telescope and fiddled with some parts. "I'm ready," he told Carolyn.

She counted down. He started an exposure looking to the west: the Trojans were now setting, and dawn was approaching. When a giant asteroid moving at nine miles per second hits the ground, he explained, a compressional wave rips through the asteroid, transforming it into a liquescent mass that tries to splash but turns into a fireball. The rocks beneath it compress to one-third their normal size, and a tremendous flash of light floods out of ground zero. "The radiant heat would set buildings on fire a hundred kilometres away," he said. If the asteroid hit in mid-Atlantic, the spreading tsunamis would obliterate many cities on the ocean's rim.

The impact of an asteroid a mile and a half across would squeeze out a ring of crushed, molten, and vaporized rock known as the cone of ejecta. The leading edge of the cone of ejecta expands at hypersonic velocity upward into the atmosphere, like a blossoming flower. It superheats the atmosphere into a bubble of gas mixed with molten and vaporized rock. The bubble bursts through the top of the atmosphere into outer space, and the vaporized rock condenses into droplets of glass, which keep on going. Thumb-size pieces of glass would soar on ballistic suborbital trajectories halfway around the earth, to reenter the atmosphere at a zone of convergence opposite the point of impact, causing firestorms of hot glass to rain down over a region the size of Australia. The blast might inject enough dust into the atmosphere to blot out the sun over the entire planet, and perhaps cause a temporary winter. Nitrogen oxides (the burned atmosphere) would turn into nitric-acid rains. Gene suspects that during the last million years a spike of powerful impacts has occurred—as many as thirty major events, including perhaps ten continental impacts. This suggests the possibility that our species came of age during a mild comet shower. *Homo sapiens*, then, has already survived several natural versions of nuclear war—with the important difference that impact holocausts do not leave radioactive fallout in their wakes. [. . .]

On May 5, 1986, on Palomar Mountain, Carolyn found a strange minor planet in a pair of films that she and Gene had taken a few days before—an asteroid moving slowly and going the wrong way, against the flow of the main belt. She telephoned Brian Marsden at the Minor Planet Center and gave him a set of rough coordinates. He gave the asteroid a temporary name: 1986 JK. The Shoemakers took more photographs of 1986 JK on following nights and continued to report the object's changing positions to the Minor Planet Center. On May 13th, Brian Marsden informed the Shoemakers and the rest of the world (via international telex) that "1986 JK appears to be an Apollo object approaching the earth." Astronomers rushed for their telescopes. The asteroid's slow motion was an illusion: the Shoemakers had been watching it head toward the earth. JK reversed its apparent motion, seemed to accelerate wildly, and passed the earth on June 1st, at a distance of two million six hundred thousand miles—one of the closest asteroidal misses on record. By the standards of normal planetary motions in the solar system, the apparition of 1986 JK was somewhat akin to having the hair on one's head parted by a bullet from a .30-30 rifle. JK was travelling on a long ellipse, like the orbit of a comet. It was a faint object, and seemed to be tiny, but when radio-astronomers using a dish antenna at the Goldstone station of NASA's Deep-Space Network bounced a radar signal off as it went by, they received a clear echo, which suggests that JK is several miles across and faint only because it is as black as a piece of charcoal. It is an Apollo object, a Mars-crosser, and a Jupiter-crosser: it could make a hole in Mars or the earth, or it could vanish into the storms of Jupiter. At intervals of fourteen years, marking time like a metronome, it ticks past the earth. Around Independence Day in the year 2000 it will be back in our skies.

PRIMO LEVI

Carbon, from
The Periodic Table (1975)

Primo Levi (1919–1987) studied chemistry at the University of Turin, but in 1938, Italy barred Jews from academic life. Levi eventually joined a partisan band in the mountains and fought against Mussolini until he was captured and sent to Auschwitz. Because his chemical training made him useful to the Nazis, he was put to work in a lab. When the Soviets arrived, Levi was taken prisoner because, as an Italian, he was an enemy. He was held in a camp in Belorussia but eventually was able to return to Italy. His Auschwitz memoir, *If This Is a Man* (1947), is one of the earliest works of survival literature. Although it led to a successful writing career, Levi continued to work in chemistry as well. In short, Levi was a world-class writer who was also a scientist, or perhaps he was a scientist who was also a world-class writer. This ambivalence permeates his unusual book *The Periodic Table*, in which each chapter is devoted to a single element. The selection here is the final chapter in that book.

This is not a chemical treatise: my presumption does not reach so far—"*ma voix est foible, et même un peu profane.*" Nor is it an autobiography, save in the partial and symbolic limits in which every piece of writing is autobiographical, indeed every human work; but it is in some fashion a history.

It is—or would have liked to be—a micro-history, the history of a trade and its defeats, victories, and miseries, such as everyone wants to tell when he feels close to concluding the arc of his career, and art ceases to be long. Having reached this point in life, what chemist, facing the Periodic Table, or the monumental indices of Beilstein or Landolt, does not perceive scattered among them the sad tatters, or trophies, of his own professional past? He only has to leaf through any treatise and memories rise up in bunches: there is among us he who has tied his destiny, indelibly, to bromine or to propylene, or the –NCO group, or glutamic acid; and every chemistry student, faced by almost any treatise, should be aware that on one of those pages, per-

haps in a single line, formula, or word, his future is written in indeci-
pherable characters, which, however, will become clear "afterward":
after success, error, or guilt, victory or defeat. Every no longer young
chemist, turning again to the *verhängnisvoll* page in that same treatise,
is struck by love or disgust, delights or despairs.

So it happens, therefore, that every element says something to
someone (something different to each) like the mountain valleys or
beaches visited in youth. One must perhaps make an exception for
carbon, because it says everything to everyone, that is, it is not specific, in the same way that Adam is not specific as an ancestor—unless
one discovers today (why not?) the chemist-stylite who has dedicated
his life to graphite or the diamond. And yet it is exactly to this carbon
that I have an old debt, contracted during what for me were decisive
days. To carbon, the element of life, my first literary dream was
turned, insistently dreamed in an hour and a place when my life was
not worth much: yes, I wanted to tell the story of an atom of carbon.

Is it right to speak of a "particular" atom of carbon? For the
chemist there exist some doubts, because until 1970 he did not have
the techniques permitting him to see, or in any event isolate, a single atom; no doubts exist for the narrator, who therefore sets out
to narrate.

Our character lies for hundreds of millions of years, bound to
three atoms of oxygen and one of calcium, in the form of limestone:
it already has a very long cosmic history behind it, but we shall
ignore it. For it time does not exist, or exists only in the form of sluggish variations in temperature, daily or seasonal, if, for the good fortune of this tale, its position is not too far from the earth's surface. Its
existence, whose monotony cannot be thought of without horror, is a
pitiless alternation of hots and colds, that is, of oscillations (always of
equal frequency) a trifle more restricted and a trifle more ample: an
imprisonment, for this potentially living personage, worthy of the
Catholic Hell. To it, until this moment, the present tense is suited,
which is that of description, rather than the past tense, which is that
of narration—it is congealed in an eternal present, barely scratched
by the moderate quivers of thermal agitation.

But, precisely for the good fortune of the narrator, whose story
could otherwise have come to an end, the limestone rock ledge of
which the atom forms a part lies on the surface. It lies within reach of

man and his pickax (all honor to the pickax and its modern equivalents; they are still the most important intermediaries in the millenial dialogue between the elements and man): at any moment—which I, the narrator, decide out of pure caprice to be the year 1840—a blow of the pickax detached it and sent it on its way to the lime kiln, plunging it into the world of things that change. It was roasted until it separated from the calcium, which remained so to speak with its feet on the ground and went to meet a less brilliant destiny, which we shall not narrate. Still firmly clinging to two of its three former oxygen companions, it issued from the chimney and took the path of the air. Its story, which once was immobile, now turned tumultuous.

It was caught by the wind, flung down on the earth, lifted ten kilometers high. It was breathed in by a falcon, descending into its precipitous lungs, but did not penetrate its rich blood and was expelled. It dissolved three times in the water of the sea, once in the water of a cascading torrent, and again was expelled. It traveled with the wind for eight years: now high, now low, on the sea and among the clouds, over forests, deserts, and limitless expanses of ice; then it stumbled into capture and the organic adventure.

Carbon, in fact, is a singular element: it is the only element that can bind itself in long stable chains without a great expense of energy, and for life on earth (the only one we know so far) precisely long chains are required. Therefore carbon is the key element of living substance: but its promotion, its entry into the living world, is not easy and must follow an obligatory, intricate path, which has been clarified (and not yet definitively) only in recent years. If the elaboration of carbon were not a common daily occurrence, on the scale of billions of tons a week, wherever the green of a leaf appears, it would by full right deserve to be called a miracle.

The atom we are speaking of, accompanied by its two satellites which maintained it in a gaseous state, was therefore borne by the wind along a row of vines in the year 1848. It had the good fortune to brush against a leaf, penetrate it, and be nailed there by a ray of the sun. If my language here becomes imprecise and allusive, it is not only because of my ignorance: this decisive event, this instantaneous work *a tre*—of the carbon dioxide, the light, and the vegetal greenery—has not yet been described in definitive terms, and perhaps it will not be for a long time to come, so different is it from that other "organic" chemistry which is the cumbersome, slow, and ponderous work of man: and yet this refined, minute, and quick-witted chemistry was "invented" two or three billion years ago by our silent sisters, the plants, which do not experiment and do not discuss, and whose temperature is identical to that of the environment in which they live. If to comprehend is the same as forming an image, we will never form an image of a happening whose scale is a millionth of a millimeter, whose rhythm is a millionth of a second, and whose protagonists are in their essence invisible. Every verbal description must be inadequate, and one will be as good as the next, so let us settle for the following description.

Our atom of carbon enters the leaf, colliding with other innumerable (but here useless) molecules of nitrogen and oxygen. It adheres to a large and complicated molecule that activates it, and simultaneously receives the decisive message from the sky, in the flashing form of a packet of solar light: in an instant, like an insect caught by a spider, it is separated from its oxygen, combined with hydrogen and (one thinks) phosphorus, and finally inserted in a chain, whether long or short does not matter, but it is the chain of life. All this happens swiftly, in silence, at the temperature and pressure of the atmosphere, and gratis: dear colleagues, when we learn to do likewise we will be *sicut Deus,* and we will have also solved the problem of hunger in the world.

But there is more and worse, to our shame and that of our art. Carbon dioxide, that is, the aerial form of the carbon of which we have up till now spoken: this gas which constitutes the raw material of life, the permanent store upon which all that grows draws, and the ultimate destiny of all flesh, is not one of the principal components of air but rather a ridiculous remnant, an "impurity," thirty times less abundant than argon, which nobody even notices. The air contains 0.03 percent; if Italy was air, the only Italians fit to build life would be, for example, the fifteen thousand inhabitants of Milazzo in the province of Messina. This, on the human scale, is ironic acrobatics, a juggler's trick, an incomprehensible display of omnipotence-arrogance, since from this ever renewed impurity of the air we come, we animals and we plants, and we the human species, with our four billion discordant opinions, our millenniums of history, our wars and shames, nobility and pride. In any event, our very presence on the planet becomes laughable in geometric terms: if all of humanity, about 250 million tons, were distributed in a layer of homogeneous thickness on all the emergent lands, the "stature of man" would not be visible to the naked eye; the thickness one would obtain would be around sixteen thousandths of a millimeter.

Now our atom is inserted: it is part of a structure, in an architectural sense; it has become related and tied to five companions so identical with it that only the fiction of the story permits me to distinguish them. It is a beautiful ring-shaped structure, an almost regular hexagon, which however is subjected to complicated exchanges and balances with the water in which it is dissolved; because by now it is dissolved in water, indeed in the sap of the vine, and this, to remain dissolved, is both the obligation and the privilege of all substances that are destined (I was about to say "wish") to change. And if then anyone really wanted to find out why a ring, and why a hexagon, and why soluble in water, well, he need not worry: these are among the not many questions to which our doctrine can reply with a persuasive discourse, accessible to everyone, but out of place here.

It has entered to form part of a molecule of glucose, just to speak plainly: a fate that is neither fish, flesh, nor fowl, which is intermediary, which prepares it for its first contact with the animal world but does not authorize it to take on a higher responsibility: that of becoming part of a proteic edifice. Hence it travels, at the slow pace of vegetal juices, from the leaf through the pedicel and by the shoot to the trunk, and from here descends to the almost ripe bunch of grapes. What then follows is the province of the winemakers: we are only interested in pinpointing the fact that it escaped (to our advantage, since we would not know how to put it in words) the alcoholic fermentation, and reached the wine without changing its nature.

It is the destiny of wine to be drunk, and it is the destiny of glucose to be oxidized. But it was not oxidized immediately: its drinker kept it in his liver for more than a week, well curled up and tranquil, as a reserve aliment for a sudden effort; an effort that he was forced to make the following Sunday, pursuing a bolting horse. Farewell to the hexagonal structure: in the space of a few instants the skein was unwound and became glucose again, and this was dragged by the bloodstream all the way to a minute muscle fiber in the thigh, and here brutally split into two molecules of lactic acid, the grim harbinger of fatigue: only later, some minutes after, the panting of the lungs was able to supply the oxygen necessary to quietly oxidize the latter. So a new molecule of carbon dioxide returned to the atmosphere, and a parcel of the energy that the sun had handed to the vine-shoot passed from the state of chemical energy to that of mechanical energy, and thereafter settled down in the slothful condition of heat, warming up imperceptibly the air moved by the running and the blood of the runner. "Such is life," although rarely is it described in this manner: an inserting itself, a drawing off to its advantage, a parasitizing of the downward course of energy, from its noble solar form to the degraded one of low-temperature heat. In this downward course, which leads to equilibrium and thus death, life draws a bend and nests in it.

Our atom is again carbon dioxide, for which we apologize: this too is an obligatory passage; one can imagine and invent others, but on earth that's the way it is. Once again the wind, which this time travels far; sails over the Apennines and the Adriatic, Greece, the Aegean, and Cyprus: we are over Lebanon, and the dance is repeated. The atom we are concerned with is now trapped in a structure that promises to last for a long time: it is the venerable trunk of a cedar, one of the last; it is passed again through the stages we have already described, and the glucose of which it is a part belongs, like the bead of a rosary, to a long chain of cellulose. This is no longer the hallucinatory and geological fixity of rock, this is no longer millions of years, but we can easily speak of centuries because the cedar is a tree of great longevity. It is our whim to abandon it for a year or five hundred years: let us say that after twenty years (we are in 1868) a wood worm has taken an interest in it. It has dug its tunnel between the trunk and the bark, with the obstinate and blind voracity of its race; as it drills it grows, and its tunnel grows with it. There it has swallowed and provided a setting for the subject of this story; then it has formed a pupa, and in the spring it has come out in the shape of an ugly gray moth which is now drying in the sun, confused and dazzled by the splendor of the day. Our atom is in one of the insect's thousand eyes, contributing to the summary and crude vision with which it orients itself in space. The insect is fecundated, lays its eggs, and dies: the small cadaver lies in the undergrowth of the woods, it is emptied of its fluids, but the chitin carapace resists for a long time, almost indestructible. The snow and sun return above it without injuring it: it is buried by the dead leaves and the loam, it has become a slough, a "thing," but the death of atoms, unlike ours, is never irrevocable. Here are at work the omnipresent, untiring, and invisible gravediggers of the undergrowth, the microorganisms of the humus. The carapace, with its eyes by now blind, has slowly disintegrated, and the ex-drinker, ex-cedar, ex-wood worm has once again taken wing.

We will let it fly three times around the world, until 1960, and in justification of so long an interval in respect to the human measure we will point out that it is, however, much shorter than the average: which, we understand, is two hundred years. Every two hundred years, every atom of carbon that is not congealed in materials by now stable (such as, precisely, limestone, or coal, or diamond, or certain plastics) enters and reenters the cycle of life, through the narrow door of photosynthesis. Do other doors exist? Yes, some syntheses created by man; they are a title of nobility for man-the-maker, but until now their quantitative importance is negligible. They are doors still much narrower than that of the vegetal greenery; knowingly or not, man has not tried until now to compete with nature on this terrain, that is, he has not striven to draw from the carbon dioxide in the air the carbon that is necessary to nourish him, clothe him, warm him, and for the hundred other more sophisticated needs of modern life. He has not done it because he has not needed to: he has found, and is still finding (but for how many more decades?) gigantic reserves of carbon already organicized, or at least reduced. Besides the vegetable and animal worlds, these reserves are constituted by deposits of coal and petroleum: but these too are the inheritance of photosynthetic activity carried out in distant epochs, so that one can well affirm that photosynthesis is not only the sole path by which carbon becomes living matter, but also the sole path by which the sun's energy becomes chemically usable.

It is possible to demonstrate that this completely arbitrary story is nevertheless true. I could tell innumerable other stories, and they would all be true: all literally true, in the nature of the transitions, in their order and data. The number of atoms is so great that one could always be found whose story coincides with any capriciously invented story. I could recount an endless number of stories about carbon atoms that become colors or perfumes in flowers; of others which, from tiny algae to small crustaceans to fish, gradually return as carbon dioxide to the waters of the sea, in a perpetual, frightening round-dance of life and death, in which every devourer is immediately devoured; of others which instead attain a decorous semi-eternity in the yellowed pages of some archival document, or the canvas of a famous painter; or those to which fell the privilege of forming part of a grain of pollen and left their fossil imprint in the rocks for our curiosity; of others still that descended to become part of the mysterious shape-messengers of the human seed, and participated in the subtle process of division, duplication, and fusion from which each of us is born. Instead, I will tell just one more story, the most secret, and I will tell it with the humility and restraint of him who knows from the start that his theme is desperate, his means feeble, and the trade of clothing facts in words is bound by its very nature to fail.

It is again among us, in a glass of milk. It is inserted in a very complex, long chain, yet such that almost all of its links are acceptable to the human body. It is swallowed; and since every living structure harbors a savage distrust toward every contribution of any material of living origin, the chain is meticulously broken apart and the fragments, one by one, are accepted or rejected. One, the one that concerns us, crosses the intestinal threshold and enters the bloodstream: it migrates, knocks at the door of a nerve cell, enters, and supplants the carbon which was part of it. This cell belongs to a brain, and it is my brain, the brain of the *me* who is writing; and the cell in question, and within it the atom in question, is in charge of my writing, in a gigantic minuscule game which nobody has yet described. It is that which at this instant, issuing out of a labyrinthine tangle of yeses and nos, makes my hand run along a certain path on the paper, mark it with these volutes that are signs: a double snap, up and down, between two levels of energy, guides this hand of mine to impress on the paper this dot, here, this one.

Acknowledgments

Allport, Gordon W.: Excerpt from *Personality: A Psychological Interpretation* by Gordon W. Allport, copyright 1937 by Holt, Rinehart and Winston and renewed 1965 by Gordon W. Allport, reprinted by permission of the publisher.

Asimov, Isaac: From *Asimov on Chemistry* by Isaac Asimov. Copyright © 1974 by Isaac Asimov. Used by permission of Doubleday, a division of Bantam Doubleday Dell Publishing Group, Inc.

Bacon, Francis: From *Novum Organum* by Francis Bacon (1620).

Bartlett, Frederic: From *Remembering* © 1932 by Frederic Bartlett. Reprinted with the permission of Cambridge University Press.

Bateson, William: From *Mendel's Principles of Heredity*, 2nd edition by William Bateson (1909).

Bolles, Edmund Blair: From *A Second Way of Knowing: The Riddle of Human Perception*. Copyright © 1991 by Edmund Blair Bolles. Reprinted by permission of the author.

Boyle, Robert: From *The Sceptical Chymist; Or Chymico-Physical Doubts & Paradoxes, Touching the Spagyrist's Principles Commonly Call'd Hypostatical, As They are Wont to be Propos'd and Defended by the Generality of Alchymists. Whereunto is Præmised Part of Another Discourse Relating to the Same Subject* by Robert Boyle (1661).

Butterfield, Herbert: Reprinted with the permission of Simon & Schuster from *The Origins of Modern Science, 1300–1800* by Herbert Butterfield. Copyright © 1957 by G. Bell & Sons, Ltd.

Cannon, Annie J.: From *The Universe of Stars*, eds. Harlow Shapeley and Cecilia H. Payne, © 1926 by Harvard Observatory. Reprinted by permission of Harvard College Observatory, Cambridge, MA.

Carson, Rachel: From *The Sea Around Us* by Rachel Carson. Copyright 1951 by Rachel Carson. Used by permission of Oxford University Press, Inc.

Chomsky, Noam: From "The Case Against B. F. Skinner" by Noam Chomsky in *The New York Review of Books*. Copyright © 1971, NYREV, Inc. Reprinted by permission of the author.

Curie, Marie: From *Pierre Curie* by Marie Curie, translated by Charlotte and Vernon Kellogg (1923).

Darwin, Charles: From *Journal of Researches into the Geology and Natural History of the Various Countries Visited by H.M.S. Beagle, 1832–36* by Charles Darwin (1839).

Davies, Paul: From *Other Worlds: A Portrait of Nature in Rebellion: Space, Superspace, and the Quantum Universe* by Paul Davies. Copyright ©

1980 by Paul Davies. Original U.S. edition published by Simon and Schuster. Reprinted with the permission of the author and Orion Publishing Group, Ltd.

Duncan, Robert Kennedy: From "Radio-Activity: A New Property of Matter," by Robert Kennedy Duncan, *Harper's Magazine* (1902).

Eddington, Arthur: Reprinted from *Stars and Atoms* by A. S. Eddington (1927) by permission of Oxford University Press.

Einstein, Albert: From *Relativity: The Special and General Theory* by Albert Einstein, translated by Robert W. Lawson. Copyright © 1961 by The Estate of Albert Einstein. Reprinted by permission of Crown Publishers, Inc.

Eiseley, Loren C.: From *The Immense Journey* by Loren Eiseley. © 1953 by Loren Eiseley. Reprinted by permission of Random House, Inc.

Feynman, Richard: From *The Character of Physical Law* by Richard Feynman. Copyright © 1965 by Richard Feynman. Published by The MIT Press, Cambridge, MA. Reprinted by permission of the publisher.

Galilei, Galileo: From *The Sidereal Messenger* (1610) by Galileo Galilei. Translated by Edward Stafford Carlos.

Galilei, Galileo: From *Dialogue Concerning the Two Chief World Systems: The Ptolemain and Copernican* by Galileo Galilei, 2nd revised edition, translated and edited by Drake, Stillman. Copyright © 1942, 1962, 1967 by the Regents of the University of California. Reprinted by permission of University of California Press.

Galilei, Galileo: From *Two New Sciences* (1638) by Galileo Galilei. Translated by Henry Crew and Alfonso de Salvio.

Gould, Stephen Jay: From *The Mismeasure of Man* by Stephen Jay Gould. Copyright © by Stephen Jay Gould. Reprinted by permission of W. W. Norton & Company, Inc.

Haldane, J. B. S.: "Food Control in Insect Societies" from *Possible Worlds and Other Essays* by J. B. S. Haldane. Copyright © 1928 by J. B. S. Haldane. Copyright renewed © 1966 by J. B. S. Haldane. Reprinted by permission of HarperCollins Publishers, Inc.

Harrison, Edward: Reprinted by permission of the publisher from *Darkness at Night: A Riddle of the Universe* by Edward Harrison, Cambridge, MA: Harvard University Press. Copyright © 1987 by Edward Harrison.

Heisenberg, Werner: Excerpts from *The Physicist's Conception of Nature* by Werner Heisenberg, copyright © 1955 by Rowohit Teschenbach Verlag, GmbH, English translation by Arnold J. Pomerans, copyright © 1958 by Hutchinson & Co. Ltd. and renewed 1986 by Century Hutchinson Publishing Group Limited, reprinted by permission of Harcourt Brace & Company.

Helmholtz, Hermann von: From a public lecture, "On the Conservation of Force" (1862). Translator unknown.

Herodotus: From *The Histories* by Herodotus, translated by Aubrey de Sélincourt. Copyright © 1954 by the Estate of Aubrey de Sélincourt. Reprinted with the permission of Penguin Books, Ltd.

Hoyle, Fred: From *The Nature of the Universe* by Fred Hoyle. Reprinted by permission of Curtis Brown, Ltd. Copyright © 1950 by Fred Hoyle. Published by Harper & Bros., 1993.

Huxley, Julian: Originally published in *The Yale Review*, 1943. Copyright © Yale University. Reprinted by permission of Blackwell Publishers, Oxford, U.K. and Malden, MA.

Huxley, Thomas H.: From a review of *On the Origin of Species* published in 1859 and reprinted in *Darwiniana*, 1893.

Jeans, Sir James: From *The Universe Around Us* by Sir James Jeans. Copyright © 1929 by Sir James Jeans. Reprinted by permission of Cambridge University Press.

Kepler, Johannes: From *Conversation with the Sidereal Messenger Recently Sent to Mankind by Galileo Galilei, Mathematician of Padua* (Prague, 1610). Translated by Edward Rosen.

Lavoisier, Antoine-Laurent: From *The Elements of Chemistry, in a New Systematic Order, Containing All the Modern Discoveries* (1789). Translated by Robert Kerr.

Leonardo da Vinci: From *The Notebooks of Leonardo*, translated by Edward MacCurdy.

Levi, Primo: From *The Periodic Table* by Primo Levi, translated by Raymond Rosenthal. Translation copyright © 1984 by Schocken Books, Inc. Reprinted by permission of Schocken Books, published by Pantheon Books, a division of Random House, Inc.

Lucretius: From *The Nature of Things: A New Translation* by Lucretius, translated by Frank O. Copley. Translation copyright © 1977 by W. W. Norton & Company, Inc. Reprinted by permission of W. W. Norton & Company, Inc.

Mach, Ernst: From *Popular Scientific Lectures*, 3rd edition (1898) by Ernst Mach. Translated by Thomas J. McCormack.

Maxwell, James Clerk: From a lecture given by James Clark Maxwell before the British Association and reprinted in *Nature*, Sept. 25, 1873.

McPhee, John: From *Basin and Range* by John McPhee. Copyright © 1981 by John McPhee. Reprinted with the permission of Farrar, Straus, and Giroux.

Newton, Isaac and Robert Hooke: From *Letters to Henry Oldenburg*, 1672.

Oppenheimer, J. Robert: Reprinted with the permission of Simon & Schuster from *Science and the Common Understanding* by J. Robert Oppenheimer. Copyright © 1953 by J. Robert Oppenheimer.

Pagels, Heinz: Reprinted with the permission of Simon & Schuster from *The Cosmic Code: Quantum Physics as the Language of Nature* by Heinz R. Pagels. Copyright © 1982 by Heinz Pagels.

Pavlov, I. P.: From *Conditioned Reflexes: An Investigation of the Physiological Activity of the Cerebral Cortex* by I. P. Pavlov (1924).

Piaget, Jean: From *Judgment and Reasoning in the Child* by Jean Piaget. Published by Routledge and Kegan Paul in 1928 and Littlefield Adams in 1976. Reprinted by permission of Littlefield Adams.

Popper, Karl: Reprinted by permission of Open Court Trade & Academic Books, a division of Carus Publishing Company, Peru, IL, from *The Philosophy of Karl Popper* (The Library of Living Philosophers). Copyright © 1974.

Preston, Richard: From *First Light* by Richard Preston. Copyright © 1987 by Richard Preston. Paperback edition published by Random House, 1997. Reprinted with permission of the author and Janklow & Nesbit Associates.

Russell, Bertrand: From *The ABC of Relativity* by Bertrand Russell. Copyright © Bertrand Russell Peace Foundation. Reprinted by permission of Routledge Ltd.

Rutherford, Ernest: From *The Transmutation of the Atom*, a lecture by Ernest Rutherford aired by the British Broadcasting Company on 11 October 1933 and based on a lecture given before the British Association.

Sagan, Carl: From *The Cosmic Connection: An Extraterrestrial Perspective*. Copyright © 1973 by Carl Sagan and Jerome Agel. For further information: Jerome Agel, 2 Peter Cooper Road, New York City, 10010.

Saussure, Horace de: From *Voyages dans les Alpes* (1796), translator unknown.

Schaller, George: From *Stones of Silence* by George B. Schaller. Copyright © 1979, 1980 by George B. Schaller. Used by permission of Viking Penguin, a division of Penguin Books USA Inc.

Smoot, George: From *Wrinkles in Time* by George Smoot and Keay Davidson. Copyright © 1993 by George Smoot and Keay Davidson. Reprinted by permission of William Morrow & Company, Inc.

Sullivan, Walter: From *Continents in Motion: The New Earth Debate*, First Edition, by Walter Sullivan. Copyright © 1974 by Walter Sullivan. Reprinted with the permission of The Estate of Walter Sullivan.

Voltaire: From *Letters Concerning the English Nation* (1733). Translated by John Lockman. Punctuation and spelling have been modernized.

Wallace, Alfred: From "On the Tendency of Varieties to Depart Indefinitely from the Original Type." Read before the Linnean Society on July 1, 1858.

Watson, James: Reprinted with the permission of Scribner, a division of Simon & Schuster, from *The Double Helix* by James D. Watson. Copyright © 1968 by James D. Watson.

Watson, John B.: From "The New Science of Animal Behavior," by John B. Watson, *Harper's* (1909).

Wegener, Alfred: From *The Origin of Continents and Oceans* by Alfred Wegener. Translation by John Biram © 1966. Reprinted by permission of Dover Books.

Wolf, Fred Alan: Excerpt from *Taking the Quantum Leap: The New Physics for Nonscientists* by Fred Alan Wolf. Copyright © 1981 by Fred Alan Wolf. Preface, Chapter 15 copyright © by Fred Alan Wolf. Reprinted by permission of HarperCollins Publishers, Inc.

Young, Louise B.: From *The Blue Planet* by Louise B. Young. Copyright © 1983 by Louise B. Young. By permission of Little, Brown & Company.

Index

A page number followed by an italic letter *n* indicates a footnote.